Unmanned Aerial Vehicle (UAV)-Enabled Wireless Communications and Networking

Unmanned Aerial Vehicle (UAV)-Enabled Wireless Communications and Networking

Editor

Margot Deruyck

MDPI • Basel • Beijing • Wuhan • Barcelona • Belgrade • Manchester • Tokyo • Cluj • Tianjin

Editor
Margot Deruyck
Ghent University – IMEC
Belgium

Editorial Office
MDPI
St. Alban-Anlage 66
4052 Basel, Switzerland

This is a reprint of articles from the Special Issue published online in the open access journal *Sensors* (ISSN 1424-8220) (available at: https://www.mdpi.com/journal/sensors/special_issues/U_AV).

For citation purposes, cite each article independently as indicated on the article page online and as indicated below:

LastName, A.A.; LastName, B.B.; LastName, C.C. Article Title. *Journal Name* **Year**, *Volume Number*, Page Range.

ISBN 978-3-0365-4663-6 (Hbk)
ISBN 978-3-0365-4664-3 (PDF)

Cover image courtesy of Dose Media

Contents

About the Editor

Margot Deruyck

Margot Deruyck (Assistant Professor) is currently working with the WAVES (Wireless, Acoustics, Environment & Expert Systems—Department of Information Technology) research group of Ghent University—IMEC. She obtained her Ph.D. in Computer Science Engineering from that same university in 2015. She is a Post-Doctoral Fellow of the FWO-V (Research Foundation—Flanders) and is an IEEE Member. Her scientific work focuses on green wireless networks (minimal energy consumption and/or exposure for human beings), UAV-based network design, and Internet of Animals. She is the Science Communication Coordinator of the EU COST CA20120 INTERACT Action.

sensors

MDPI

Editorial

Editorial: Special Issue "Unmanned Aerial Vehicle (UAV)-Enabled Wireless Communications and Networking"

Margot Deruyck

Department of Information Technology, IMEC-Ghent University-WAVES, 9052 Ghent, Belgium; margot.deruyck@ugent.be

Citation: Deruyck, M. Editorial: Special Issue "Unmanned Aerial Vehicle (UAV)-Enabled Wireless Communications and Networking". *Sensors* **2022**, *22*, 4458. https://doi.org/10.3390/s22124458

Received: 20 May 2022
Accepted: 31 May 2022
Published: 13 June 2022

In the last decade, the behavior of mobile data users has completely changed. We are no longer limited to our own house if we want to share pictures, make a video call, stream music or video, or game. In fact, nowadays, we can do all those things in the park, bus, or train, actually wherever we are. With the upcoming next generation standard—the 5th Generation or 5G—not only will significantly higher data rates will be offered, but also an ever larger spectrum of applications and services to the mobile users and platforms, all with heterogeneous requirements. Besides serving users, 5G will also bring many opportunities to the industry. This means that 5G will be serving a massive density of nodes which can either be human-held or machine-type; a tendency we currently already see in the emerging Internet of Things (IoT) and its applications. The above-mentioned evolution will of course also have its repercussions on the network itself. The future wireless network will be characterized by a high degree of flexibility compared with the past, allowing them to adapt smoothly, autonomously, and efficiently to the quickly changing traffic demand evolutions both in time and space. The networks of today are designed to cope with average or peak traffic predictions and the offered capacity on a local scale is highly dependent on the density of the infrastructure equipment in a considered area. Adding mobile infrastructure or UABSs (Unmanned Aerial Base Stations)—i.e., a base station mounted on a UAV (Unmanned Aerial Vehicle) or drone—would provide a huge added value to the network.

One of the key challenge of designing such a UAV-based network is a proper allocation of both the UAV's location and network resources. Ref. [1] proposed a fast and practical algorithm for UAV relay networks to provide the optimal solution for the number of transmit time slots and the UAV relay location in a sequential manner. The transmit power at the base station and the UAV was determined in advanced based on the availability of channel state information (CSI). Simulation results demonstrated that the proposed algorithms can significantly reduce the computational effort and complexity to determine the optimal UAV location. Ref. [2] formulated the uplink resource allocation problem for a NOMA (Non-Orthogonal Multiple Access)-IoT based UAV network with the objective of maximizing the system throughput while minimizing the delay of IoT applications. Power allocation was investigated to achieve fairness between the users. The algorithm achieved a 31.8% delay improvement, 99.7% reliability increase, and 50.8% fairness enhancement when compared with the maximum channel quality indicator CQI) algorithm, allowing the algorithm to be used for URLLC (Ultra-Reliable Low-Latency Communication) applications. Due to the large-scale deployment of UAVs in UAV-based networks, also the shortage of spectrum is a realistic threat for the roll-out of these networks. Ref. [3] proposed a cooperative spectrum sensing scheme for cognitive UAV networks based on a CHMM (Continuous Hidden Markov Model) with a novel SNR (Signal-to-Noise Ratio) estimation method.

Another typical use case for UAV-aided networks—besides offering an extra degree of flexibility in the network—is to extend the network's coverage in isolated or rural areas. Ref. [4] comprehensively surveyed the user of three platforms to deliver broadband services to remote and low-income areas: UAVs, APs (Altitude Platforms), and LEO (Low

Earth Orbit) satellites. The UAVs are considered as a noteworthy solution since their efficient maneuverability can solve rural coverage issues or not-spots. Ref. [5] investigated six different routing algorithms for this kind of applications. A time-dependent variant of the Dijkstra's algorithm which determines the fastest route by taking into account the time when the message reaches the node and the time allocated for data transfer was developed for a DTN (Delay Tolerant Network) of drones flying in an aerial area divided either into squares or into equilateral triangles. Their simulation results showed a better performance of the proposed time-dependent Dijkstra algorithm compared with state-of-the art routing protocols. One step further than extending the coverage of an existing network is to provide wireless communication in areas hit by a natural disaster. Ref. [6] jointly addressed the transmission power selection, data-rate maximization, and interference mitigation problems for such a scenario. Considering all these conflicting parameters, the problem was investigated as a budget-constrained multi-player multi-armed bandit problem. The whole process was carried out in a decentralized manner, where no information was exchanged between UAVs. To achieve this, two power-budget-aware algorithms were proposed to realize the selection of the transmission power value efficiently. Both algorithms showed outstanding performance over random power value selection in terms of achievable data rate. UAVs have also showed their added value for the enhancement of wireless communication in millimeter-wave bands. Typically, antenna arrays have been employed for this purpose. However, many beam-forming methods for improving communication quality are based on channel estimation, which are resource-intensive due to the complexity of channel estimation in practice. Ref. [7] formulated a MIMO (Massive Input Massive Output) blind beamforming problem at the receivers for UAV-assisted communications in which channel estimation was omitted to save resources. An analytical method, called ACMA (Analytical Constant Modulus Algorithm), was introduced relying only on data received by the antenna. ACMA could achieve good signal recovery accuracy, a reasonable sum rate, and acceptable complexity.

Many challenges related to the design of UAV networks in general, such as, for example, spectrum sharing/coexistence, are computationally very challenging and hence an excellent case to apply machine learning and AI (Artificial Intelligence) on. Ref. [8] proposed a multi-agent DQL (Deep Q-learning)-based transmission power control algorithm to minimize the outage probability while satisfying the interference requirement of an interfered system. To deal with the potential risk of action-value overestimation from the DQL, they developed even a DDQL (Double DQL). The proposed DQL power control algorithm performed equal or close to the optimal exhaustive search algorithm for varying positions of the interfered system. With a similar performance by the DDQL power control yields the same performance, the authors concluded that the actional value overestimation did not adversely affect the quality of the learned power control policy. Besides using machine learning to actual develop the network, it can also be applied in properly predicting the user demand. Ref. [9] proposed a UAV positioning algorithm with the objective of extending an existing network and balancing the traffic load. To properly predict the users and their movement, the performance of various state-of-the-art machine learning algorithms was investigated. Random Forest and Gradient Boosting presented both the best performance, with Random Forest having a better prediction and training time.

The use of UAVs in wireless communication networks—whether it is to provide extra flexibility, extend the coverage or cover isolated, rural, or post-disasters areas—makes the network more vulnerable for security issues. Ref. [10] proposed a security model applying cooperative friendly jamming using artificial noise and drone mobility to prevent eavesdropping and improve security. Ref. [11] developed two dynamic selection techniques which identify the most effective classifier for the detection of GPS (Global Positioning System) spoofing attacks.

It might be clear that using UAVs for wireless communication and networks can provide a tremendous added value. The scenarios here presented are only a fraction of what we can achieve by UAV-aided communication. A nice overview of what is possible was provided in ref. [12], along with the recent advances and challenges that are brought by the recent research trends.

Funding: This research received no external funding.

Conflicts of Interest: The author declare no conflict of interest.

References

1. Han, S.I.; Baek, J. Optimal UAV Deployment and Resource Management in UAV Relay Networks. *Sensors* **2021**, *21*, 6878. [CrossRef] [PubMed]
2. Karem, R.; Ahmed, M.; Newagy, F. Resource Allocation in Uplink NOMA-IoT Based UAV for URLLC Applications. *Sensors* **2022**, *22*, 1566. [CrossRef] [PubMed]
3. Feng, Y.; Xu, W.; Zhang, Z.; Wang, F. Continuous Hidden Markov Model Based Spectrum Sensing with Estimated SNR for Cognitive UAV Networks. *Sensors* **2022**, *22*, 2620. [CrossRef] [PubMed]
4. Cabrera-Castellanos, D.F.; Aragón-Zavala, A.; Castañón-Ávila, G. Closing Connectivity Gap: An Overview of Mobile Coverage Solutions for Not-Spots in Rural Zones. *Sensors* **2021**, *21*, 8037. [CrossRef] [PubMed]
5. Deaconu, A.M.; Udroiu, R.; Nanau, C.-Ş. Algorithms for Delivery of Data by Drones in an Isolated Area Divided into Squares. *Sensors* **2021**, *21*, 5472. [CrossRef] [PubMed]
6. Amrallah, A.; Mohamed, E.M.; Tran, G.K.; Sakaguchi, K. Enhanced Dynamic Spectrum Access in UAV Wireless Networks for Post-Disaster Area Surveillance System: A Multi-Player Multi-Armed Bandit Approach. *Sensors* **2021**, *21*, 7855. [CrossRef] [PubMed]
7. Liu, P.; Fan, K.; Chen, Y. Analytical Blind Beamforming for a Multi-Antenna UAV Base-Station Receiver in Millimeter-Wave Bands. *Sensors* **2021**, *21*, 6561. [CrossRef] [PubMed]
8. Jo, S.; Yang, W.; Choi, H.K.; Noh, E.; Jo, H.-S.; Park, J. Deep Q-Learning-Based Transmission Power Control of a High Altitude Platform Station with Spectrum Sharing. *Sensors* **2022**, *22*, 1630. [CrossRef] [PubMed]
9. Oliveira, F.; Luís, M.; Sargento, S. Machine Learning for the Dynamic Positioning of UAVs for Extended Connectivity. *Sensors* **2021**, *21*, 4618. [CrossRef] [PubMed]
10. Jeon, G.-H.; Lee, J.-H.; Sung, Y.-S.; Park, H.-J.; Lee, Y.-J.; Yun, S.-W.; Lee, I.-G. Cooperative Friendly Jamming Techniques for Drone-Based Mobile Secure Zone. *Sensors* **2022**, *22*, 865. [CrossRef] [PubMed]
11. Talaei Khoei, T.; Ismail, S.; Kaabouch, N. Dynamic Selection Techniques for Detecting GPS Spoofing Attacks on UAVs. *Sensors* **2022**, *22*, 662. [CrossRef] [PubMed]
12. Abdelmaboud, A. The Internet of Drones: Requirements, Taxonomy, Recent Advances, and Challenges of Research Trends. *Sensors* **2021**, *21*, 5718. [CrossRef] [PubMed]

sensors

MDPI

Article

Continuous Hidden Markov Model Based Spectrum Sensing with Estimated SNR for Cognitive UAV Networks

Yuqing Feng [1], Wenjun Xu [2,*], Zhi Zhang [1] and Fengyu Wang [3]

1. The State Key Laboratory of Networking and Switching Technology, Beijing University of Posts and Telecommunications, Beijing 100876, China; fyq@bupt.edu.cn (Y.F.); zhangzhi@bupt.edu.cn (Z.Z.)
2. The Key Lab of Universal Wireless Communications, Ministry of Education, Beijing University of Posts and Telecommunications, Beijing 100876, China
3. School of Artificial Intelligence, Beijing University of Posts and Telecommunications, Beijing 100876, China; fengyu.wang@bupt.edu.cn

* Correspondence: wjxu@bupt.edu.cn

Abstract: In this paper, to enhance the spectrum utilization in cognitive unmanned aerial vehicle networks (CUAVNs), we propose a cooperative spectrum sensing scheme based on a continuous hidden Markov model (CHMM) with a novel signal-to-noise ratio (SNR) estimation method. First, to exploit the Markov property in the spectrum state, we model the spectrum states and the corresponding fusion values as a hidden Markov model. A spectrum prediction is obtained by combining the parameters of CHMM and a preliminary sensing result (obtained from a clustered heterogeneous two-stage-fusion scheme), and this prediction can further guide the sensing detection procedure. Then, we analyze the detection performance of the proposed scheme by deriving its closed-formed expressions. Furthermore, considering imperfect SNR estimation in practical applications, we design a novel SNR estimation scheme which is inspired by the reconstruction of the signal on graphs to enhance the proposed CHMM-based sensing scheme with practical SNR estimation. Simulation results demonstrate the proposed CHMM-based cooperative spectrum sensing scheme outperforms the ones without CHMM, and the CHMM-based sensing scheme with the proposed SNR estimator can outperform the existing algorithm considerably.

Keywords: cognitive UAV networks; clustered two-stage-fusion cooperative spectrum sensing; continuous hidden Markov model; SNR estimation

Citation: Feng, Y.; Xu, W.; Zhang, Z.; Wang, F. Continuous Hidden Markov Model Based Spectrum Sensing with Estimated SNR for Cognitive UAV Networks. *Sensors* **2022**, *22*, 2620. https://doi.org/10.3390/s22072620

Academic Editor: Omprakash Kaiwartya

Received: 10 January 2022
Accepted: 22 March 2022
Published: 29 March 2022

Publisher's Note: MDPI stays neutral with regard to jurisdictional claims in published maps and institutional affiliations.

1. Introduction

With the advantage of high flexibility and low deployment cost, unmanned aerial vehicles (UAVs) have been widely used in military communications, weather monitoring, emergency rescue [1] and some other UAV-assisted Internet of Things (IoT) applications [2]. The large-scale deployment of UAVs has exacerbated the shortage of spectrum resources. However, the existing spectrum allocation strategies cannot effectively use the scarce spectrum resources, which becomes the bottleneck for enhancing the communication performance of UAVs [3]. Cognitive radio (CR) is proposed to solve the problem, which improves the spectrum efficiency by perceiving spectrum holes and providing secondary UAVs with opportunities to reuse idle spectrum. Thus, CR can further guide the spectrum utilization in cognitive unmanned aerial vehicle networks (CUAVNs), including the resource allocation for low-latency communications [4], high-quality services with limited resources [5], maximum achievable throughput [6], and optimal power allocation. To enable CUAVNs, accurate spectrum sensing attaches great importance.

With the development of UAVs, spectrum sensing in unmanned aerial vehicle networks (UAVNs) has attracted attention from both academia and industry. The detection performance is enhanced by using multiple secondary UAVs in [7,8]. Authors in [3,9]

consider the combination of UAVs and terrestrial communication equipment for air-ground-integrated spectrum sensing. The heterogeneity of spatial information is further taken into account by using the 3D [10], so as to improve the spectrum sensing detection performance. The above studies introduce information from multiple users and spatial dimensions to improve the spectrum utilization of CUAVNs. However, they do not consider the temporal information of the spectrum, which affects the detection performance of UAVNs. Exploring the temporal correlation of spectrum states by focusing on the Markov property is an effective and novel idea to enhance the detection probability [11].

Another concern is that knowledge of the signal-to-noise ratio (SNR) is required before spectrum sensing [12], e.g., energy detection and cyclostationary feature detection. Therein, it is assumed that SNR is perfectly known. However, UAVs need to move across a large area in lots of applications, and the links between two UAVs that are far apart frequently break and reestablish [13], which degrades the sensing performance and the spectrum utilization. Besides, large flying areas, unstable links, and dynamic network topologies lead to variable SNRs in CUAVNs, which makes the assumption of pre-known SNRs no longer applicable, bringing new challenges for sensing in UAVNs.

To overcome these problems, we propose a continuous hidden Markov model (CHMM) based sensing scheme with a novel space smoothing second- and fourth-order moments (SS-M2M4) SNR estimator. Note that the true states of the unauthorized spectrum are not observable, however, the sensing results originating from the unobservable spectrum states can be easily obtained. Therefore, a hidden Markov chain model fits well with the spectrum sensing procedure. To fully exploit the temporal information and further enhance the detection performance, we combine the spectrum sensing scheme with CHMM. To the best of our knowledge, our work is the first to consider the continuous hidden Markov model in non-centralized CUANs. Moreover, we further provide theoretical analysis of the detection probability and the false alarm probability, while previous works mainly focus on numerical simulations. Besides, for generalized estimation and higher accuracy, based on a general SNR estimation method of a second- and fourth- moments (M2M4) estimator [14], we propose a SS-M2M4 SNR estimator. The SNR smoothness of the neighboring CUAVs has been taken into account, enlightened by the smoothness of the graph signal [15]. Compared to the widely-used M2M4 estimator, our proposed SNR estimator achieves more accurate estimation. With the SNR provided by the proposed estimator, our CHMM-based spectrum sensing scheme can achieve a high detection probability. Our contributions are summarized as follows:

- We propose a spectrum sensing scheme based on a continuous hidden Markov model in CUAVNs to obtain better sensing performance.
- We derive the closed-form detection probability and false alarm probability expressions of the proposed CHMM-based spectrum sensing.
- Considering the fading similarity within the neighboring CUAVs in practical applications, we propose an SNR estimation scheme based on signal smoothness, which reduces the SNR estimation error effectively, and further enhances the performance of CHMM-based cooperative spectrum sensing.

The remainder of this paper is organized as follows. Section 2 discusses the related work. Section 3 introduces the system model. In Section 4, the CHMM based spectrum sensing scheme is proposed. In Section 5, the closed-form detection probability and false alarm probability expressions of the proposed method are derived. In Section 6, we design a novel SNR estimator to enhance the proposed CHMM-based spectrum sensing scheme with a more accurate SNR. The simulation results are presented to evaluate the proposed algorithms in Section 7. Section 8 concludes the work and discusses the possible future work.

2. Related Work

2.1. Hidden Markov Model (HMM) Based Spectrum Sensing

Hidden Markov model (HMM) based spectrum sensing means using the HMM to model the spectrum sensing procedure. As the HMM has a hidden layer and an observable layer, it fits well with the sensing procedure whose spectrum states are unknown but the receiver can be obtained. Exploring the Markov property of the spectrum states is an effective way to enhance the sensing performance [11]. Compared with deep learning based spectrum sensing [16,17], CHMM-based spectrum sensing has a stronger interpretability and smaller delay, and the initial probability distribution can be used to calculate the initial spectrum utilization. In addition, the obtained parameters can also be used for digital twinning of the communication system. It has been validated that the spectrum state can be modeled as a Markov chain by analyzing real-world measurements [18]. A hidden Markov model based scheme [19] is proposed to predict the arrival of the primary user (PU). Authors in [20] evaluate the reliability of HMM based cooperative spectrum sensing in cognitive radio networks, in the presence of random malfunctioning of secondary user nodes participating in the process. Occupancy prediction schemes based on a discrete hidden Markov model (DHMM) and a continuous hidden Markov model are investigated in [21–24], respectively. Authors in [21,22] adopt DHMM to model the spectrum sensing procedure, which do not make full use of the information obtained by the receiver. Authors in [23,24] use CHMM to model the sensing procedure of CUAVNs. However, they do not consider the dynamicity of UAVs, which is an important characteristic of the UAV networks [25]. Besides, centralized spectrum sensing methods in [23,24] do not work well in CUAVNs since the global fusion center is usually unreachable to secondary CUAVs. Thus, in this paper, we propose a CHMM-based spectrum sensing method to enhance the spectrum utilization in CUANs.

2.2. SNR Estimation

SNR estimation refers to the calculation of SNR by using signal information [14]. Various algorithms require SNR estimation for optimal performance if the SNR is not constant, such as linear diversity combining techniques and Viterbi algorithms with soft-decision [14]. Note that the knowledge of SNR is also required for typical commonly-used spectrum sensing [12]. However, in practical UAV applications, it is difficult to obtain accurate SNR. Conventionally, SNR estimators require knowledge of the signal or the channel, such as the maximum likelihood (ML) SNR estimator [26] and particle swarm optimization (PSO) SNR estimator, based on parameters of hardwares or channels [27]. In addition, there exist estimators designed for specific signals or specific feature spectrum sensing methods, such as the SNR estimator [28] for M-ary amplitude phase shift keying (M-APSK) modulated signals, the SNR estimator for signal with Polar code [29], and the estimator for eigenvalue-based spectrum detectors [30]. Lacking the prior knowledge of the signal/channel and design for specific signals/sensing methods make it hard to generally adopt the above estimators to various CUAVNs. Besides, some scholars have paid attention to deep learning based SNR estimation methods, such as the convolutional neural networks (CNN)—long short term memory (LSTM) based SNR estimators [31] and the CNN-based SNR estimators designed for UAVNs [32]. However, in the spectrum sensing of CUAVNs, the deep learning based methods are too complicated and take more time, leading to the sensing term being missed. Taking all the above into consideration, we design a generalized and low-complexity SNR estimator named SS-M2M4. Based on the proposed SNR estimator, the CHMM-based spectrum sensing method can further enhance detection performance.

3. System Model

In order to detect whether the authorized spectrum of PU is occupied, a spectrum sensing method is adopted by multiple UAVs [3,33]. In this paper, we propose a CHMM-based spectrum sensing method with consideration of the imperfect SNR estimation to enhance the detection probability. The overview of this unified scheme is shown in Figure 1. First, to obtain the single-time spectrum sensing fusion result, we adopt the max-min distance clustering algorithm [34] and heterogeneous two-stage-fusion spectrum sensing scheme [35], which are described in Sections 3.1 and 3.2, respectively. Then, to fully take advantage of the temporal correlation of the spectrum states, we propose a CHMM-based spectrum sensing method, which will be introduced in Section 4. Considering the imperfect SNR estimation in practical applications, we propose a novel SNR estimator shown in Section 6 to offer the sensing scheme more desirable SNRs. Finally, combining the CHMM-based spectrum sensing and the proposed SNR estimator, we propose the unified CHMM-based spectrum sensing scheme with advanced SNR estimator.

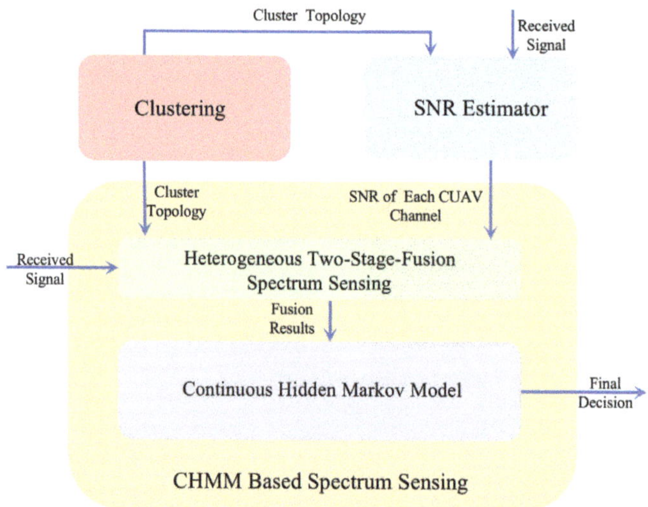

Figure 1. The unified framework of the CHMM-based spectrum sensing scheme with the SS-M2M4 estimator.

3.1. Clustering Method

Considering the fusion delay brought by distributed cooperative spectrum sensing (DCSS), and to further provide the proposed SS-M2M4 SNR estimator with graph topology, a max-min distance clustering algorithm [34] is adopted. It divides CUAVs with similar position and mobility into the same cluster, and selects the one with the highest trust value in each cluster as the cluster head. Thus, we can adopt intra-cluster centralized cooperative spectrum sensing. Due to the similar locations, the UAVs in the same cluster have similar SNRs, which facilitates the SNR estimation. The cluster heads are less than the number of the CUAVs, so the fusion delay will be reduced and, therefore, we adopt inter-cluster distributed cooperative spectrum sensing.

3.2. Heterogeneous Two-Stage-Fusion Spectrum Sensing

To further improve the sensing performance in CUAVNs, a heterogeneous cooperative spectrum sensing scheme is employed with the clustering outcome, as shown in Figure 2. A symbol table of the notations is shown in Table 1.

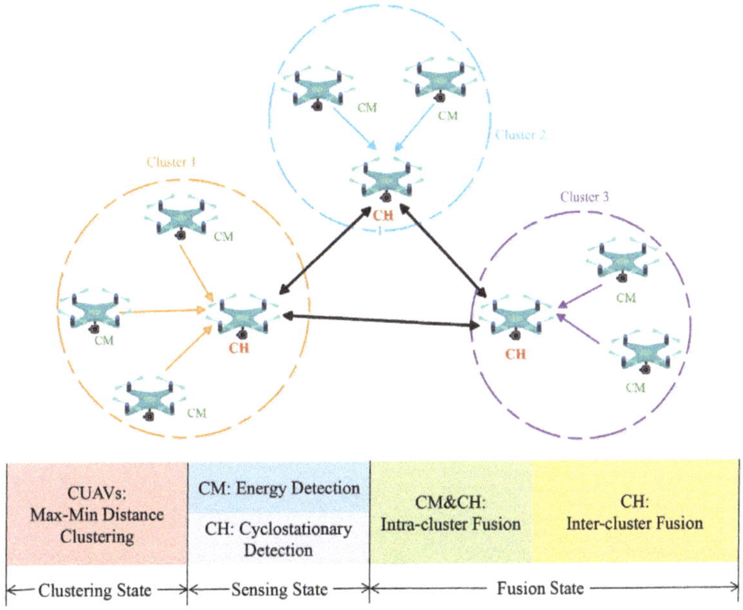

Figure 2. Heterogeneous Two-Stage-Fusion.

Table 1. Notations.

Notation	Description	Notatiom	Description
a	the amplitude of the transmitted signal	T_i	the likelihood ratio test statistic of the ith cluster
f_c	the carrier frequency of the transmitted signal	T_{ij}	the cyclostationary detection statistics or energy detection statistics
φ	the carrier phase offset of the transmitted signal	ω_{ij}	weight of $T_{E_{ij}}$
M	the number of sampling point	ρ_i	weight of T_{C_i}
α	the cyclic frequency of the received signal	γ_{C_i}	the SNR of the ith cluster head
T_C	the test statistic for the first-order cyclostationary detection	$\gamma_{E_{ij}}$	the SNR of the jth node in the ith cluster
T_E	the test statistic for energy detector	L	the Laplacian matrix of the cluster heads topology
T_{ic}	the statistics of the ith cluster	α_l	step size
T_{C_i}	the cyclostationary detection statistics of the ith cluster head	d	maximum node degree of the cluster graph
$T_{E_{ij}}$	the energy detection statistics of the jth the node in the ith cluster	δ_{h_i}	the weight of the ith cluster head
K	the number of CUAVs in the ith cluster	T^*	the final consensus
K_E	the number of cluster members in the ith cluster	W	the consensus weight matrix

3.2.1. Heterogeneous Spectrum Sensing Based on Clustering

To be specific, in the sensing state, a heterogeneous cooperative spectrum sensing scheme [36] is used according to the clustering result. Here, the "heterogeneous" means two different detection schemes: energy detection and cyclostationary detection [36]. Cluster heads adopt cyclostationary detection, while other secondary UAVs in the same cluster adopt energy detection. Cyclostationary feature detection is adopted to the cluster head, since it has great detection accuracy and can maintain good performance even in environments with low SNR. The cluster members (CMs) adopt energy detection, since it is simple to implement and does not require prior knowledge of the channels. Similarly to [36], we assume that the transmitted signal of the primary user is a sinusoidal signal, in which the n-th sample is expressed as

$$s(n) = ae^{j2\pi f_c n+\varphi}. \tag{1}$$

where a is the amplitude, f_c is the carrier frequency, and φ is the carrier phase offset. The received signal of the cluster heads is written as

$$y_C(n) = \begin{cases} \omega(n) & H_0 \\ s(n)+\omega(n) & H_1 \end{cases}, \tag{2}$$

where $\omega(n)$ denotes the additive Gaussian noise with zero mean and unit variance. H_0 indicates that the spectrum is absent, while H_1 indicates that the spectrum is occupied. The test statistic for the first-order cyclostationary detection [36] is expressed as

$$T_C = |\frac{1}{M}\sum_{n=1}^{M} y_{C(n)}e^{-j2\pi\alpha n}|, \tag{3}$$

where α is the cyclic frequency, M denotes the number of sampling point. The received signal $y_E(n)$ of the CUAV cluster member is modeled as the same as $y_C(n)$

$$y_E(n) = \begin{cases} \omega(n), & H_0 \\ s(n)+\omega(n), & H_1 \end{cases}. \tag{4}$$

The test statistic for the energy detector [36] is expressed as

$$T_E = \frac{1}{M}\sum_{n=1}^{M} |y_E(n)|^2, \tag{5}$$

where M denotes the number of sampling points. Let $T_{ic} = [T_{C_i}, T_{E_{i1}}, T_{E_{i2}}, \ldots, T_{E_{iK_E}}]$ denotes the vector of the statistics from the ith cluster, T_{C_i} is the cyclostationary detection statistics of the ith cluster head, and its distribution [36] can be expressed as

$$T_{C_i} \sim \begin{cases} \mathcal{FN}(\sqrt{\frac{2}{\pi M}}, \frac{\pi-2}{\pi M}), & H_0 \\ \mathcal{FN}(\zeta_i, 2\gamma_{C_i}+\frac{1}{M}-\zeta_i^2), & H_1 \end{cases}, \tag{6}$$

where $\zeta_i = \sqrt{\frac{2}{\pi M}}e^{-M\gamma_{C_i}} + \sqrt{2\gamma_{C_i}}(1-2Q(\sqrt{2M\gamma_{C_i}}))$, γ_{C_i} denotes the SNR of the ith cluster head, and $\mathcal{FN}$ is the folded normal distribution [37]. $T_{E_{ij}}$ is the energy detection statistics of the jth the node in the ith cluster. When M is large enough, $T_{E_{ij}}$ approximatively follows normal distribution according to the central limit theorem [38], but slightly different to [38], our $T_{E_{ij}}$ is $1/\mathrm{M}$ times the energy detection statistics in [38], thus the our SNR is

M times the SNR in [38], and we assume the energy of the received signal is 1. Thus, the distribution of $T_{E_{ij}}$ can be expressed as

$$T_{E_{ij}} \sim \begin{cases} \mathcal{N}(\gamma_{E_{ij}}{}^{-1}, \frac{2\gamma_{E_{ij}}{}^{-2}}{M}), & H_0 \\ \mathcal{N}(1+\gamma_{E_{ij}}^{-1}, \frac{2(\gamma_{E_{ij}}^{-2}+2\gamma_{E_{ij}}{}^{-1})}{M}), & H_1 \end{cases}, \tag{7}$$

where $\gamma_{E_{ij}}$ is the SNR of the jth node in the ith cluster.

3.2.2. Two-Stage-Fusion

In the fusion duration, we adopt a two-stage-fusion scheme [34], which includes the intra-cluster fusion stage and the inter-cluster fusion stage. The CUAV cluster members adopt energy detection, and we apply high-accuracy centralized soft fusion in the intra-cluster since the performance of energy detection is somewhat not precise. In the intra-cluster fusion stage, according to the assumption that the observations are independent, we can attain the likelihood ratio test (LRT) [36] of the ith cluster

$$\begin{aligned} T_i &= \prod_{j=1}^{K} \frac{P[T_{ij}|H_1]}{P[T_{ij}|H_0]} \\ &= \prod_{j=1}^{K_E} \frac{P[T_{E_{ij}}|H_1]}{P[T_{E_{ij}}|H_0]} \times \frac{P[T_{Ci}|H_1]}{P[T_{Ci}|H_0]}, \end{aligned} \tag{8}$$

there are K CUAVs, K_E CMs in the ith cluster, where $K_E = K-1$. T_{ij} denotes cyclostationary detection statistics or energy detection statistics, $P[T_{ij}|H_1]$ and $P[T_{ij}|H_0]$ represent the probability density under hypotheses H_1 and H_0, respectively. According to the $T_{E_{ij}}$, T_{C_i}, the LRT in Equation (8) can be simplified [36] as

$$T_i = \sum_{j=1}^{K_E} \omega_{ij} T_{E_{ij}} + \rho_i T_{Ci}. \tag{9}$$

where $\omega_{ij} = \frac{\gamma_{E_{ij}}}{2(1+\gamma_{E_{ij}})}$, and $\rho_i = \sqrt{2M^2\gamma_{C_i}}$.

As for the inter-cluster fusion stage, considering the good sensing performance of cyclostationary detection and the large distance between the cluster heads, distributed consensus-based fusion [38] is performed. Each cluster head communicates with its neighbouring cluster heads to exchange information, and the exchange process is iteratively done. The initial information (which is the intra-cluster fusion result) of the ith cluster head is denoted as $T_i(0)$. Then, according to the network topology, these cluster heads repeatedly iterate until $T_i(k)$ covers a common value. The consensus-based scheme [38] is

$$T_i(k+1) = WT_i(k). \tag{10}$$

where $W = I - \alpha\Delta^{-1}L$, and L is the Laplacian matrix of the cluster heads topology. α is the step size, and it satisfies $0 < \alpha < d^{-1}$, d is maximum node degree of the graph. $\Delta = diag\{\delta_{h_1}, \delta_{h_2}, \ldots, \delta_{h_n}\}$, where δ_{hi} [38] is the weight according to the channel condition of the ith cluster head, and it satisfies $\delta_{h_i} \geq 1$. The cluster heads communicate with their own neighbors, then a final consensus [38] is reached as

$$T^* = \lim_{k\to\infty} T(k) = \frac{\sum_{i=1}^{n} \delta_{h_i} T_i(0)}{\sum_{h_{i=1}}^{n} \delta_{h_i}}. \tag{11}$$

4. Continuous Hidden Markov Model Based Spectrum Sensing

To fully take advantage of the temporal correlation of the spectrum states, we propose a CHMM-based spectrum sensing scheme as shown in Figure 3. Firstly, hidden HMM and CHMM are introduced, and the suitability regarding the spectrum states as a continuous hidden Markov model are analyzed. Then, the model is trained with the forward-back algorithm and Baum–Welch algorithm. The prediction obtained according to CHMM model is used to assist the sensing.

4.1. Hidden Markov Model

The hidden Markov model is a double stochastic process with a hidden layer and an observable layer. The hidden process is an unobservable Markov chain, which can be obtained through the observed states.

The true states of the spectrum are not observable but the sensing results can be easily obtained. Therefore, the hidden Markov chain model fits well with the PU spectrum state. For HMM, there are three basic problems that need to be solved, i.e., the evaluation problem that computes the probability of the observed fusion result sequence, the learning problem that adjusts the model parameters to maximize the probability of the observed sequence, and the predication problem that calculates the most likely hidden spectrum state sequence according to the observation sequence and model parameters.

In order to avoid the distortion caused by the discretization of continuous variables in the cluster heads, we consider the continuous HMM, which replaces the discrete observation states with continuous characteristics. With more specific spectrum information, we can obtain better detection performance.

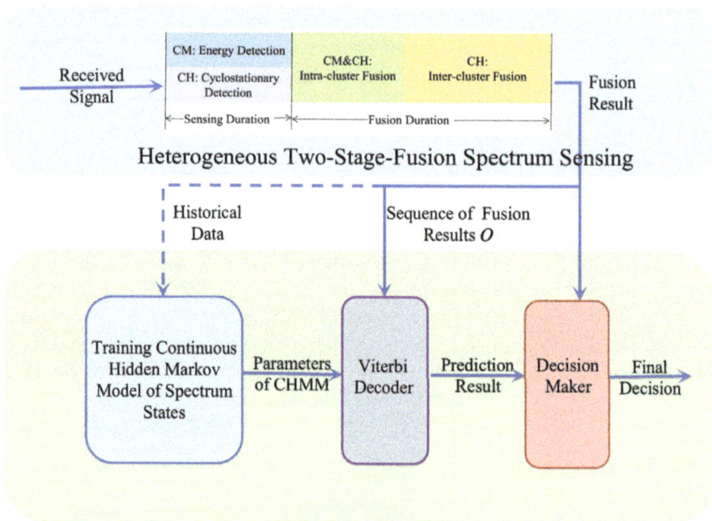

Figure 3. CHMM-based Spectrum Sensing.

4.2. Continuous Hidden Markov Model of Spectrum States

The PU spectrum state at time instant t is given by x_t, and it can be 0 or 1, where 0 represents spectrum absence, 1 denotes spectrum occupancy. The sequence of the PU states $X = (x_1, x_2, \ldots, x_t)$ can be seen as the hidden Markov chain. o_t is the fusion value of the heterogeneous two-fusion-stage spectrum sensing at time instant t, and $O = (o_1, o_2, \ldots, o_t)$ is the observable layer. The hidden Markov chain and the observable layer constitute a continuous hidden Markov model, which can be formulated as $\lambda = (\pi, A, \mu, \Sigma, C)$ [39], where π represents the initial probability vector of the hidden spectrum state, A is the transition matrix of the two states. The continuous hidden Markov model can be represented in Figure 4. μ, Σ, C are the parameters of the observation probability distribution. The Gaussian mixture model (GMM) is used to model the probability, as the Gaussian process has good adaptability in dealing with complex regression problems and classification. Thus, the observation probability in state i according to the GMM can be written as

$$b_i(o) = \sum_{m=1}^{M} C_{im} N(o, \mu_{im}, \Sigma_{im}), i = 0, 1, \tag{12}$$

which is composed of M Gaussian mixtures. i denotes the spectrum state. C_{im} is the proportion of the mth mixture coefficient in state i, o denotes the fusion result calculated by the clustered heterogeneous two-fusion-stage scheme, μ_{im} and Σ_{im} represent the mean and the covariance of mth mixture in state i, respectively.

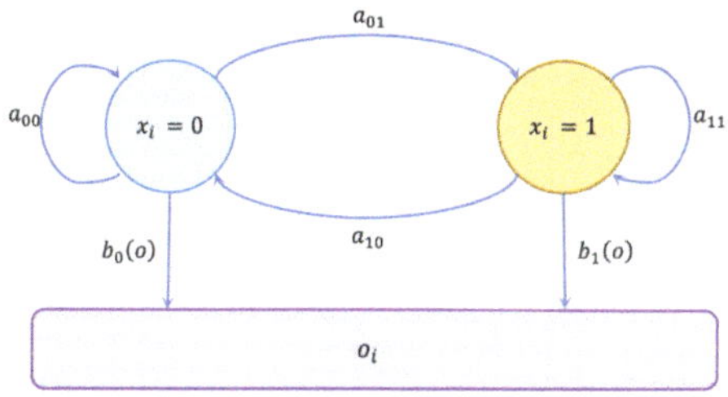

Figure 4. Continuous hidden Markov model.

To employ the continuous hidden Markov model to cognitive UAV networks, the forward-backward algorithm and Baum–Welch algorithm are utilized to solve the evaluation problem and the training problem, respectively. As for the prediction problem, the Viterbi algorithm is utilized.

4.3. *Evaluation Process and Learning Process of Continuous Hidden Markov Model*

For the evaluation problem, the forward-back algorithm is used to calculate the probability of the observed fusion value sequence, and it can be divided into two parts: the forward algorithm and backward algorithm. For a given λ and the spectrum state at time t, the forward quantity $\alpha_t(i)$ is the joint probability of sequence O from the initial time to time t, and the state in S_i at time t, $\beta_t(i)$ is the joint probability of sequence from time $t+1$ to the final time and the state in S_i at time t. The forward and backward quantities [39] are defined as

$$\alpha_t(i) = P(o_1, o_2, \ldots, o_t, q_t = S_i|\lambda), \tag{13}$$

$$\beta_t(i) = P(o_{t+1}, o_{t+2}, \ldots, o_T|q_t = S_i, \lambda), \tag{14}$$

with the initializations

$$\alpha_1(i) = \pi_i b_i(o_1), i = 0, 1, \tag{15}$$

$$\beta_T(i) = 1, i = 0, 1. \tag{16}$$

The forward and backward quantities can be calculated by

$$\alpha_t(j) = [\sum_{i=1}^{2} \alpha_t(i) a_{i,j}] b_j(o_t), t = 1, 2, \ldots, T, j = 0, 1, \tag{17}$$

$$\beta_t(i) = \sum_{j=1}^{2} a_{i,j} b_j(o_{t+1}) \beta_{t+1}(j), t = T-1, T-2, \ldots, 1, i = 0, 1. \tag{18}$$

Then, combining the forward and backward algorithm, we can get the forward-backward algorithm, then the probability of the observed fusion result sequence O with the given model parameters λ is obtained [39] as

$$P(O|\lambda) = \sum_{i=1}^{2} \alpha_t(i)\beta_t(i). \tag{19}$$

To solve the learning problem, the Baum–Welch algorithm is adopted, which is one of the expectation–maximization algorithms and uses the forward-backward algorithm in each expectation process. Before using this algorithm, we need to define three parameters: $\gamma_t(i)$, $\xi_t(i,j)$ and $\gamma_t(j,m)$. $\gamma_t(i)$ denotes the probability of the ith spectrum state at time t, $\xi_t(i,j)$ denotes the probability that at time t the spectrum state is S_i and at time $t+1$ the spectrum state is S_j, $\gamma_t(j,m)$ denotes the probability of the mth Gaussian mixture of state S_j at time t . $\gamma_t(i)$, $\xi_t(i,j)$ and $\gamma_t(j,m)$ can be calculated [39] as follows:

$$\gamma_t(i) = \frac{\alpha_t(i)\beta_t(i)}{\sum_{i=1}^{2} \alpha_t(i)\beta_t(i)}, \tag{20}$$

$$\xi_t(i,j) = \frac{\alpha_t(i)a_{ij}b_j(o_{t+1})\beta_{t+1}(j)}{\sum_{i=1}^{2}\sum_{j=1}^{2} \alpha_t(i)a_{ij}b_j(o_{t+1})\beta_{t+1}(j)}, \tag{21}$$

$$\gamma_t(j,m) = \frac{\alpha_t(j)\beta_t(j)}{\sum_{i=1}^{2} \alpha_t(i)\beta_t(i)} \cdot \frac{c_{j,m}N(o_t,\mu_{j,m},\Sigma_{j,m})}{\sum_{n=1}^{2} N(o_t,\mu_{j,m},\Sigma_{j,m})}. \tag{22}$$

When enough training data are provided, that is, the sequences of fusion results obtained from the heterogeneous two-stage-fusion sensing scheme and corresponding spectrum states, the Baum–Welch algorithm offers a way to train the model, outputting good CHMM parameters. In specific, initial model parameters are first selected according to the spectrum condition. Second, $\alpha_t(i)$, $\beta_{t+1}(j)$, $\gamma_t(i)$, $\xi_t(i,j)$ and $\gamma_t(j,m)$ are calculated. Third, the parameters are updated according to Appendix A. The forward-backward procedure and the updating procedure are repeated until the probability of the observation sequence $P(O|\lambda)$ satisfies the convergence condition or the increments of parameters are less than threshold. Finally, we can obtain the trained model parameters $\lambda = (\pi, A, \mu, \Sigma, C)$.

4.4. Predication of Spectrum State with CHMM

In this section, we adopt the Viterbi algorithm to solve the predication problem. With the learned model parameters and the observed sequence of fusion results, we can calculate the joint probability of the observed sequence. The real spectrum state sequence is calculated [39] as

$$\begin{aligned} P(O,X|\lambda) &= P(O|X,\lambda)P(X,\lambda) \\ &= \pi_{x_1}b_{x_1}(o_1)a_{x_1x_2}b_{x_2}(o_2)a_{x_{T-1}x_T}b_{x_T}(o_T), \end{aligned} \tag{23}$$

where x_t denotes the spectrum state at time t, $a_{x_tx_{t+1}}$ represents the transition probability from state x_t to x_{t+1}, $b_{x_t}(o_t)$ denotes the observation probability of o_t, when the real spectrum state is x_t. Then, select the sequence with the maximum probability as the prediction sequence

$$X = \underset{allX}{\text{argmax}}\, P(O,X|\lambda). \tag{24}$$

Next, we can get the prediction x_T and the prediction sequence X, and compare the predication result X with the real state sequence to get the prediction accuracy P_r. When the prediction is "busy", P_r can denote the probability that the spectrum is really occupied at time T, $1 - P_r$ can denote the probability that the spectrum is really absent at time T.

Similarly, when the prediction is "idle", the probability of occupancy is $1 - P_r$, and the probability of absence is P_r.

Combining the prediction accuracy of cluster heads with the fusion result of the detectors, we can get a new false alarm probability and detection probability. There are mainly two kinds of predictions: busy and idle. When the prediction is "busy", we multiply the fusion result T by ϵ $(\epsilon > 1)$. Similarly, when the prediction result is "idle", we multiply the fusion value by η $(\eta < 1)$. Then we obtain the final fusion statistic adjusted by the prediction. After that, we compare the final fusion statistic with the threshold. When the fusion statistic is larger than the threshold, the decision is "busy" and vice versa. Thus, the final detection probability P_D can be calculated as

$$P_D = P_r \cdot P_d(\epsilon T) + (1 - P_r) \cdot P_d(\eta T). \tag{25}$$

5. Analysis of Detection Performance for CHMM-Based Spectrum Sensing

Under H_1 means the spectrum is occupied. When the sampling points are large enough, $T_{E_{ij}}$ approximately follows normal distribution, as does T_{C_j} (which will be explained in Section 7.1). Thus, when the spectrum is occupied, the fusion result can be calculated as

$$\begin{aligned} T &= \sum_{i=1}^{n} \delta_{h_i} \left(\sum_{j=1}^{K_{iE}} \omega_{ij} T_{E_{ij}} + \rho_i T_{Ci}\right) \\ &= \sum_{i=1}^{n} \sum_{j=1}^{K_{iE}} \delta_{h_i} \omega_{ij} T_{E_{ij}} + \sum_{i=1}^{n} \delta_{h_i} \rho_i T_{Ci}, \end{aligned} \tag{26}$$

where n is the number of clusters, δ_{hi} is the weight according to the channel condition of the ith cluster head, K_{iE} is the number of cluster members (CMs) in the ith cluster. According to [36], the weight can be simplified as $\omega_{ij} = \gamma_{E_{ij}} / 2(1 + \gamma_{E_{ij}})$, and $\rho_i = \sqrt{2M^2 \gamma_{C_i}}$. The fusion result T is an approximately normally distributed random variable with mean μ_T and variance ${\sigma_T}^2$, the mean and the variance are expressed as

$$\begin{aligned} \mu_T &= \sum_{i=1}^{n} \sum_{j=1}^{K_{iE}} \delta_{h_i} \omega_{ij} (1 + {\gamma_{E_{ij}}}^{-1}) + \sum_{i=1}^{n} \delta_{h_i} \rho_i \sqrt{2\gamma_{C_i}} \\ &= \sum_{i=1}^{n} \sum_{j=1}^{K_{iE}} \frac{\delta_{h_i}}{2} + \sum_{i=1}^{n} 2M \delta_{h_i} \gamma_{Ci}, \end{aligned} \tag{27}$$

and

$${\sigma_T}^2 = \sum_{i=1}^{n} \sum_{j=1}^{K_{iE}} \frac{{\delta_{h_i}}^2 (1 + 2\gamma_{E_{ij}})}{2M(1 + \gamma_{E_{ij}})^2} + \sum_{i=1}^{n} 2M {\delta_{h_i}}^2 {\gamma_{Ci}}^2. \tag{28}$$

Then we can obtain the final detection probability P_D of the CUAVNs,

$$P_D = P_r \cdot Q\left(\frac{\frac{\lambda_T}{\epsilon} - \mu_T}{\sigma_T}\right) + (1 - P_r) \cdot Q\left(\frac{\frac{\lambda_T}{\eta} - \mu_T}{\sigma_T}\right), \tag{29}$$

where λ_T is the threshold.

Under H_0, when M is large enough, $T_{E_{ij}}$ approximately follows normal distribution, and the probability density $T_{C_{ij}}$ can be approximately represented as

$$f(T_{C,i}) = \frac{2}{\sqrt{2\pi} \sigma_{T_{C,i}}} \exp\left\{ -\frac{{T_{C,i}}^2}{2{\sigma_{T_{C,i}}}^2} \right\}. \tag{30}$$

The probability density of the fusion result T can be represented as

$$f(T) = \frac{n}{\sqrt{2\pi\sigma^2}} \exp\left\{-\frac{(T-\gamma)^2}{2\sigma^2}\right\} \times erfc\left(\frac{\sigma_C(T-\gamma)}{\sqrt{2\sigma_E{}^2\sigma^2}}\right), \tag{31}$$

where $\gamma = \sum_{i=1}^{n}\sum_{j=1}^{K_E} \delta_{h_i}/\{2(1+\gamma_{E_{ij}})\}$, $\sigma^2 = \sigma_E{}^2 + \sigma_C{}^2$, $\sigma_C{}^2 = \sum_{i=1}^{n} 2M\delta_{h_i}{}^2\gamma_{C_i}$, $\sigma_E{}^2 = \sum_{i=1}^{n}\sum_{j=1}^{K_E} \delta_{h_i}{}^2/\{2M(1+\gamma_{E_{ij}})^2\}$. The proof of $f(T)$ is shown in Appendix B.

The false alarm probability can be calculated as follows, where $F(T) = \int_0^T f(x)dx$,

$$\begin{aligned} P_F &= P_r \cdot (1 - F(\eta T)) + (1 - P_r) \cdot (1 - F(\epsilon T)) \\ &= 1 - P_r F(\eta T) - (1 - P_r) F(\epsilon T). \end{aligned} \tag{32}$$

6. CHMM-Based Spectrum Sensing with Practical SNR Estimation

The above work is based on the assumption that the SNR is perfectly known, i.e., we use a perfectly-known SNR when calculating the fusion weight, detection probability and false alarm probability. In this section, we further consider the scenario that the SNR is not perfectly known, and design a novel SNR estimation algorithm: space smoothing-based M2M4 (SS-M2M4), which modifies M2M4 with spatial smoothness, a technique that is used in the field of signal reconstruction in graphs. This algorithm can provide a more accurate SNR for the proposed CHMM-based spectrum sensing method.

6.1. Typical SNR Estimator

M2M4 is one of the most widely used blind estimators [14]. M2 and M4 represent the second and the fourth moment of y_n, respectively, where y_n refers to the samples of the received signal. M2 and M4 can be calculated as follows [14]

$$M_2 = E\{y_n y_n^*\}, \tag{33}$$

$$M_4 = E\{(y_n y_n^*)^2\}. \tag{34}$$

Then with second-order moments and fourth-order moments, we can estimate SNR as follows,

$$\hat{\rho} = \frac{\sqrt{2M_2^2 - M_4}}{M_2 - \sqrt{2M_2^2 - M_4}}. \tag{35}$$

In practice, the second and fourth moments are usually calculated by their own time averages:

$$M_2 = \frac{1}{M}\sum_{n=0}^{M} y_n{}^2, \tag{36}$$

$$M_4 = \frac{1}{M}\sum_{n=0}^{M} y_n{}^4. \tag{37}$$

6.2. SS-M2M4 SNR Estimation Algorithm

In CUAVNs, the CUAVs within the same cluster are close to each other, and thus, their large-scale fading and shadow fading are generally similar [40]. Therefore, their SNRs are correlated. However, the M2M4 algorithm estimates these SNRs individually, ignoring the spatial correlation among the SNRs of those neighbouring users. CUAVNs have their own topology, and the SNR estimation problem can be naturally modeled as graph signal processing problems. Therefore, we propose a novel SNR estimation algorithm as shown in Algorithm 1, which considers the smoothness between neighbors [15]. Firstly, the M2M4 algorithm is applied to calculate the SNR of each CUAV, and the estimation result can be represented as $\mathbf{p}_0 = \{p_{01}, p_{02}, \ldots, p_{0N}\}$. After that, to ensure that the secondary CUAVs that are close to each other are in the same cluster, the max–min distance clustering algorithm described in Section 3.1 is adopted. Based on the clustering result, we consider each cluster as a new graph, and then correct the original SNR to get the final estimated SNR $\hat{\mathbf{p}} = \{\hat{p}_1, \hat{p}_2, \ldots, \hat{p}_N\}$ by Equation (42).

The problem of estimating $\hat{\mathbf{p}}$ from the original SNR $\mathbf{p}_0$ can be modeled as the following optimization problem,

$$\min_{\hat{\mathbf{p}}} \quad \frac{1}{2} \| \mathbf{p}_0 - \hat{\mathbf{p}} \|_2^2 + \frac{\rho}{2} \| \mathbf{H}\hat{\mathbf{p}} \|^2, \tag{38}$$

where $\mathbf{H}$ is a high-pass graph filter, and ρ is the regularization parameter. The first term penalizes the error of the estimated graph signal, the second term encourages the smoothness of the estimated SNR. Similar to [15], we set $\mathbf{H} = \mathbf{L}^{1/2}$, where $\mathbf{L}$ is the Laplacian matrix, so that $\| \mathbf{H}\hat{\mathbf{p}} \|_2^2 = \hat{\mathbf{p}}^T \mathbf{L} \hat{\mathbf{p}}$. The smoothness of the estimated graph signal can be characterized by the graph Laplacian quadratic form

$$S(\hat{\mathbf{p}}) = \hat{\mathbf{p}}^T \mathbf{L} \hat{\mathbf{p}}. \tag{39}$$

It can be written as

$$S(\hat{\mathbf{p}}) = \sum_{(i,j \in \varepsilon)} W_{ij} (\hat{p}_j - \hat{p}_i)^2, \tag{40}$$

where ε is the set of edges. The smaller the function value $S(\hat{\mathbf{p}})$ is, the smoother the SNR difference of the cluster, especially when neighboring CUAVs connected by an edge with a large weight have similar values [15]. We can then represent the optimization problem as

$$\min_{\hat{\mathbf{p}}} \quad \frac{1}{2} \| \mathbf{p}_0 - \hat{\mathbf{p}} \|_2^2 + \frac{\rho}{2} \hat{\mathbf{p}}^T \mathbf{L} \hat{\mathbf{p}}. \tag{41}$$

To get the best estimated SNRs, we take the derivation of the formula, and get the optimal solution as

$$\hat{\mathbf{p}} = (\mathbf{1} + \rho \mathbf{L})^{-1} \mathbf{p}_0. \tag{42}$$

Algorithm 1 Space smoothing-based M2M4 SNR estimation algorithm

1: **Input:** The received signal sample of each CUAV $\{y_{i1}, y_{i2}, \ldots, y_{in}\}$.

2: **Output:** The SNR of all the N CUAVs $\hat{\mathbf{p}} = \{p_1, p_2, \ldots, p_N\}$.

Step 1: M2M4 SNR Estimation

3: Calculate SNR of all the CUAVs with the traditional M2M4 algorithm in Section 6.1, obtain $\{p_{01}, p_{02}, \ldots, p_{0N}\}$.

Step 2: Space Smoothing

4: Adopt the Max-Min distance clustering algorithm, obtain the clustering result.

5: With the clustering result, calculate the SNR of CUAVs in each cluster by Equation (34), then obtain the final estimated SNR $\hat{\mathbf{p}} = \{\hat{p}_1, \hat{p}_2, \ldots, \hat{p}_N\}$.

6.3. CHMM-Based Spectrum Sensing with SS-M2M4 SNR Estimation

The estimated SNRs can provide better SNR information for the CHMM-based spectrum sensing procedure, and thus we can obtain more accurate fusion weight in a heterogenous two-stage-fusion stage. Therefore, the CHMM-based spectrum sensing scheme with estimated SNRs achieves a higher detection probability, and the utilization of spectrum sensing can be further enhanced.

The proposed continuous hidden Markov model based spectrum sensing with space smoothing M2M4 SNR estimator can achieve good detection performance in CUAVNs, since it has some advantages over the existing HMM-based spectrum sensing method and the existing SNR estimators. The proposed method can either achieve better performance or be more suitable for UAV applications. The characteristics of these existed methods compared to the proposed method are summarized in Table 2.

Table 2. The characteristics of the existing methods.

Methods	Characteristics	Examples
DHMM-based spectrum sensing	Errors caused by quantization degrade detection performance	Suguna et al. [21], Eltom et al. [22]
CHMM-based spectrum sensing	Centralized spectrum sensing and lack of dynamicity, not suitable for CUAVNs	Halaseh et al. [23], Cheng et al. [24]
Deep learning-based spectrum sensing	Poor interpretability and missing of the sensing term caused by the large delay	Xie et al. [16], Xie et al. [17]
SNR estimator with pre-knowledge	Hard to get pre-knowledge of channel in UAVs applications	Raza et al. [26], Manesh et al. [27]
SNR estimator designed for specific signals	Not generalized for UAVs applications	He et al. [28], Chen et al. [29]
Deep learning-based SNR estimator	Missing of the sensing term caused by the long estimation time	Ngo et al. [31], Yang et al. [32]

7. Evaluation and Numerical Results

In this section, the validity of the approximation of folded normal distribution is firstly presented. Then, the detection performance of the proposed CHMM-based spectrum sensing scheme is evaluated by comparing with the non-CHMM ones. Next, we verify the effectiveness of the proposed SS-M2M4 estimator and further demonstrate the performance of the unified CHMM-based spectrum sensing with the SS-M2M4 estimator. Finally, we further consider mutipath effects in CUAVNs, and verify the effectiveness of the proposed scheme under the Rice channel.

7.1. Approximation of Folded Normal Distribution

The cluster heads adopt cyclostationary detection to sense the primary spectrum. Under the hypotheses H_1, the cyclostationary detection statistic x follows the folded normal distribution, which can be represented as

$$f(x) = \frac{1}{\sqrt{2\pi\frac{1}{M}}}\exp\left\{\frac{(x-\sqrt{2\gamma})^2}{2\frac{1}{M}}\right\} + \frac{1}{\sqrt{2\pi\frac{1}{M}}}\exp\left\{\frac{(x+\sqrt{2\gamma})^2}{2\frac{1}{M}}\right\}, x > 0. \quad (43)$$

Figure 5 shows the probability density function of the cyclostationary feature under hypotheses H_1 when $SNR = -15$ dB, $M = 2048$. As shown in Figure 5, the folded normal distribution mainly coincides with the normal distribution. When x is less than 0.0025, which is already 10.8σ away from the mean, the two distributions start diverging. In fact, the two terms of folded normal distribution $f(x)$ can be seen as two normal distributions with opposite means and the same variance. In spectrum sensing, when the channel is occupied, the mean $\sqrt{2\gamma}$ is away from 0, and it is much larger than the variance. Thus, when $x > 0$, the second term of $f(x)$ contributes little to the folded normal distribution. Therefore, in CUAVNs, folded normal distribution can be seen as its first term, that is, a normal distribution in the positive axis.

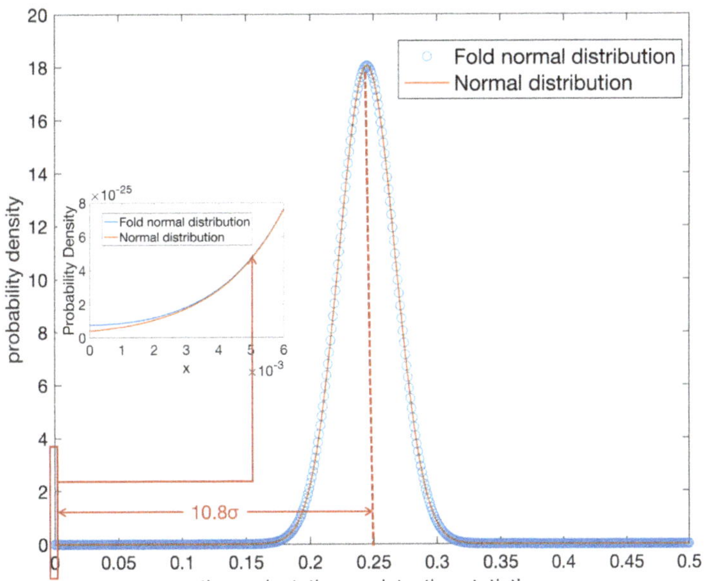

Figure 5. Approximation of folded distribution.

7.2. *Performance of CHMM-Based Spectrum Sensing with Perfect SNR Estimation*

In the simulations, the number of secondary CUAVs is set to 20. The CUAVs move according to the random walk mobility model [34], in which the maximum velocity of the CUAVs is 36km/h, and the sensing time is 20 μs. The range of the SNR in our CUAVNs is set as $[-15, -3]$ dB according to [32,33,41]. Locations of secondary nodes lead to different SNRs, we assume that the maximum distance from the transmitter to the secondary UAV is about 1.5 to 2 times the minimum [1]. According to the large-scale fading calculation formula [40], assuming that the fading coefficient is 2, we can obtain 5 dB as a SNR range in our simulation, in other words, the maximum SNR difference of secondary users is within the range of 5 dB. We assume that all secondary CUAVs experience additive white Gaussian noise.

According to [42], the spectrum utilization rate below 3G (The Federal Communications Commission (FCC) of the United States and the European Union (EU) have set 2.4 GHz and 5.8 GHz as the band of civil UAVs. The EU also allocates 433 MHz and 863–870 MHz to UAVs. Similarly, China has set 840.5–845 MHz, 1438–1444 MHz and 2408–2440 MHz as that for UAVs. Compared to 5.8 GHz, more UAVs work on lower bands, below 3 GHz, thus the dilemma of spectrum scarcity is more serious on the below-3G bands) in Berkeley is about 0.3, therefore, we set the initial distribution of the spectrum state distribution as $\pi = (0.7, 0.3)^T$, that is to say, the probability of spectrum presence is 0.3, and the probability of spectrum absence is 0.7. We assume a 1st order Markov chain, and the spectrum state at time t is known, where the distribution is either $(1,0)^T$ or $(0,1)^T$. As stated before, the spectrum utilization is around 0.3, the transition probability of absence to presence is thus set as 0.25, and in the same way, the transition probability of presence to presence is set as 0.35. Therefore, the transition matrix is $A = [0.75, 0.25; 0.65, 0.35]$. Next, we use MATLAB to generate a spectrum state sequence with a length of 8000 under the parameters above. According to each single spectrum state (hidden state) and different clustering results, the simulated energy detection statistics $T_{E_{ij}}$ and cyclostationary statistics $T_{C,i}$ are obtained, respectively. Then corresponding observation values o_t are calculated according to the two-step-fusion method.

Figures 6 and 7 show the received operating curve (ROC) of the CHMM-based sensing method and non-CHMM-based ones under the AWGN channel with 20 CUAVs. In Figure 6, the soft–soft represents the heterogeneous two-stage sensing scheme represented in Section 2.2, in which both the intra-cluster fusion stage and the inter-cluster fusion stage adopt a soft combining rule [43]. It can be observed from Figure 6 that the soft–soft heterogeneous sensing scheme with CHMM predication outperforms the non-predication one (soft–soft). Due to CHMM avoiding the distortion caused by the discretization of DHMM, the proposed CHMM-based sensing scheme can further improve the detection probability compared with the DHMM-based soft–soft scheme, which adopts DHMM to model the sensing procedure. What stands out in Figure 6 is the achieved high detection probability of 0.91 when false alarm probability is around 0.1. A higher P_D with small P_F indicates that the proposed algorithm can offer the secondary CUAVs more opportunities to access the spectrum and maintain tolerable interruption to the primary user.

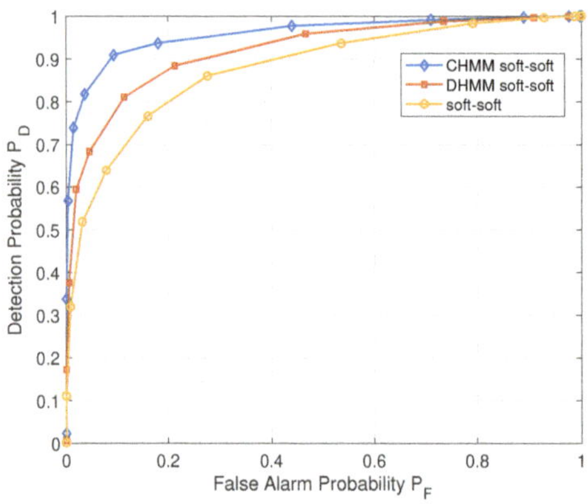

Figure 6. ROC of CHMM soft–soft scheme (the proposed CHMM-based heterogeneous two-stage fusion sensing scheme), DHMM soft–soft scheme and non-CHMM soft–soft scheme.

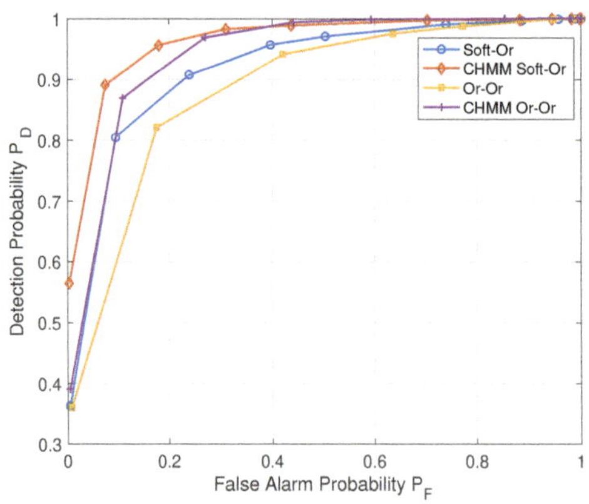

Figure 7. ROC of soft–or, or–or scheme and CHMM soft–or, or–or scheme.

In addition, we also simulate the soft–or and or–or schemes [36] to verify the universality of the CHMM model for spectrum sensing in Figure 7, where soft–or means a soft combining rule at the intra-cluster fusion stage and/or a combining rule [43] at the inter-cluster fusion stage. Similarly, or–or means an or combining rule at both the intra-cluster fusion stage and the inter-cluster fusion stage. As shown in Figure 7, in addition to the good detection performance offered by the heterogeneous soft–soft scheme, the proposed CHMM can also achieve obvious improvement when it is implemented into the other two fusion schemes: or–or and soft–or. Then, with the help of a better sensing performance, the spectrum efficiency and throughput can be further improved.

7.3. Performance of CHMM-Based Spectrum Sensing with SS-M2M4 SNR Estimator

In this section, the simulation evaluation of our proposed SS-M2M4 in comparison with the existing M2M4 algorithm is presented. The simulation results with SNR in $[-15, -3]$ are shown in Figure 8. Note that the SNR = -15 dB represents the SNR range from $[-15, -10]$. Similarly, -3 dB represents the range $[-3, 2]$. MSE is used to measure the effect of SNR estimation, and the calculation scheme of MSE is given as

$$MSE = \frac{1}{N}\sum_{i=1}^{N}(p_i - \hat{p}_i)^2, \tag{44}$$

where $\hat{p}_i$ represents the estimated SNR of the ith CUAV, p_i represents the actual SNR of the ith CUAV. Figure 8 shows that the proposed scheme offers a good SNR error decrease. It can be seen that Figure 8 also shows that the proposed scheme can improve the performance of SNR estimation up to 3dB (when a 2.3×10^{-3} MSE is required, the M2M4 estimator needs the actual SNR to be -6 dB, but SS-M2M4 only needs it to be -9 dB). An accurate SNR estimator can further help the CUAV get a better detection performance. Consequently, it is reasonable to expect good detection performance when the CHMM-based spectrum sensing adopts the novel SNR estimation algorithm.

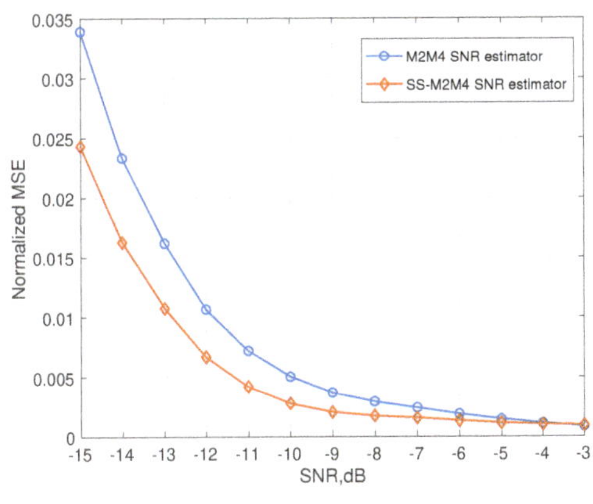

Figure 8. MSE of SS-M2M4 estimator and M2M4 estimator.

Figure 9 shows the CHMM-based spectrum sensing with the SS-M2M4 estimator significantly outperforming the original ones, which either do not use the CHMM model (spectrum sensing + SS-M2M4 SNR estimator) or employ the M2M4 estimator (CHMM spectrum sensing + M2M4 SNR estimator), or neither (spectrum sensing + M2M4 SNR estimator). Specifically, when the false alarm is 0.1, the unified scheme achieves a detection probability of 0.95, while the CHMM-based spectrum sensing with the original SNR estimator (CHMM spectrum sensing + M2M4 SNR estimator) can only achieve around 0.82. Besides, the unified scheme can further enhance the detection performance compared with the DHMM-based spectrum sensing with the proposed SS-M2M4 (DHMM spectrum sensing + SS-M2M4 SNR estimator), and the DHMM-based spectrum sensing with the proposed SNR estimator can also outperform the DHMM-based spectrum sensing with the original SNR estimator (DHMM spectrum sensing + M2M4 SNR estimator). In other words, the proposed SNR estimator has a universality to a different spectrum sensing method and the unified scheme can utilize the spectrum more efficiently and maintain a tiny interference to the primary user.

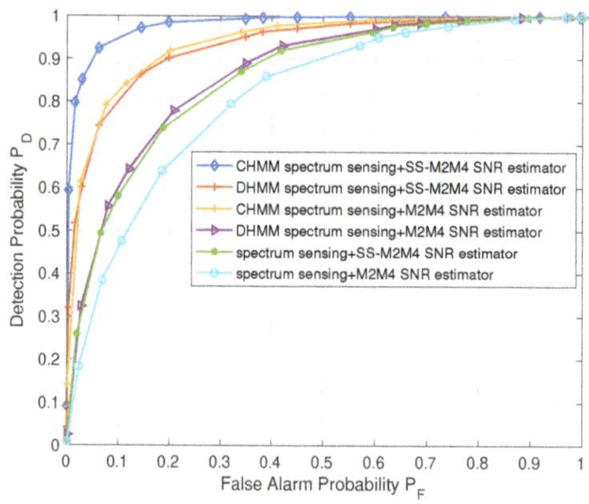

Figure 9. Detection probability versus false alarm probability.

7.4. Performance of CHMM-Based Spectrum Sensing with SS-M2M4 SNR Estimator under Rice Channel

In the above simulation, we assume the AWGN channel and verify the effectiveness of the proposed scheme under the AWGN channel. In this section, we have further considered the Rice channel when mutipath effects have been introduced in UAV applications [44]. The simulation of the proposed scheme under the Rice channel is shown in Figure 10, according to [44], we set the Rician factor as K = 10. Figure 10 shows that even under the Rice channel, the proposed method (Rice CHMM sensing+SS-M2M4 estimator) can obtain better performance compared with DHMM-based spectrum sensing with the proposed SS-M2M4 SNR estimator (Rice DHMM sensing + SS-M2M4 estimator), CHMM-based spectrum sensing with M2M4 SNR estimator (Rice CHMM sensing+M2M4 estimator), and other methods. Although the detection probability under the Rice channel is lower than that under the AWGN channel, the proposed scheme can also achieve better performance than other schemes under the Rice channel, in other words, our scheme is effective in CUANs.

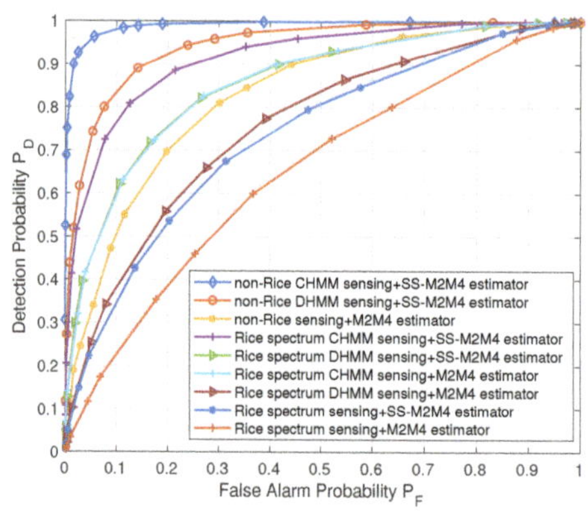

Figure 10. Detection probability versus false alarm probability under the Rice channel.

8. Conclusions

In this paper, we consider modeling primary user states as a Markov chain, and propose a spectrum sensing scheme based on a continuous hidden Markov model with perfect SNR estimation. We derive the detection probability and false alarm probability of the heterogenous-fusion clustered CUAVNs. Taking the similarity of the SNRs of CUAVs in the same cluster into account, we propose a space smoothing based SNR estimator for sensing in CUAVNs to offer a more accurate SNR to the proposed sensing method. Simulation results show that the unified CHMM-based sensing scheme with the proposed SNR estimator enhances the sensing performance considerably.

The work can be further extended in the following aspects in our future research. First, in the proposed CHMM based spectrum sensing method, the hidden spectrum chain is modeled by first-order Markov chain, and it has not made full use of the existing historical information. In the future research, the high-order Markov chain can be used to model the spectrum sensing process to further improve the accuracy of prediction. Second, in this paper, we considered cognitive UAV networks with a single primary user. Recently, cognitive UAV networks with multi primary users [33] have been proposed to enhance the spectrum utilization. Thus, we can explore our work in multi-PUs CUAVNs in the future.

Author Contributions: Conceptualization and methodology, Y.F. and W.X.; software and validation, Y.F.; formal analysis, Y.F.; investigation, Y.F., W.X. and Z.Z.; resources, Y.F.; data curation, Y.F.; writing—original draft preparation, Y.F.; writing—review and editing, Y.F. and F.W.; visualization,Y.F. and F.W.; supervision, W.X.; project administration, Z.Z.; funding acquisition, W.X. All authors have read and agreed to the published version of the manuscript.

Funding: This work was supported in part by the National Key Research and Development Program of China under Grant 2019YFC1511302; and in part by the National Natural Science Foundation of China under Grant 61871057.

Institutional Review Board Statement: Not applicable.

Informed Consent Statement: Not applicable.

Data Availability Statement: Not applicable.

Conflicts of Interest: The authors declare no conflict of interest.

Appendix A. Formulas of the Updated CHMM Parameters

The parameters of the CHMM [39] $\lambda = (\pi, A, \mu, \Sigma, C)$ can be updated with parameters $\gamma_t(i)$, $\xi_t(i,j)$ and $\gamma_t(j,m)$ as follows

$$\pi_i = \gamma_t(i), \tag{A1}$$

$$a_{ij} = \frac{\sum_{t=1}^{T-1} \delta_t(i,j)}{\sum_{t=1}^{T-1} \gamma_t(i)}, \tag{A2}$$

$$c_{j,m} = \frac{\sum_{t=1}^{T} \gamma_t(j,m)}{\sum_{t=1}^{T} \sum_{n=1}^{2} \gamma_t(j,n)}, \tag{A3}$$

$$\mu_{j,m} = \frac{\sum_{t=1}^{T} \gamma_t(j,m) o_t}{\sum_{t=1}^{T} \gamma_t(j,m)}, \tag{A4}$$

$$\Sigma_{j,m} = \frac{\sum_{t=1}^{T} \gamma_t(j,m)(o_t - \mu_{j,m})(o_t - \mu_{j,m})^T}{\sum_{t=1}^{T} \gamma_t(j,m)}. \tag{A5}$$

Appendix B. Proof of Distribution of the Fusion Value

In this section, we derive the distribution of the fusion result T, where $T = T_E + T_C$. To obtain the fusion result T, we first add the energy detection result T_{Eij} and the cyclostationary detection result T_{C_i}, respectively, and then calculate the summation of the two different detection results. When M is large enough, the sum of the energy detection T_E follows normal distribution written as

$$f(T_E) = \frac{1}{\sqrt{2\pi\sigma_E{}^2}} e^{-\frac{(T_E-\gamma)^2}{2\sigma_E{}^2}}, \tag{A6}$$

where mean $\gamma = \sum_{i=1}^{n}\sum_{j=1}^{K_E}\delta_{h_i}/\{2(1+\gamma_{E_{ij}})\}$, variance $\sigma_E{}^2 = \sum_{i=1}^{n}\sum_{j=1}^{K_E}\delta_{h_i}{}^2/\{2M(1+\gamma_{E_{ij}})^2\}$. To obtain the distribution which the sum of cyclostationary detection follows, we do the following derivation. For simplicity, x denotes one cyclostationary detection result T_{C_i}, y denotes another cyclostationary detection result T_{C_j}, and the distribution of x and y can be represent as

$$f(x) = \frac{2}{\sqrt{2\pi}\sigma} e^{-\frac{x^2}{2\sigma^2}}, \tag{A7}$$

$$f(y) = \frac{2}{\sqrt{2\pi}\sigma} e^{-\frac{y^2}{2\sigma^2}}, \tag{A8}$$

with the same mean 0 and the same variance $\sigma^2 = \frac{2}{M}$, $z = ax + by$ is defined as the sum of x and y, we can derive the distribution of z as

$$\begin{aligned}
f(z) &= \int_0^{+\infty} \frac{1}{a} f_x\left(\frac{z-by}{a}\right) f_y(y)dy \\
&= \int_0^{+\infty} \frac{1}{a}\frac{2}{\sqrt{2\pi}\sigma} \exp\left\{-\frac{\left(\frac{z-by}{a}\right)^2}{2\sigma^2}\right\} \frac{2}{\sqrt{2\pi}\sigma} \exp\left\{-\frac{y^2}{2\sigma^2}\right\}dy \\
&= \left(\frac{2}{\sqrt{2a\pi}\sigma}\right)^2 \exp\left\{-\frac{\frac{z^2}{a^2+b^2}}{2\sigma^2}\right\} \int_0^{+\infty} \exp\left\{-\frac{\left(\sqrt{a^2+b^2}y - \frac{bz}{\sqrt{a^2+b^2}}\right)^2}{2a^2\sigma^2}\right\}dy \\
&= \left(\frac{2}{\sqrt{2a\pi}\sigma}\right)^2 \int_{-\frac{bz}{\sqrt{2}a\sqrt{a^2+b^2}\sigma}}^{+\infty} \exp\left\{-t^2\right\}dt \frac{\sqrt{2}a\sigma}{\sqrt{a^2+b^2}} \exp\left\{-\frac{z^2}{2(a^2+b^2)\sigma^2}\right\} \\
&= \left(\frac{2}{\sqrt{2a\pi}\sigma}\right)^2 \frac{\sqrt{2}a\sigma}{\sqrt{a^2+b^2}} \exp\left\{-\frac{z^2}{2(a^2+b^2)\sigma^2}\right\} \int_{-\infty}^{+\infty} \exp\left\{-t^2\right\}dt \\
&= \frac{4}{\sqrt{2\pi(a^2+b^2)\sigma^2}} \exp\left\{-\frac{z^2}{2(a^2+b^2)\sigma^2}\right\},
\end{aligned} \tag{A9}$$

where $t = \frac{\sqrt{a^2+b^2}y - \frac{bz}{\sqrt{a^2+b^2}}}{\sqrt{2}a\sigma}$, $y = \frac{\sqrt{2}a\sigma t + \frac{bz}{\sqrt{a^2+b^2}}}{\sqrt{a^2+b^2}}$. From the above proof, we can conclude that the distribution of z is similar with x and y, but there exist some differences: the coefficient has doubled, the variance is the sum of σ_x and σ_y. Thus T_C follows the following distribution

$$f(T_C) = \frac{2n}{\sqrt{2\pi\sigma_C{}^2}} \exp\left\{-\frac{T_C{}^2}{2\sigma_C{}^2}\right\}. \tag{A10}$$

After obtaining the distribution of cyclostationary detection, we can further derive the distribution of the fusion result T as follows,

$$
\begin{aligned}
f(T) &= \int_0^{+\infty} f_{T_E}(T-T_C) f_{T_C}(T_C) dT_C \\
&= \frac{2n}{\sqrt{2\pi\sigma_E^2}\sqrt{2\pi\sigma_C^2}} \int_0^{+\infty} \exp\left\{-\frac{T_C^2}{2\sigma_C^2}\right\} \exp\{-\frac{(T-T_C-\gamma)^2}{2\sigma_E^2}\} dT_C \\
&= \frac{2n}{2\pi\sqrt{\sigma_E^2\sigma_C^2}} \exp\left\{-\frac{(T-\gamma)^2}{2\sigma^2}\right\} \int_0^{+\infty} \exp\left\{\frac{-\left[\sigma T_C - \frac{\sigma_C^2(T-\gamma)}{\sigma}\right]^2}{2\sigma_E^2\sigma_C^2}\right\} dT_C \\
&= \frac{2n}{2\pi\sqrt{\sigma_E^2\sigma_C^2}} \exp\left\{-\frac{(T-\gamma)^2}{2\sigma^2}\right\} \frac{2\sqrt{\sigma_E^2\sigma_C^2}}{\sigma} \int_0^{+\infty} \exp\left\{-t^2\right\} dt \\
&= \frac{n}{\sqrt{2\pi\sigma^2}} \exp\left\{-\frac{(T-\gamma)^2}{2\sigma^2}\right\} erfc\left(\frac{\sigma_C(T-\gamma)}{\sqrt{2\sigma_E^2\sigma^2}}\right),
\end{aligned}
\quad (A11)
$$

where n is number of the clusters, $t = \left[\sqrt{\sigma_E^2 + \sigma_C^2} T_C - \frac{\sigma_C^2(T-\gamma)}{\sqrt{\sigma_E^2+\sigma_C^2}}\right] / \sqrt{2\sigma_E^2\sigma_C^2}$, $\sigma_C^2 = \sum_{i=1}^{n} 2M\delta_{h_i}^2 \gamma_{C_i}$, $\sigma^2 = \sigma_E^2 + \sigma_C^2$. Finally, the distribution of the fusion result T under hypotheses H_0 can be obtained as

$$
f(T) = \frac{n}{\sqrt{2\pi\sigma^2}} \exp\left\{-\frac{(T-\gamma)^2}{2\sigma^2}\right\} erfc\left(\frac{\sigma_C(T-\gamma)}{\sqrt{2\sigma_E^2\sigma^2}}\right). \quad (A12)
$$

References

1. Mozaffari, M.; Saad, W.; Bennis, M.; Nam, Y.-H.; Debbah, M. A tutorial on UAVs for wireless networks: Applications, challenges, and open problems. *IEEE Commun. Surv. Tutor.* **2019**, *21*, 2334–2360. [CrossRef]
2. Kumar, K.; Kumar, S.; Kaiwartya, O.; Sikandar, A.; Kharel, R.; Mauri, J.L. Internet of unmanned aerial vehicles: QoS provisioning in aerial ad-hoc networks. *Sensors* **2020**, *20*, 3160. [CrossRef]
3. Shen, F.; Ding, G.; Wang, Z.; Wu, Q. UAV-based 3D spectrum sensing in spectrum- heterogeneous networks. *IEEE Trans. Veh. Technol.* **2019**, *68*, 5711–5722. [CrossRef]
4. Lu, Y.; Huang, X.; Zhang, K.; Maharjan, S.; Zhang, Y. Low-latency federated learning and blockchain for edge association in digital twin empowered 6G networks. *IEEE Trans. Ind. Inform.* **2020**, *17*, 5098–5107. [CrossRef]
5. Lu, Y.; Huang, X.; Zhang, K.; Maharjan, S.; Zhang, Y. Communication-efficient federated learning and permissioned blockchain for digital twin edge networks. *IEEE Internet Things J.* **2021**, *8*, 2276–2288. [CrossRef]
6. Liu, X.; Zheng, K.; Chi, K.; Zhu, Y.-H. Cooperative spectrum sensing optimization in energy-harvesting cognitive radio networks. *IEEE Trans. Wirel. Commun.* **2020**, *19*, 7663–7676. [CrossRef]
7. Chen, X.; Feng, Z.; Wei, Z.; Gao, F.; Yuan, X. Performance of joint sensing-communication cooperative sensing UAV network. *IEEE Trans. Veh. Technol.* **2020**, *69*, 15545–15556. [CrossRef]
8. Chen, H.; Wang, Z.; Zhang, L. Collaborative spectrum sensing for illegal drone detection: A deep learning-based image classification perspective. *China Commun.* **2020**, *17*, 81–92. [CrossRef]
9. Chen, H.; Kung, H.T.; Vlah, D.; Hague, D.; Muccio, M.; Poland, B. Collaborative compressive spectrum sensing in a UAV environment. In Proceedings of the 2011 Military Communications Conference, Baltimore, MD, USA, 7–10 November 2011; pp. 142–148.
10. Shang, B.; Liu, L.; Rao, R.M.; Marojevic, V.; Reed, J.H. 3D spectrum sharing for hybrid D2D and UAV networks. *IEEE Trans. Commun.* **2020**, *68*, 5375–5389. [CrossRef]
11. Zhao, Y.; Hong, Z.; Luo, Y.; Wang, G.; Pu, L. Prediction-based spectrum management in cognitive radio networks. *IEEE Syst. J.* **2018**, *12*, 3303–3314. [CrossRef]
12. Tandra, R.; Sahai, A. SNR walls for signal detection. *IEEE J. Sel. Top. Signal Process.* **2008**, *2*, 4–17. [CrossRef]
13. Xu, W.; Li, X.; Lee, C.; Pan, M.; Feng, Z. Joint sensing duration adaptation, user matching, and power allocation for cognitive OFDM-NOMA systems. *IEEE Trans. Wirel. Commun.* **2018**, *17*, 1269–1282. [CrossRef]
14. Pauluzzi, D.R.; Beaulieu, N.C. A comparison of SNR estimation techniques in the AWGN channel. In Proceedings of the IEEE Pacific Rim Conference on Communications, Computers, and Signal Processing, Victoria, BC, Canada, 17–19 May 1995; pp. 36–39.
15. Qiu, K.; Mao, X.; Shen, X.; Wang, X.; Li, T.; Gu, Y. Time-varying graph signal reconstruction. *IEEE J. Sel. Top. Signal Process.* **2017**, 11, 870–883 [CrossRef]
16. Xie, J.; Fang, J.; Liu, C.; Li, X. Deep learning-based spectrum sensing in cognitive radio: A CNN-LSTM approach. *IEEE Commun. Lett.* **2020**, *24*, 196–2200. [CrossRef]

17. Xie, J.; Fang, J.; Liu, C.; Yang, L. Unsupervised deep spectrum sensing: A variational auto-encoder based approach. *IEEE Trans. Veh. Technol.* **2020**, *69*, 5307–5319. [CrossRef]
18. Ghosh, C.; Cordeiro, C.; Agrawal, D.P.; Rao, M.B. Markov chain existence and hidden Markov models in spectrum sensing. In Proceedings of the 2009 IEEE International Conference on Pervasive Computing and Communications, Galveston, TX, USA, 9–13 March 2009; pp. 1–6.
19. Santana, G.M.D.; Cristo, R.S.; Diguet, J.-P.; Dezan, C.; Diana, O.P.M.; Kalinka, B.R.L.J.C. A Case Study of Primary User Arrival Prediction Using the Energy Detector and the Hidden Markov Model in Cognitive Radio Networks. In Proceedings of the 2019 IEEE Symposium on Computers and Communications (ISCC), Barcelona, Spain, 29 June–3 July 2019; pp. 1195–1198.
20. Soumya, D.; Tamaghna, A. Faulty node detection in HMM-based cooperative spectrum sensing for cognitive radio networks. *Comput. J.* **2018**, *61*, 1469–1478.
21. Suguna, R.; Rathinasabapathy, V. Spectrum sensing in cognitive radio enabled wireless sensor networks using discrete Markov model. In Proceedings of the 2018 International Conference on Smart Systems and Inventive Technology (ICSSIT), Tirunelveli, India, 13–14 December 2018; pp. 442–447.
22. Eltom, H.; Kandeepan, S.; Liang, Y.-C.; Evans, R.J. Cooperative soft fusion for HMM-based spectrum occupancy prediction. *IEEE Commun. Lett.* **2018**, *22*, 2144–2147. [CrossRef]
23. Halaseh, R.A.; Dahlhaus, D. Continuous hidden Markov model based interference-aware cognitive radio spectrum occupancy prediction. In Proceedings of the 2016 IEEE 27th Annual International Symposium on Personal, Indoor, and Mobile Radio Communications (PIMRC), Valencia, Spain, 4–8 September 2016; pp. 1–6.
24. Cheng, H.; Mark, B.L.; Ephraim, Y.; Chen, C.-H. Multiband spectrum sensing with non-exponential channel occupancy times. In Proceedings of the ICC 2021—IEEE International Conference on Communications, Montreal, QC, Canada, 14–23 June 2021; pp. 1–6.
25. Kumar, K.; Kumar, S.; Kaiwartya, O.; Kashyap, P.K.; Lloret, J.; Song, H. Drone assisted flying ad-Hoc networks: mobility and service oriented modeling using Neuro-fuzzy. *Ad Hoc Netw.* **2020**, *106*, 10242. [CrossRef]
26. Raza, M.A.; Hussain, A. Maximum likelihood SNR estimation of hyper cubic signals over Gaussian channel. *IEEE Commun. Lett.* **2016**, *20*, 45–48. [CrossRef]
27. Manesh, M.R.; Quadri, A.; Subramaniam, S.; Kaabouch, N. An optimized SNR estimation technique using particle swarm optimization algorithm. In Proceedings of the 2017 IEEE 7th Annual Computing and Communication Workshop and Conference (CCWC), Las Vegas, NV, USA, 9–11 January 2017; pp. 1–6.
28. He, R.; Yang, D.; Wang, H.; Kuang, J.; Wen, X. A code-aided and moment-based joint SNR estimation for M-APSK over AWGN Channels. In Proceedings of the 2017 IEEE 85th Vehicular Technology Conference (VTC Spring), Sydney, NSW, Australia, 4–7 June 2017; pp. 1–6.
29. Chen, C.; Wu, J.; Wang, C.; Miao, X.; Bu, X. On a new SNR estimation approach with Polar codes. In Proceedings of the 2021 IEEE 4th Advanced Information Management, Communicates, Electronic and Automation Control Conference (IMCEC), Chongqing, China, 18–20 June 2021; pp. 727–732.
30. Sharma, S.K.; Chatzinotas, S.; Ottersten, B. SNR estimation for multi-dimensional cognitive receiver under correlated channel/noise. *IEEE Trans. Wirel. Commun.* **2013**, *12*, 6392–6405. [CrossRef]
31. Ngo, T.; Kelley, B.; Rad, P. Deep learning based prediction of signal-to-noise ratio (SNR) for LTE and 5G systems. In Proceedings of the 2020 8th International Conference on Wireless Networks and Mobile Communications (WINCOM), Reims, France, 27–29 October 2020; pp. 1–6.
32. Yang, Y.; Jing, X.; Mu, J.; Gao, H. SNR estimation of UAV control signal based on convolutional neural network. In Proceedings of the 2021 International Wireless Communications and Mobile Computing (IWCMC), Harbin, China, 28 June–2 July 2021; pp. 780–784.
33. Shang, B.; Marojevic, V.; Yi, Y.; Abdalla, A.S.; Liu, L. Spectrum sharing for UAV communications: spatial spectrum sensing and open issues. *IEEE Veh. Technol. Mag.* **2020**, *12*, 104–112. [CrossRef]
34. Nie, R.; Xu, W.; Zhang, Z.; Zhang, P.; Pan, M.; Lin, J. Max-Min distance clustering based distributed cooperative spectrum sensing in cognitive UAV networks. In Proceedings of the ICC 2019—2019 IEEE International Conference on Communications (ICC), Shanghai, China, 20–24 May 2019; pp. 1–6.
35. Nie, R. Research on Spectrum Sensing Scheme in UAV Networks. Master's Thesis, Beijing University of Posts and Telecommunications, Beijing, China, 2020.
36. Ejaz, W.; Hattab, G.; Cherif, N.; Ibnkahla, M.; Abdelkefi, F.; Siala, M. Cooperative spectrum sensing with heterogeneous devices: Hard combining versus soft combining. *IEEE Syst. J.* **2018**, *12*, 981–992. [CrossRef]
37. Tsagris, M.; Beneki, C.; Hassani, H. On the folded normal distribution. *Mathematics* **2014**, *2*, 12–28. [CrossRef]
38. Zhang, W.; Guo, Y.; Liu, H.; Chen, Y.; Wang, Z.; Mitola, J., III. Distributed consensus-based weight design for cooperative spectrum sensing. *IEEE Trans. Parallel Distrib. Syst.* **2015**, *26*, 54–64. [CrossRef]
39. Rabiner, L.R. A tutorial on hidden Markov models and selected applications in speech recognition. *Proc. IEEE* **1989**, *77*, 257–286. [CrossRef]
40. Goldsmith, A. *Wireless Communications*; Cambridge University Press: Cambridge, UK, 2005.

41. Tan, Z.; Nguyen, A.H.T.; Khong, A.W.H. An efficient dilated convolutional neural network for UAV noise reduction at low input SNR. In Proceedings of the 2019 Asia-Pacific Signal and Information Processing Association Annual Summit and Conference (APSIPA ASC), Lanzhou, China, 18–21 November 2019; pp. 1885–1892.
42. Yang, J. Spatial Channel Characterization for Cognitive Radios. Master's Thesis, University of California, Berkeley, CA, USA, 2004.
43. Sharma, G.; Sharma, R. Performance comparison of hard and soft fusion techniques for energy efficient CSS in cognitive radio. In Proceedings of the 2018 International Conference on Advanced Computation and Telecommunication (ICACAT), Bhopal, India, 28–29 December 2018; pp. 1–4.
44. Goddemeier, N.; Wietfeld, C. Investigation of air-to-air channel characteristics and a UAV specific extension to the rice model. In Proceedings of the 2015 IEEE Globecom Workshops (GC Wkshps), San Diego, CA, USA, 6–10 December 2015; pp. 1–5.

sensors

MDPI

Article

Deep Q-Learning-Based Transmission Power Control of a High Altitude Platform Station with Spectrum Sharing

Seongjun Jo [1], Wooyeol Yang [1], Haing Kun Choi [2], Eonsu Noh [3], Han-Shin Jo [1,*] and Jaedon Park [3,*]

1 Department of Electronic Engineering, Hanbat National University, Daejeon 34158, Korea; 30211176@edu.hanbat.ac.kr (S.J.); 30211173@edu.hanbat.ac.kr (W.Y.)
2 TnB Radio Tech., Seoul 08504, Korea; radioeng@tnbrt.com
3 Agency for Defense Development, Daejeon 34186, Korea; nes@add.re.kr
* Correspondence: hsjo@hanbat.ac.kr (H.-S.J.); jaedon2@add.re.kr (J.P.)

Abstract: A High Altitude Platform Station (HAPS) can facilitate high-speed data communication over wide areas using high-power line-of-sight communication; however, it can significantly interfere with existing systems. Given spectrum sharing with existing systems, the HAPS transmission power must be adjusted to satisfy the interference requirement for incumbent protection. However, excessive transmission power reduction can lead to severe degradation of the HAPS coverage. To solve this problem, we propose a multi-agent Deep Q-learning (DQL)-based transmission power control algorithm to minimize the outage probability of the HAPS downlink while satisfying the interference requirement of an interfered system. In addition, a double DQL (DDQL) is developed to prevent the potential risk of action-value overestimation from the DQL. With a proper state, reward, and training process, all agents cooperatively learn a power control policy for achieving a near-optimal solution. The proposed DQL power control algorithm performs equal or close to the optimal exhaustive search algorithm for varying positions of the interfered system. The proposed DQL and DDQL power control yields the same performance, which indicates that the actional value overestimation does not adversely affect the quality of the learned policy.

Keywords: Deep Q-learning (DQL); Double Deep Q-learning (DDQL); dynamic spectrum sharing; High Altitude Platform Station (HAPS); cellular communications; power control; interference management

Citation: Jo, S.; Yang, W.; Choi, H.K.; Noh, E.; Jo, H.-S.; Park, J. Deep Q-Learning-Based Transmission Power Control of a High Altitude Platform Station with Spectrum Sharing. *Sensors* **2022**, *22*, 1630. https://doi.org/10.3390/s22041630

Academic Editor: Margot Deruyck

Received: 12 January 2022
Accepted: 18 February 2022
Published: 19 February 2022

1. Introduction

A High Altitude Platform Station (HAPS) is a network node operating in the stratosphere at an altitude of approximately 20 km. The International Telecommunication Union (ITU) defines a HAPS in Article 1.66A as "A station on an object at an altitude of 20 to 50 km and a specified, nominal, fixed point relative to the Earth". Various studies have been performed on HAPS in recent years, and the commercial applications of HAPS have significantly increased [1]. In addition, the HAPS has potential as a significant component of wireless network architectures [2]. It is also an essential component of next-generation wireless networks, with considerable potential as a wireless access platform for future wireless communication systems [3–5].

Because the HAPS is located at high altitudes ranging from 20 to 50 km, the HAPS-to-ground propagation generally experiences lower path loss and a higher line-of-sight probability than typical ground-to-ground propagation. Thus, the HAPS can provide a high data rate for wide coverage; however, it is likely to interfere with various other terrestrial services, e.g., fixed, mobile, and radiolocation. The World Radiocommunication Conference 2019 (WRC-19) adopted a HAPS as the IMT Base Station (HIBS) in the frequency bands below 2.7 GHz previously identified for IMT by Resolution 247 [6], which addresses the potential interference of HAPS with an existing service. In such a situation, if the existing service is not safe from HAPS interference, the two systems cannot coexist. Therefore, the HAPS transmitter is requested to reduce its transmission power to satisfy the interference–to–noise ratio (INR) requirement for protecting the receiver of the existing service. However, if the HAPS transmission power is excessively reduced, the signal–to–interference–plus–noise ratio (SINR) of the HAPS downlink decreases; thus, the outage probability may exceed the desired level. Herein, a HAPS transmission power control algorithm is proposed that aims to minimize the outage probability of the HAPS downlink while satisfying the INR requirement for protecting incumbents.

1.1. Related Works

Studies have been performed on improving the performance of HAPS. In [7], resource allocation for an Orthogonal Frequency Division Multiple Access (OFDMA)-based HAPS system that uses multicasting in the downlink to maximize the number of user terminals by maximizing the radio resources was studied. The authors of [8] proposed a wireless channel allocation algorithm for a HAPS 5G massive multiple-input multiple-output (MIMO) communication system based on reinforcement learning. Combining Q-learning and backpropagation neural networks allows the algorithm to learn intelligently for varying channel load and block conditions. In [9], a criterion for determining the minimum distance in a mobile user access system was derived, and a channel allocation approach based on predicted changes in the number of users and the call volume was proposed.

Additionally, spectrum sharing studies on HAPS have been performed. In [10], a spectrum sharing study was conducted to share a fixed service using a HAPS with other services in the 31/28-GHz band. Interference mitigation techniques were introduced, e.g., increasing the minimum operational elevation angle or improving the antenna radiation pattern to facilitate sharing with other services. In addition, the possibility of dynamic channel allocation was analyzed. In [11], sharing between a HAPS and a fixed service in the 5.8-GHz band was investigated using a coexistence methodology based on a spectrum emission mask.

In contrast to previous studies in which HAPS communication improvement and spectrum sharing were dealt with separately, in the present study, a combination of spectrum sharing with other systems and HAPS downlink coverage improvement is considered. In this regard, this study is more advanced than previous HAPS-related studies.

Deep Q-learning (DQL) is a reinforcement learning algorithm that applies deep neural networks to reinforcement learning to solve complex problems in the real world. DQL is widely used in various fields, including UAV, drone, and HAPS. In [12], the optimal UAV-BS trajectory was presented using a DQL for optimal placement of UAVs, and the author of [13] used a DQL to determine the optimal link between two UAV nodes. In [14], a DQL is used to find the optimal flight parameters for the collision-free trajectory of the UAV. In [15], two-hop communication was considered to optimize the drone base station trajectory and improve network performance, and a DQL was used to solve the joint two-hop communication scenario. In [16], a DQL was used for multiple-HAPS coordination for communications area coverage. A Double Deep Q-learning (DDQL) is an algorithm developed to prevent the overestimation of a DQL and shows better performance than the DQL in various fields [17].

1.2. Contributions

The contributions of the present study are as follows. (1) For the first time, a multiagent DQL was used to improve the HAPS outage performance and solve the problem of spectrum sharing with existing services. (2) We defined the power control optimization problem to minimize the outage probability of the HAPS downlink under the interference constraint for protecting the existing system. The state and reward for the training agent were designed to consider the objective function and constraints of the optimization problem. (3) Because the HAPS has a multicell structure, the number of power combinations increases exponentially as the number of cells (N_{cell}) and power levels increase linearly.

Thus, the optimal exhaustive search method requires an impractically long computation time to solve the multicell power optimization problem. The proposed DQL algorithm performs comparably to an optimal exhaustive search with a feasible computation time. (4) Even for varying positions of the interfered system, the proposed DQL produces a proper power control policy, maintaining stable performance. (5) Comparing the proposed DQL algorithm with the DDQL algorithm shows no performance degradation due to overestimation in the proposed DQL. The remainder of this paper is organized as follows.

Section 2 presents the system model, including the system deployment model, HAPS model, interfered system model, and path loss model. In Section 3, the downlink SINR and INR are calculated. In Section 4, a DQL-based HAPS power control algorithm is proposed. Section 5 presents the simulation results, and Section 6 concludes the paper.

2. System Model

2.1. System Deployment Model

HAPS communication networks are assumed to consist of a single HAPS, multiple ground user equipment (*UE*) devices (referred to as *UE*s hereinafter), and a ground interfered receiver. The HAPS, *UE*, and interfered receiver are distributed in the three-dimensional Cartesian coordinate system, as shown in Figure 1. The coordinates of the HAPS antenna and the interfered receiver antenna are (0, 0, h_{HAPS}) and (X, Y, h_V), respectively. The N_{UE} *UE* devices with an antenna height of h_{UE} are uniformly distributed within the circular HAPS area.

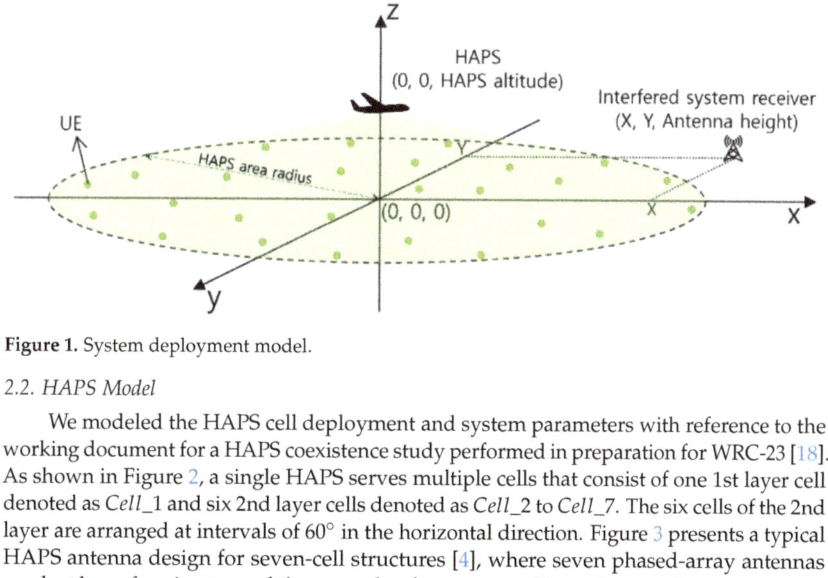

Figure 1. System deployment model.

2.2. HAPS Model

We modeled the HAPS cell deployment and system parameters with reference to the working document for a HAPS coexistence study performed in preparation for WRC-23 [18]. As shown in Figure 2, a single HAPS serves multiple cells that consist of one 1st layer cell denoted as $Cell\_1$ and six 2nd layer cells denoted as $Cell\_2$ to $Cell\_7$. The six cells of the 2nd layer are arranged at intervals of 60° in the horizontal direction. Figure 3 presents a typical HAPS antenna design for seven-cell structures [4], where seven phased-array antennas conduct beamforming toward the ground to form seven cells, as shown in Figure 2. The 1st layer cell has an antenna tilt of 90°, i.e., perpendicular to the ground; the 2nd layer cell has an antenna tilt of 23°.

The antenna pattern of the HAPS was designed using the antenna gain formula presented in Recommendation ITU-R M.2101 [19]. The transmitting antenna gain is calculated as the sum of the gain of a single element and the beamforming gain of a multi-antenna array. The single element antenna gain is determined by the azimuth angle (ϕ) and the elevation angle (θ) between the transmitter and receiver and is calculated as follows:

$$A_E(\phi,\theta) = G_{E,max} - min\{-[A_{E,H}(\phi) + A_{E,v}(\theta)], A_m\}, \quad (1)$$

where $G_{E,max}$ represents the maximum antenna gain of a single element, $A_{E,H}(\phi)$ represents the horizontal radiation pattern calculated using Equation (2), and $A_{E,v}(\theta)$ represents the vertical radiation pattern calculated using Equation (3).

$$A_{E,H}(\phi) = -min\left[12\left(\frac{\phi}{\phi_{3\text{dB}}}\right)^2, A_m\right] \tag{2}$$

Here, $\phi_{3\text{dB}}$ represents the horizontal 3 dB beamwidth of a single element, and A_m represents the front-to-back ratio.

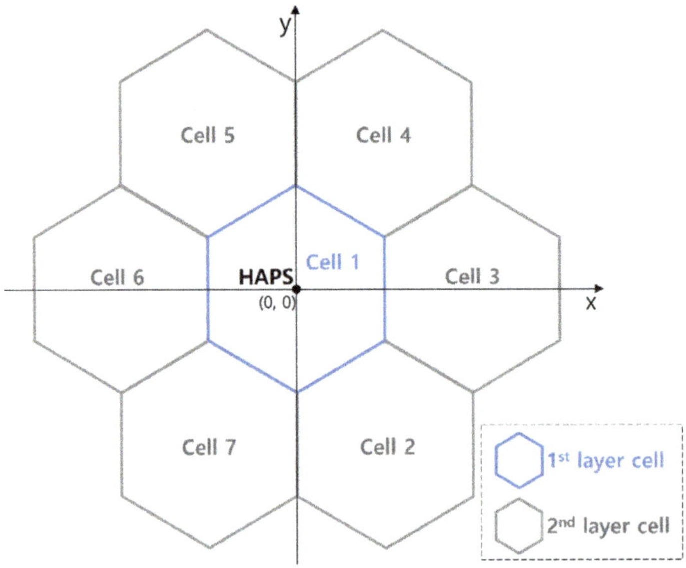

Figure 2. HAPS seven-cell layout.

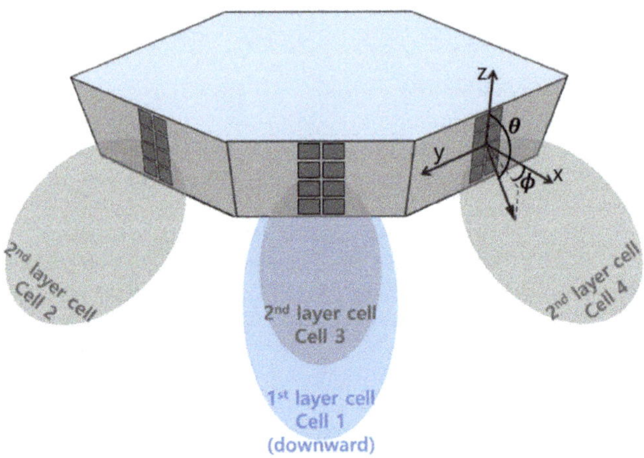

Figure 3. Typical antenna structure for multi-cell HAPS communication.

$$A_{E,V}(\theta) = -min\left[12\left(\frac{\theta - 90}{\theta_{3dB}}\right)^2, SLA_v\right] \quad (3)$$

Here, θ_{3dB} represents the vertical 3 dB bandwidth of a single element, and SLA_v represents the front-to-back ratio.

The transmitting antenna gain of the HAPS is calculated using the antenna arrangement and spacing, as well as the target beamforming direction. The gain for beam i is calculated as follows:

$$A_{A,Beami}(\theta, \phi) = A_E(\theta, \phi) + 10\log_{10}\left(\left|\sum_{m=1}^{N_H} \sum_{n=1}^{N_V} w_{i,n,m} \cdot v_{n,m}\right|^2\right), \quad (4)$$

where N_H and N_V represent the number of antennas in the horizontal and vertical directions, respectively. $v_{n,m}$ is the superposition vector that overlaps the beams of the antenna elements, which is calculated using Equation (5), and $w_{i,n,m}$ is the weight that directs the antenna element in the beamforming direction, which is calculated using Equation (6).

$$\begin{aligned} &n = 1, 2, \ldots N_V; m = 1, 2, \ldots N_H \\ &v_{n,m} = exp\left(\sqrt{-1} \cdot 2\pi\left((n-1) \cdot \tfrac{d_V}{\lambda} \cdot cos(\theta) + (m-1) \cdot \tfrac{d_H}{\lambda} \cdot sin(\theta) \cdot sin(\phi)\right)\right) \end{aligned} \quad (5)$$

Here, d_H and d_V represent the intervals between the horizontal and vertical antenna arrays, respectively, and λ represents the wavelength.

$$\begin{aligned} w_{i,n,m} = \tfrac{1}{\sqrt{N_H N_V}} \quad & exp(\sqrt{-1} \\ & \cdot 2\pi\left((n-1) \cdot \tfrac{d_V}{\lambda} \cdot sin(\theta_{i,etilt}) - (m-1) \cdot \tfrac{d_H}{\lambda} \cdot cos(\theta_{i,etilt})\right. \\ & \cdot sin(\phi_{i,escan}))) \end{aligned} \quad (6)$$

Here, $\phi_{i,escan}$ and $\theta_{i,etilt}$ represent the ϕ and θ of the main beam direction, respectively.

The 1st layer cell of the HAPS uses a 2 × 2 antenna array, and the 2nd layer cell uses a 4 × 2 antenna array. Figure 4 shows the antenna pattern of the 1st layer cell, and Figure 5 shows the antenna pattern of the 2nd layer cell.

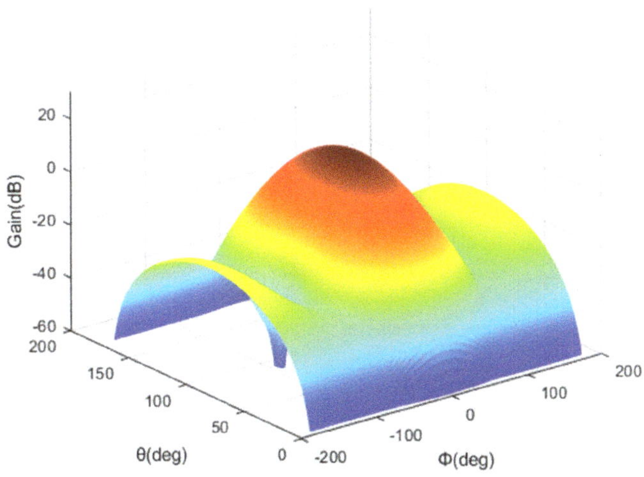

Figure 4. 1st layer cell antenna pattern.

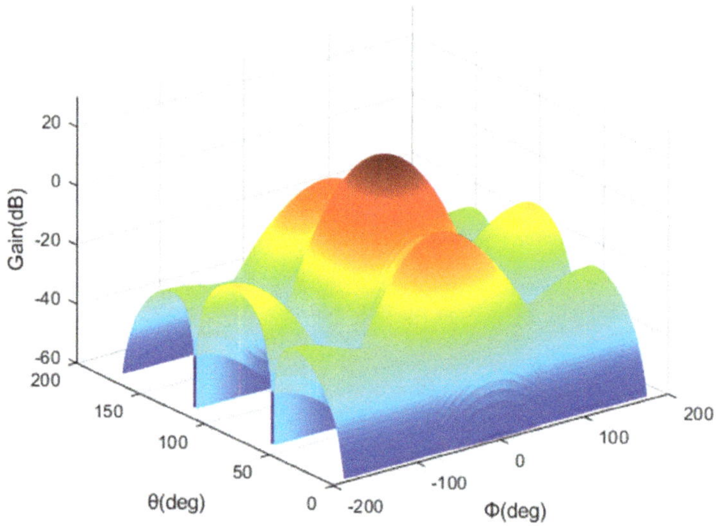

Figure 5. 2nd layer cell antenna pattern.

2.3. Interfered System Model

Various interfered systems, e.g., fixed, mobile, and radiolocation services, can be considered for the interference scenario involving a HAPS. We adopted a ground IMT base station (BS) for the interfered system, referring to the potential interference scenario [6]. The antenna pattern of the interfered system was applied by referring to Recommendation ITU-R F.1336 [20]. The receiving antenna gain is calculated as follows:

$$G(\phi,\theta) = G_0 + G_{hr}(x_h) + R{\cdot}G_{vr}(x_v), \quad (7)$$

where G_0 represents the maximum gain in the azimuth plane; $G_{hr}(x_h)$ represents the relative reference antenna gain in the azimuth plane in the normalized direction of (x_h, 0), which is calculated using Equation (8); and $G_{vr}(x_v)$ represents the relative reference antenna gain in the elevation plane in the normalized direction of (0, x_v), which is calculated using Equation (9). R represents the horizontal gain compression ratio when the azimuth angle is shifted from 0° to ϕ, which is calculated using Equation (10).

$$\begin{aligned} &G_{hr}(x_h) = -12x_h^2 && for\ x_h \le 0.5 \\ &G_{hr}(x_h) = -12x_h^{(2-k_h)} - \lambda_{\text{kh}} && for\ 0.5 < x_h \\ &G_{hr}(x_h) \ge G_{180} \end{aligned} \quad (8)$$

$$\begin{aligned} &G_{\text{vr}}(x_v) = -12x_v^2 && for\ x_v < x_k \\ &G_{\text{vr}}(x_v) = -15 + 10\log(x_v^{-1.5} + k_v) && for\ x_k \le x_v < 4 \\ &G_{\text{vr}}(x_v) = -\lambda_{\text{kv}} - 3 - C\log(x_v) && for\ 4 \le x_v < 90/\theta_3 \\ &G_{\text{vr}}(x_v) = G_{180} && for\ x_v \ge 90/\theta_3 \end{aligned} \quad (9)$$

$$R = \frac{G_{hr}(x_h) - G_{hr}(180°/\phi_3)}{G_{hr}(0) - G_{hr}(180°/\phi_3)} \quad (10)$$

Here, x_h and λ_{kh} are given by Equations (11) and (12), respectively; ϕ_3 represents the 3 dB beamwidth in the azimuth plane; and k_h is an azimuth pattern adjustment factor based on the leaked power. The relative minimum gain G_{180} was calculated using Equation (13).

$$x_h = |\phi\,|/\phi_3 \quad (11)$$

$$\lambda_{\mathrm{kh}} = 3\left(1 - 0.5^{-k_h}\right) \tag{12}$$

$$G_{180} = -15 + 10\log(1 + 8k_a) - 15\log\left(\frac{180^\circ}{\theta_3}\right) \tag{13}$$

Returning to Equation (9), x_v is given by Equation (14), and the 3-dB beamwidth in the elevation plane θ_3 is calculated using Equation (15), where G_0 represents the maximum gain in the azimuth plane. In addition, x_k is calculated using Equation (16), where k_v is an elevation pattern adjustment factor based on the leaked power. λ_{kv} was calculated using Equation (17), and the attenuation inclination factor C was calculated using Equation (18). Figure 6 shows the antenna pattern of the interfered system calculated using Equation (7), which is the pattern for a typical terrestrial BS with a broad beamwidth in the azimuth plane but a narrow beamwidth in the elevation plane.

$$x_v = |\theta\,|/\theta_3 \tag{14}$$

$$\theta_3 = 107.6 \times 10^{-0.1G_0} \tag{15}$$

$$x_k = \sqrt{1.33 - 0.33k_v} \tag{16}$$

$$\lambda_{\mathrm{kv}} = 12 - C\log(4) - 10\log\left(4^{-1.5} + k_v\right) \tag{17}$$

$$C = \frac{10\log\left(\frac{\left(\frac{180^\circ}{\theta_3}\right)^{1.5}\cdot\left(4^{-1.5}+k_v\right)}{1+8k_p}\right)}{\log\left(\frac{22.5^\circ}{\theta_3}\right)} \tag{18}$$

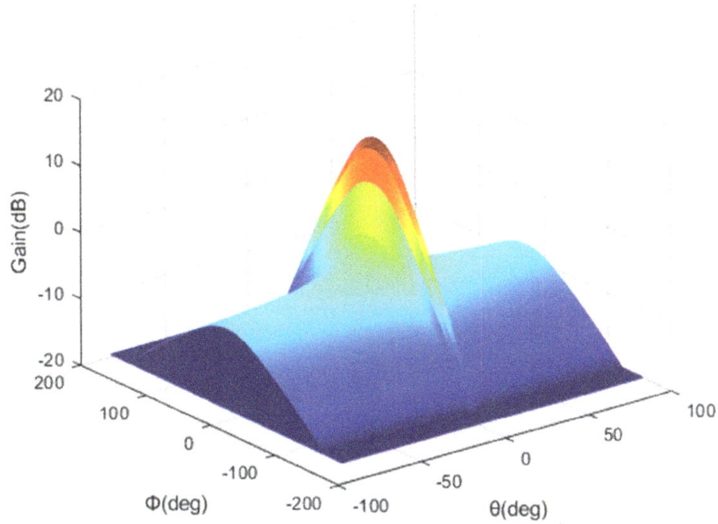

Figure 6. Interfered system antenna pattern.

2.4. Path Loss Model

The path loss model of Recommendation ITU-R P.619 [21] was applied to the working document for the HAPS coexistence study performed in preparation for WRC-23 [22]. The total path loss that occurs when the HAPS signal reaches the *UE* and the IMT BS is expressed as follows:

$$L_p = FSL + A_{xp} + A_g + A_{bs}, \tag{19}$$

where FSL represents the free-space path loss calculated using Equation (20), which occurs in a straight path from a transmitting antenna to a receiving antenna in a vacuum state, and A_{xp} is assumed to be 3 dB for depolarization attenuation. A_g represents the attenuation loss due to atmospheric gases. A_{bs} represents the resistive loss due to the spread of the antenna beam as the beam spreads attenuation. A_g and A_{bs} were calculated using the formulae in P.619.

$$FSL = 92.45 + 20\log(f \cdot d) \tag{20}$$

Here, f represents the carrier frequency (in GHz), and d represents the distance (in km) between the transmitter and receiver.

3. Calculation of Downlink SINR and INR

3.1. Calculation of Downlink SINR

The signal received by the UE from the HAPS transmission for the ith cell ($Cell\_i$) is calculated as follows:

$$S_{Cell\_i} = P_{Cell\_i} + G_{Cell\_i} + G_p + G_{r,UE} - L_p - L_{ohm}, \tag{21}$$

where P_{Cell_i} represents the HAPS transmission power for $Cell\_i$, G_{Cell_i} represents the transmitting antenna gain of $Cell\_i$, G_p represents the polarization gain, $G_{r,UE}$ represents the receiving antenna gain, and L_{ohm} represents the ohmic loss. The UE receives signals from all N_{cell} cells and considers the remaining signals (except for the strongest $Cell\ j$ signal) as interference. Equation (22) is used to calculate the signal and interference, and the receiver noise is calculated using Equation (23).

$$\begin{gathered} j = \underset{i}{\operatorname{argmax}} S_{Cell\_i} \\ S_{HAPS} = S_{Cell\_j} \\ I_{HAPS,UE} = 10\log\left(\sum_{\substack{i=1 \\ i \neq j}}^{N_{cell}} 10^{\frac{S_{Cell\_i}}{10}}\right) \end{gathered} \tag{22}$$

$$N = 10\log(k \times T \times BW) + N_f \tag{23}$$

Here, k and T represent the Boltzmann constant and noise temperature, respectively, and BW represents the channel bandwidth. N_f represents the noise figure. Finally, the downlink SINR is calculated as follows:

$$\eta = 10\log\left(\frac{10^{\frac{S_{HAPS}}{10}}}{10^{\frac{I_{HAPS,UE}}{10}} + 10^{\frac{N}{10}}}\right). \tag{24}$$

3.2. Calculation of INR

The interference power received by the interfered receiver from the HAPS transmitter servicing $Cell\ i$ is calculated as follows:

$$I_{Cell\_i} = P_{Cell\_i} + G_{Cell\_i} + G_p + G_{r,V} - L_p - L_{ohm}, \tag{25}$$

where $G_{r,V}$ represents the antenna gain of the interfered receiver. The aggregated interference power at the interfered receiver is calculated as follows:

$$I_{HAPS,V} = 10\log\left(\sum_{i=1}^{N_{cell}} 10^{\frac{I_{Cell\_i}}{10}}\right). \tag{26}$$

Finally, after converting the aggregated interference into INR form in accordance with Equation (27) and comparing it with the protection criteria (INR_{th}) of the interfered receiver, it is possible to check whether the interfered receiver is protected from the interference of the HAPS.

$$INR = I_{HAPS,V} - N \tag{27}$$

4. DQL-Based HAPS Transmission Power Control Algorithm

4.1. Problem Formulation

To satisfy the INR_{th} of the interfered system, the transmission power of the HAPS must be reduced. However, as the power of the HAPS is reduced, the η of the UE decreases, and the outage probability P_{out} increases. Thus, the objective of this study was to find a HAPS transmission power set for each cell, i.e., $\boldsymbol{P} = \{P_{Cell\_i} | i = 1, \cdots, N_{cell}\}$, that satisfies the INR_{th} of the interfered system while minimizing P_{out}. The optimization problem of the HAPS transmission power can be formulated as follows:

$$\begin{aligned} &\min_{\boldsymbol{P}} P_{out} = \frac{N_{UE,o}(\boldsymbol{P})}{N_{UE}} \\ \text{s.t.} \quad &C1: \ INR \leq INR_{th} \\ &C2: \ P_{min} \leq P_{Cell_i} \leq P_{max} \ \forall i \in \{1, \cdots, N_{cell}\}, \end{aligned} \tag{28}$$

where $N_{UE,o}(\boldsymbol{P})$ represents the number of UEs that do not satisfy the minimum required SINR η_o for a given HAPS transmission power set $\boldsymbol{P}$.

4.2. Proposed Algorithm

To control the HAPS transmission power, it is necessary to independently determine the power level of each cell. Accordingly, the total number of HAPS transmission power sets increases exponentially to $N_p^{N_{cell}}$ as the number of selectable powers N_p increases linearly. Although an exhaustive search algorithm can be used to find optimal solutions, this incurs excessive complexity and a long computation time. To solve this problem, we propose a DQL-based power optimization algorithm that can find a near-optimal $\boldsymbol{P}$ with low complexity. In the proposed DQL model, each agent functions as the power controller of a cell; accordingly, the number of agents is N_{cell}.

The agent—the subject of learning—learns a deep neural network called Deep Q Network (DQN) and selects an action using this network. DQL is an improved Q-learning method. Q-learning is a method for selecting the best action in a specific state through the Q-table of a state-action pair. As the state–action space grows in Q-learning, creating a Q-table and finding the best policy become highly complex. In addition, the use of Q-learning is limited because learning in the Q-table format becomes more complex when multiple agents are used. In contrast, a DQL is a promising way to solve the curse of dimensionality by approximating a Q function using a deep neural network instead of a Q-table. The proposed algorithm uses a method in which each agent learns a policy based on its observation and action while treating all other agents as part of the environment to solve the multiple-agent problem.

The basic DQL parameters (state, action, and reward) are presented below. Each agent learns the policy independently using the training data at each timestep t. The state space of the mth agent comprises a set of $(N_{cell} - 1)$ interferences that the agent provides to UEs located at the centers of other cells and the agent's interference to the interfered receiver, which is expressed as

$$S_t = \{I_v, \{I_{UE\_i} | i = 1, \cdots, N_{cell}, \ and \ i \neq m\}\}. \tag{29}$$

Two power sets configure the action space of an agent: $\boldsymbol{A}_1 = \{29, 31, 33, 35, 37\}$ and $\boldsymbol{A}_2 = \{26, 28, 30, 32, 34\}$ (unit: dBm). The agent of $Cell\_1$ in the 1st layer cell selects an action from $\boldsymbol{A}_1$, and the agents of the 2nd layer cell select an action from $\boldsymbol{A}_2$. All agent actions are initialized to the minimum power value to minimize the interference to the interfered receiver at the beginning of the learning process. The reward is calculated as follows. First, because the interfered receiver must be safe from HAPS interference, an agent receives a fixed r_t of -100 (deficient value) for $INR > INR_{th}$. In contrast, for $INR \le INR_{th}$, an agent receives r_t computed according to the lower 5% downlink SINR of each cell $\{\hat{\eta}_i | i = 1,\ 2,\ \cdots,\ N_{cell}\}$ and the required SINR η_o. The reward can be expressed as

$$r_t = \begin{cases} r_{1,\,t} + r_{2,\,t} & for\ INR \le INR_{th} \\ r_t = -100 & otherwise, \end{cases}, \quad (30)$$

where

$$\begin{aligned} r_{1,t} &= 10 \cdot (\Sigma(\hat{\eta}_i - \eta_o)) \quad for\ \hat{\eta}_i \ge \eta_o \\ r_{2,t} &= \Sigma(\hat{\eta}_i + \eta_o) \qquad\quad for\ \hat{\eta}_i < \eta_o. \end{aligned} \quad (31)$$

Figure 7 shows the structure of the proposed DQL-based HAPS transmission power control algorithm. Each agent learns its DQN, and one DQN consists of the main network, target network, and replay memory. The main network estimates the Q-value $Q(s, a; w)$ corresponding to the state–action pair through a deep neural network with a weight w. The main network consists of an input layer composed of seven neurons, a hidden layer consisting of 24 neurons, and an output layer consisting of five neurons. It is a fully connected network. w is updated every t in the direction that minimizes the loss function $L(w) = \mathbb{E}\left[(y_j - Q(s, a; w))^2\right]$. The target network calculates the target value $y_j = r_j + \gamma \max_{a'} \hat{Q}(s', a'; w^-)$, where γ is the discount factor; s' and a' denotes the state and action, respectively, in the next step; and $\hat{Q}(s', a'; w^-)$ is the Q-value estimated through the target network with weight w^-. The agent's transition tuple $(s_t,\ a_t,\ r_t,\ s_{t+1})$ is piled in the replay memory, from which a minibatch (size of 512 tuples) are randomly sampled at each step. The minibatch data are used to compute the target value y_j. In a DQL, learning is stabilized, and the learning performance is improved through replay memory and a separate target network [23].

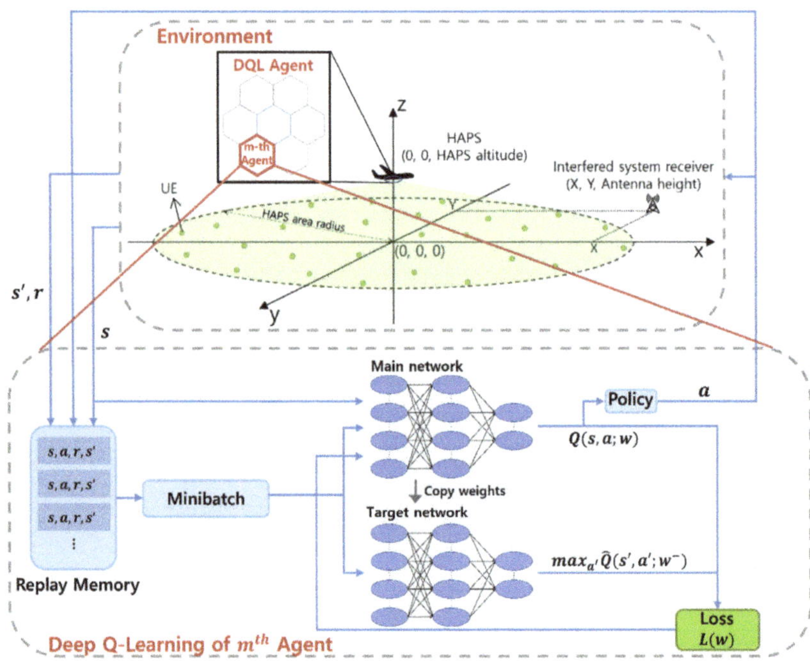

Figure 7. DQL-based HAPS power control architecture.

Algorithm 1 describes the proposed DQL-based HAPS transmission power control algorithm. For DQN training, N was set as 100,000, and the minibatch size was set as 512. M was set as 500, and T was set as 10. The Adam optimizer was used to minimize $L(w)$, and the learning rate and γ were 0.01 and 0.995, respectively. An ε-greedy policy was used to balance exploration and exploitation; ε was initially set as 1 and was reduced by 0.01 for every episode.

Algorithm 1. Training Process for the DQL-Based HAPS Power Control Algorithm

1: Initialize the replay memory $\mathcal{D}$ to capacity N
2: Initialize the Q-function with random weights w
3: Initialize the target $\hat{Q}$-function with the same weights: $w^{-} = w$
4: **for** episode = 1, M **do**
5:   Initialize action $a_0 = \min_a A$
6:   **for** timestep = 1, T **do**
7:     **if** $t = 1$
8:       Calculate s_t via Equations (21) and (25)
9:     **end if**
10:     With probability, select a random action a_t
11:     Otherwise, select $a_t = \operatorname{argmax}_a Q(s_t, a; w)$
12:     Assign the selected power to the mth cell and compute INR and η
13:     Observe the reward r_t and s_{t+1}
14:     Store the experience in (s_t, a_t, r_t, s_{t+1}) in $\mathcal{D}$
15:     Sample a random minibatch of experiences from $\mathcal{D}$
16:     Set $y_j = r_j + \gamma \max_{a'} \hat{Q}(s', a'; w^{-})$
17:     Perform optimization via $L(w)$ and update w
18:     Update the target network $\hat{Q}$ with $w^{-} = w$ every 4 steps
19:   **end for**
20: **end for**

A DDQL is a reinforcement learning algorithm to improve performance degradation due to the overestimation of the DQL. Action-value can be overestimated by the maximization step in line 16 of Algorithm 1. Therefore, the DDQL calculates the target value as $y_j = r_j + \gamma \hat{Q}\left(s', \operatorname{argmax}_{a'} Q(s', a'; w); w^{-}\right)$ to eliminate the maximization step. The DDQL-based HAPS power control algorithm proceeds the same way as Algorithm 1 except for calculating the target value.

5. Simulation Results

5.1. Simulation Configuration

The simulation was conducted using MATLAB for three positions of the interfered receiver, and the learning order of the agent was randomly set for each t. Subsequently, the simulation proceeded according to Algorithm 1. When all M episodes were finished, the simulation ended, and the set $\boldsymbol{P}_c$ composed of the power selected by each agent was calculated as the simulation result. Finally, the performance of the simulation was verified by comparing $\boldsymbol{P}_c$ with the optimal power set $\boldsymbol{P}^*$ obtained via an exhaustive search algorithm considering all $N_p^{N_{cell}}$ cases. The total elapsed time of the DQL and exhaustive search was about 7500 s and 21,000 s, respectively. The total elapsed time of the exhaustive search increased exponentially with the rise of N, but the DQL did not. Therefore, the computational efficiency of the DQL is more remarkable as the number of cells and power levels increase. In this simulation, performance comparison with the DDQL was additionally performed to check performance degradation due to overestimation of the DQL.

We applied the HAPS parameters and interfered system parameters, referring to the working document for the HAPS coexistence study performed in preparation for WRC-23 [18,24]. The simulation parameters of the two systems are presented in Tables 1 and 2, respectively.

Table 1. HAPS system parameters.

Parameter	Value
Center frequency (f)	2545 MHz
Channel bandwidth (BW)	20 MHz
Area radius	90 km
Altitude (h_{HAPS})	20 km
Number of cells (N_{cell})	7
Antenna pattern	Recommendation ITU-R M.2101
Element gain $(G_{E,max})$	8 dBi
Horizontal/vertical 3 dB beamwidth of single element	65° for both H/V
Antenna array configuration (Row × column)	2 × 2 elements (1st layer cell) 4 × 2 elements (2nd layer cell)
Ohmic losses (L_{ohm})	2 dB
Antenna tilt	90° (1st layer cell) 23° (2nd layer cell)
Antenna polarization	Linear/±45°
Number of distributed UEs (N_{UE})	1000
UE height	1.5 m
UE antenna gain	−3 dBi
Minimum required SINR (η_o)	−10 dB

Table 2. Interfered system (IMT BS) parameters.

Parameter	Value
Center frequency (f)	2545 MHz
Channel bandwidth (BW)	20 MHz
Noise figure $\left(N_f\right)$	5 dB
Antenna height (h_V)	20 m
Antenna tilt	10°
Antenna pattern	Recommendation ITU-R F.1336 (recommends 3.1) $k_a = 0.7$ $k_p = 0.7$ $k_h = 0.7$ $k_v = 0.3$ Horizontal 3 dB beamwidth: 65° Vertical 3 dB beamwidth is determined from the horizontal beamwidth equations in Recommendation ITU-R F.1336. Vertical beam widths of actual antennas may also be used when available.

Table 2. *Cont.*

Parameter	Value
Antenna polarization	Linear/±45°
Maximum antenna gain (G_0)	16 dBi
Protection criteria (INR_{th})	−6 dB

5.2. Numerical Analysis

Figure 8 shows the SINR maps obtained using $\boldsymbol{P}_{max} = \{37, 34, 34, 34, 34, 34, 34\}$ and $\boldsymbol{P}_{min} = \{29, 26, 26, 26, 26, 26, 26\}$ for all cells, that is, with no power control. We considered the three positions of the interfered receiver that do not satisfy the INR_{th} of −6 dB for the use of $\boldsymbol{P}_{max}$. In addition, the three locations were designed considering the representative interference power, which can accurately reflect the operating characteristics of the proposed power control algorithm. Interfered receiver ① was located in the main beam direction for *Cell_3* and received the highest interference from *Cell_3*. Therefore, the minimum power use of only *Cell_3* satisfied an INR_{th} of −6 dB. Interfered receiver ② was placed on the boundary between *Cell_3* and *Cell_4* and thus received equal (and the strongest) interference from these two cells. Interfered receiver ③ was located in the main beam direction for *Cell_3*, as the interfered receiver. However, the minimum power use of only *Cell_3* could not satisfy the INR_{th} of −6 dB, and at least one other cell had to use less than the maximum power.

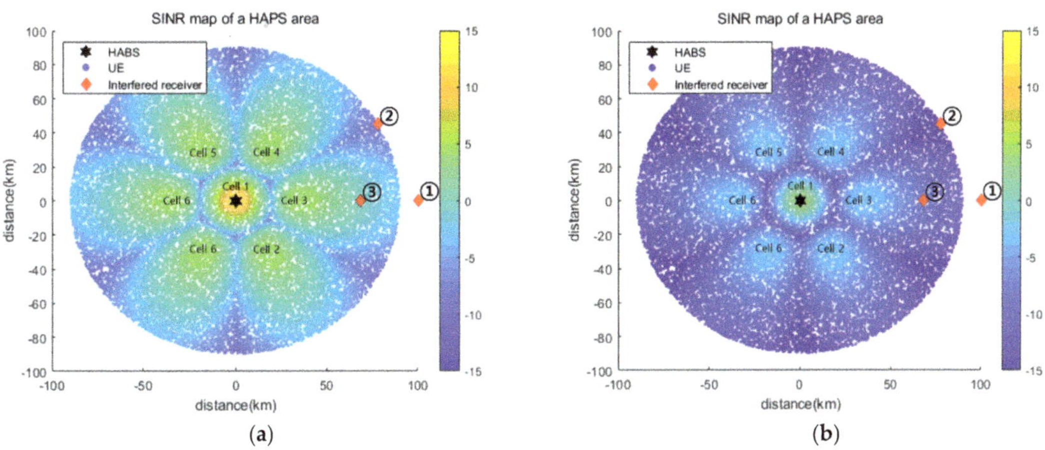

Figure 8. (**a**) SINR map for $\boldsymbol{P}_{max}$; (**b**) SINR map for $\boldsymbol{P}_{min}$.

Table 3 presents the INR and P_{out} for $\boldsymbol{P}_{max}$ and $\boldsymbol{P}_{min}$ with varying interfered receiver locations. The results confirm that the P_{out} and INR had a tradeoff relationship. The same P_{out} is shown regardless of the interference receiver position because of the absence of power control. Next, we compared the simulation results of the optimal exhaustive search and the proposed DQL-based power control algorithm for the three positions of the interfered receiver.

Table 3. INR and P_{out} for the interfered receiver locations.

Interfered Receiver	Location (km)	INR for P_{max} (dB)	INR for P_{min} (dB)	P_{out} for P_{max} (%)	P_{out} for P_{min} (%)
①	100, 0, 0.02	−3.01	−11.01	0	43.7
②	77.9, 45, 0.02	−4.08	−12.08	0	43.7
③	65.8, 0, 0.02	1.81	−6.19	0	43.7

5.2.1. Simulation Results for Interfered Receiver ①

Figure 9 shows the SINR map based on the $\boldsymbol{P_c}$ acquired using the proposed DQL-based power control algorithm for interfered receiver ①. Table 4 presents a performance comparison of the $\boldsymbol{P}^*$ values obtained via an exhaustive search and $\boldsymbol{P_c}$ and a comparison of DQL and DDQL results. As shown, $\boldsymbol{P_c}$ was equal to the optimal value $\boldsymbol{P}^*$, providing the same P_{out} and INR performance. Because the interfered receiver was located in the azimuth main beam direction of *Cell_3*, the power of *Cell_3* significantly affected the interfered receiver. Even though all other cells used the maximum power, their interference was negligible. Therefore, all the cells except for *Cell_3* used the maximum power for minimizing P_{out}, as shown in Table 4.

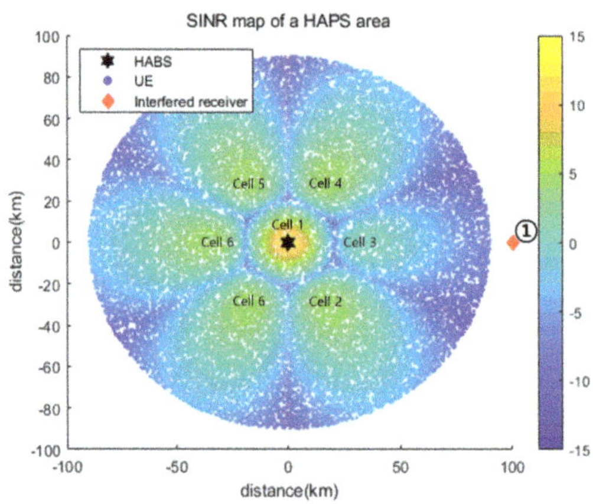

Figure 9. SINR map based on the $\boldsymbol{P_c}$ obtained using the proposed DQL-based power control algorithm for interfered receiver ①.

Table 4. Performance comparison for interfered receiver ①.

	P_{Cell_1} (dBm)	P_{Cell_2} (dBm)	P_{Cell_3} (dBm)	P_{Cell_4} (dBm)	P_{Cell_5} (dBm)	P_{Cell_6} (dBm)	P_{Cell_7} (dBm)	INR (dB)	P_{out} (%)
Optimal	37	34	30	34	34	34	34	-6.93	0.6
DQL	37	34	30	34	34	34	34	-6.93	0.6
DDQL	37	34	30	34	34	34	34	-6.93	0.6

Figure 10 presents the INR and p_{out} for each learning episode. As shown, the INR and p_{out} converged to the optimal values of the exhaustive search algorithm as the number of learning episodes increased. The INR started at −11.01 dB, which was the value for the use of $\boldsymbol{P_{min}}$, as shown in Table 3, and converged to the optimal value of −6.93 dB. Similarly, p_{out} started at 43.7% and converged to 0.6%. A large variance due to frequent exploration was observed at the beginning of the learning, but it gradually decreased and converged as the learning progressed. Figure 11 presents the cumulative and average rewards for each learning episode. As shown, the reward rapidly increased and then gradually converged at approximately 300 episodes, indicating that the proposed DQL training process allowed the agent to learn the power control algorithm quickly and stably.

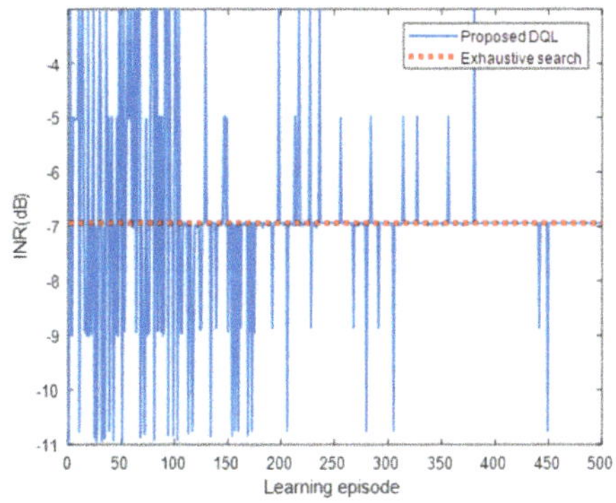

(**a**)

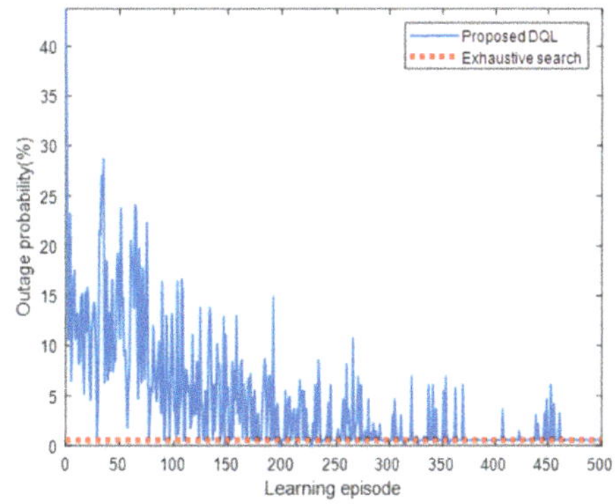

(**b**)

Figure 10. (**a**) INR and (**b**) p_{out} for each learning episode for interfered receiver ①.

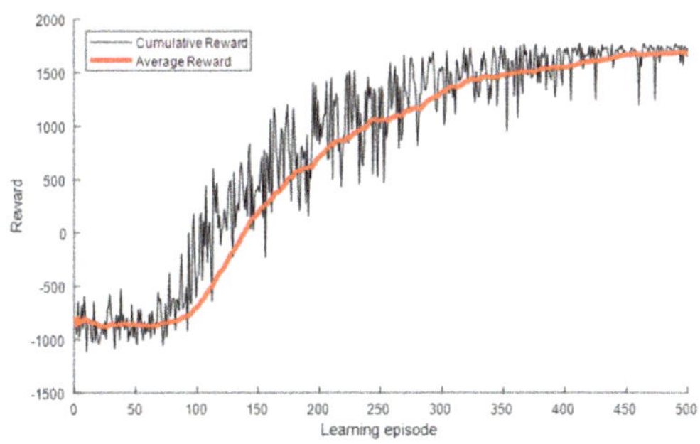

Figure 11. Reward for each learning episode for interfered receiver ①.

We compared the learning results of the DQL and DDQL. Even when the DDQL is used, the results are the same as in Table 4 and Figures 10 and 11, which shows that the overestimation of the DQL did not occur. As a result, it was confirmed that performance degradation due to overestimation did not happen, and sufficient learning is possible only with DQL.

5.2.2. Simulation Results for Interfered Receiver ②

Figure 12 shows the SINR map based on $\boldsymbol{P}_c$ acquired using the proposed DQL-based power control algorithm for interfered receiver ②. Table 5 presents a performance comparison of the $\boldsymbol{P}^*$ values obtained via an exhaustive search and $\boldsymbol{P}_c$ and a comparison of the DQL and DDQL results. As shown, $\boldsymbol{P}_c$ was equal to the optimal value $\boldsymbol{P}^*$, providing the same P_{out} and INR performance. The interfered receiver was located on the boundary between $Cell\_3$ and $Cell\_4$ and, thus, received equal (and the strongest) interference from these two cells. In addition, even though all the cells other than $Cell\_3$ and $Cell\_4$ used the maximum power, their interference was marginal. Therefore, in the optimal power control, $Cell\_3$ and $Cell\_4$ reduced the power required to satisfy the INR_{th}, whereas all the other cells used the maximum power for minimizing P_{out}, as shown in Table 5.

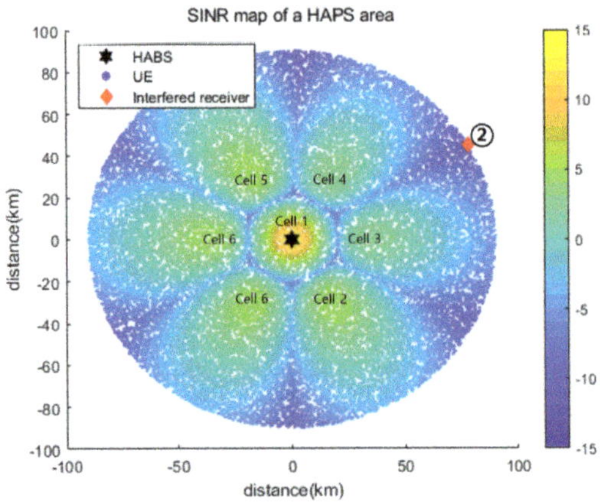

Figure 12. SINR map based on the P_c obtained using the proposed DQL-based power control algorithm for the interfered receiver ②.

Table 5. Performance comparison for interfered receiver ②.

	P_{Cell_1} (dBm)	P_{Cell_2} (dBm)	P_{Cell_3} (dBm)	P_{Cell_4} (dBm)	P_{Cell_5} (dBm)	P_{Cell_6} (dBm)	P_{Cell_7} (dBm)	INR (dB)	P_{out} (%)
Optimal	37	34	32	32	34	34	34	−6.08	0.2
DQL	37	34	32	32	34	34	34	−6.08	0.2
DDQL	37	34	32	32	34	34	34	−6.08	0.2

As shown in Figure 13, the INR and p_{out} converged to the optimal values of the exhaustive search algorithm. Similar to the case of receiver ①, as the learning progressed, the INR converged from −12.08 to −6.08 dB, and the p_{out} converged from 43.7% to 0.2%. Figure 14 shows that the reward gradually converged at approximately 300 episodes, indicating that the proposed DQL training process allowed the agent to quickly and stably learn the power control algorithm. We compared the learning results of the DQL and DDQL. Even when the DDQL was used, the results were the same as in Table 5 and Figures 13 and 14, verifying that the desired learning is attainable with the DQL only.

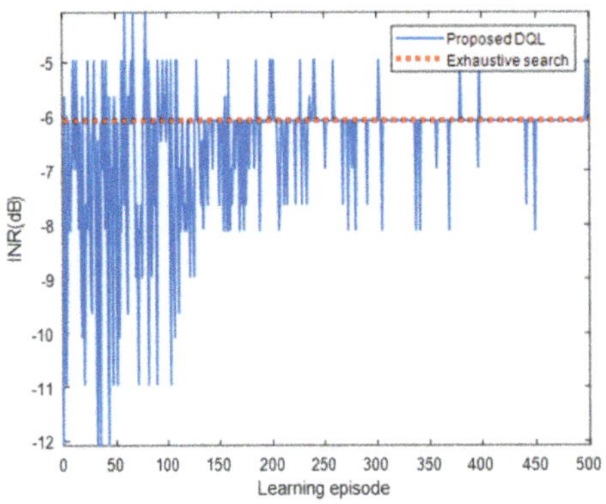

(a)

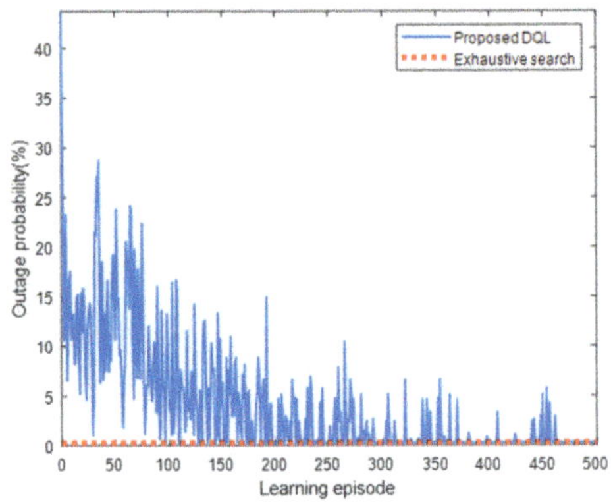

(b)

Figure 13. (a) INR and (b) p_{out} for each learning episode for interfered receiver ②.

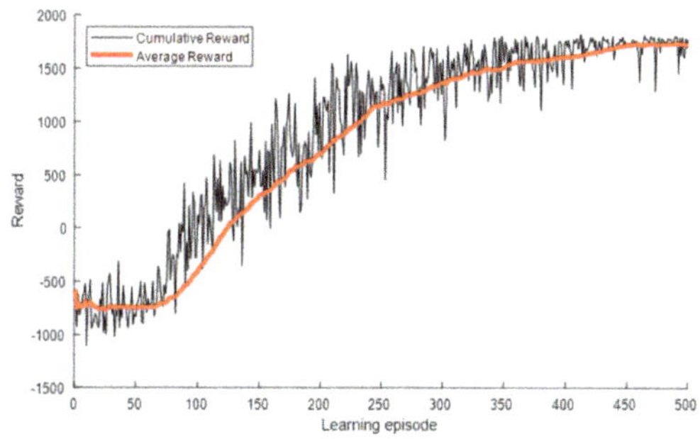

Figure 14. Reward for each learning episode for interfered receiver ②.

5.2.3. Simulation Results for Interfered Receiver ③

Figure 15 shows the SINR map based on $\boldsymbol{P}_c$ obtained using the proposed DQL-based power control algorithm for interfered receiver ③. The interfered receiver was located in the azimuth main lobe direction of *Cell_3*. It was closer to the HAPS than the receiver considered in Section 5.2.1 and was more severely affected by *Cell_3*; INR_{th} was not satisfied even for the minimum power of *Cell_3*. Thus, the optimal power control adjusted the power of *Cell_2* and *Cell_4*, which caused the second-most interference. Table 6 presents a comparison of the $\boldsymbol{P}^*$ values obtained using an exhaustive search and $\boldsymbol{P}_c$ and a comparison of the DQL and DDQL results. Although the p_{out} of $\boldsymbol{P}_c$ was 0.6% higher than that of $\boldsymbol{P}^*$, it corresponded to the third-smallest value among the 78,125 values generated by the exhaustive search algorithm. In summary, the proposed power control algorithm achieved outstanding performance close to the optimal value.

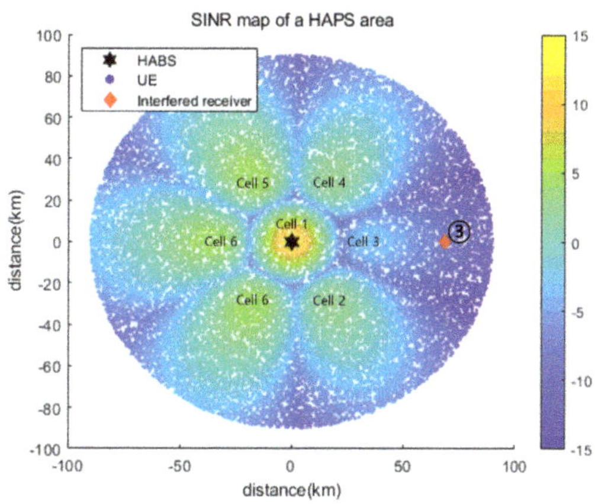

Figure 15. SINR map based on $\boldsymbol{P}_c$ obtained using the proposed DQL-based power control algorithm for interfered receiver ③.

Table 6. Performance comparison for interfered receiver ③.

	$P_{Cell\,1}$(dBm)	$P_{Cell\,2}$(dBm)	$P_{Cell\,3}$(dBm)	$P_{Cell\,4}$ (dBm)	$P_{Cell\,5}$ (dBm)	$P_{Cell\,6}$ (dBm)	$P_{Cell\,7}$ (dBm)	INR (dB)	P_{out} (%)
Optimal	37	34	26	32	34	34	34	−6.02	5.1
DQL	37	32	26	32	34	34	34	−6.06	5.7
DDQL	37	32	26	32	34	34	34	−6.06	5.7

As shown in Figure 16, the INR and p_{out} converged to the optimal values of the exhaustive search algorithm, with slight gaps. Similar to the results presented in Section 5.2.1, as the learning progressed, the INR converged from −6.19 to −6.06 dB, and the p_{out} converged from 43.7% to 5.7%. Figure 17 shows the cumulative and average rewards for each learning episode. The reward exhibited no noticeable improvement until approximately 130 episodes, after which it rapidly increased and then gradually converged at approximately 350 episodes. This is because to satisfy the INR_{th}, more agents had to take action, and the actions had to be more diverse. Nonetheless, the proposed DQL training process allowed the agent to learn the power control algorithm quickly and stably. We compared the learning results of the DQL and DDQL. Even when the DDQL was used, the results were the same as in Table 6 and Figures 16 and 17, verifying that the desired learning is attainable with the DQL only.

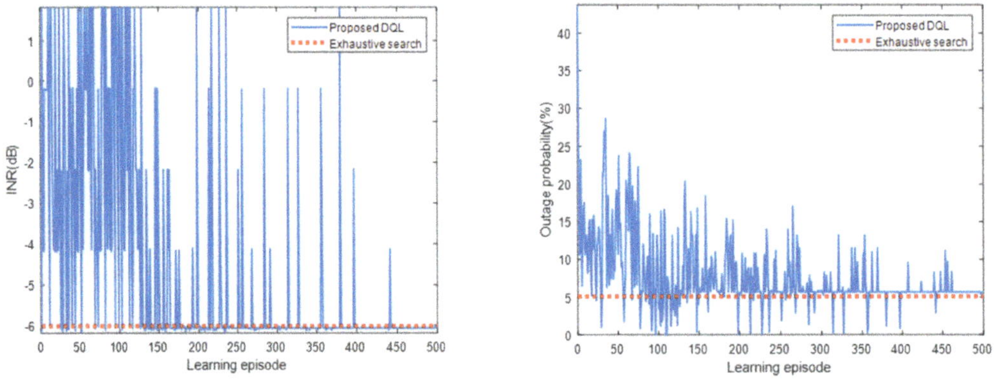

(**a**) (**b**)

Figure 16. (**a**) INR and (**b**) p_{out} for each learning episode for interfered receiver ③.

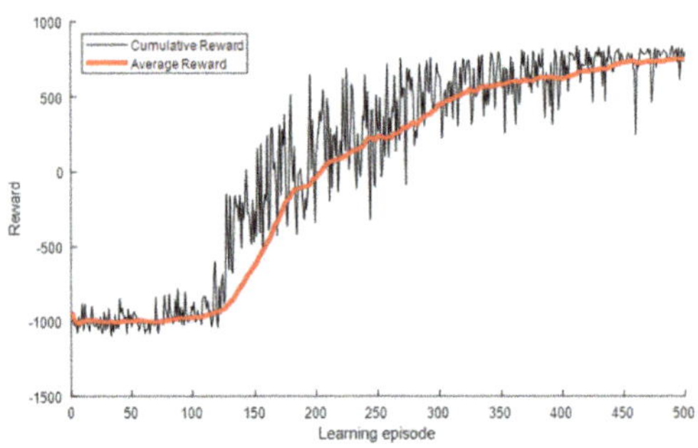

Figure 17. Reward for each learning episode for interfered receiver ③.

6. Conclusions

This paper proposed a DQL-based transmission power control algorithm for multicell HAPS communication that involved spectrum sharing with existing services. The proposed algorithm aimed to find a solution to the power control optimization problem for minimizing the outage probability of the HAPS downlink under the interference constraint to protect existing systems. We compared the solution with the optimal solution acquired using the exhaustive search algorithm. The simulation results confirmed that the proposed algorithm was comparable to the optimal exhaustive search.

Future work will include various power levels and expanding to multiple-HAPS communication in spectrum sharing with multiple interference systems. Since the increase in the power level could reveal a value-based algorithm's limit, it is preferred to apply the policy-based algorithm. Given that multiple-HAPS communication could lead to the non-stationarity problem of multiagent reinforcement learning, its solution would be worth studying.

Author Contributions: Conceptualization and methodology, S.J. and H.-S.J.; software, S.J. and W.Y.; validation, formal analysis, and investigation, S.J. and W.Y. and H.-S.J.; resources and data curation, H.K.C. and E.N.; writing—original draft preparation, S.J. and W.Y.; writing—review and editing, S.J., H.-S.J. and J.P.; visualization, W.Y. and H.K.C.; supervision, J.P.; project administration, H.-S.J. and J.P.; funding acquisition, J.P. All authors have read and agreed to the published version of the manuscript.

Funding: This work was supported by the Agency for Defense Development (ADD).

Institutional Review Board Statement: Not applicable.

Informed Consent Statement: Not applicable.

Data Availability Statement: Not applicable.

Conflicts of Interest: The authors declare no conflict of interest.

References

1. Arum, S.C.; Grace, D.; Mitchell, P.D. A review of wireless communication using high-altitude platforms for extended coverage and capacity. *Comput. Commun.* **2020**, *157*, 232–256. [CrossRef]
2. Alam, M.S.; Kurt, G.K.; Yanikomeroglu, H.; Zhu, P.; Đào, N.D. High altitude platform station based super macro base station constellations. *IEEE Commun. Mag.* **2021**, *59*, 103–109. [CrossRef]
3. Kurt, G.K.; Khoshkholgh, M.G.; Alfattani, S.; Ibrahim, A.; Darwish, T.S.; Alam, M.S.; Yanikomeroglu, H.; Yongacoglu, A. A vision and framework for the high altitude platform station (HAPS) networks of the future. *IEEE Commun. Surv. Tutor.* **2021**, *23*, 729–779. [CrossRef]
4. Hsieh, F.; Jardel, F.; Visotsky, E.; Vook, F.; Ghosh, A.; Picha, B. UAV-based Multi-cell HAPS Communication: System Design and Performance Evaluation. In Proceedings of the GLOBECOM 2020-2020 IEEE Global Communications Conference, Taipei, Taiwan, 7–11 December 2020.
5. Xing, Y.; Hsieh, F.; Ghosh, A.; Rappaport, T.S. High Altitude Platform Stations (HAPS): Architecture and System Performance. In Proceedings of the 2021 IEEE 93rd Vehicular Technology Conference (VTC2021-Spring), Online, 7–11 April 2021.
6. International Telecommunications Union (ITU). *World Radio Communication Conference 2019 (WRC-19) Final Acts*; International Telecommunications Union: Geneva, Switzerland, 2020; p. 366.
7. Ibrahim, A.; Alfa, A.S. Using Lagrangian relaxation for radio resource allocation in high altitude platforms. *IEEE Trans. Wirel. Commun.* **2015**, *14*, 5823–5835. [CrossRef]
8. Guan, M.; Wu, Z.; Cui, Y.; Cao, X.; Wang, L.; Ye, J.; Peng, B. An intelligent wireless channel allocation in HAPS 5G communication system based on reinforcement learning. *EURASIP J. Wirel. Commun. Netw.* **2019**, *2019*, 1–9. [CrossRef]
9. Guan, M.; Wang, L.; Chen, L. Channel allocation for hot spot areas in HAPS communication based on the prediction of mobile user characteristics. *Intell. Autom. Soft Comput.* **2016**, *22*, 613–620. [CrossRef]
10. Oodo, M.; Miura, R.; Hori, T.; Morisaki, T.; Kashiki, K.; Suzuki, M. Sharing and compatibility study between fixed service using high altitude platform stations (HAPS) and other services in the 31/28 GHz bands. *Wirel. Pers. Commun.* **2002**, *23*, 3–14. [CrossRef]
11. Mokayef, M.; Rahman, T.A.; Ngah, R.; Ahmed, M.Y. Spectrum sharing model for coexistence between high altitude platform system and fixed services at 5.8 GHz. *Int. J. Multimed. Ubiquitous Eng.* **2013**, *8*, 265–275. [CrossRef]
12. Lee, W.; Jeon, Y.; Kim, T.; Kim, Y.I. Deep Reinforcement Learning for UAV Trajectory Design Considering Mobile Ground Users. *Sensors* **2021**, *21*, 8239. [CrossRef] [PubMed]
13. Koushik, A.M.; Hu, F.; Kumar, S. Deep Q-Learning-Based Node Positioning for Throughput-Optimal Communications in Dynamic UAV Swarm Network. *IEEE Trans. Cogn. Commun. Netw.* **2019**, *5*, 554–566. [CrossRef]
14. Raja, G.; Anbalagan, S.; Narayanan, V.S.; Jayaram, S.; Ganapathisubramaniyan, A. Inter-UAV collision avoidance using Deep-Q-learning in flocking environment. In Proceedings of the 2019 IEEE 10th Annual Ubiquitous Computing, Electronics and Mobile Communication Conference, New York, NY, USA, 10–12 October 2019; pp. 1089–1095.
15. Fotouhi, A.; Ding, M.; Hassan, M. Deep Q-Learning for Two-Hop Communications of Drone Base Stations. *Sensors* **2021**, *21*, 1960. [CrossRef] [PubMed]
16. Anicho, O.; Charlesworth, P.B.; Baicher, G.S.; Nagar, A.K. Reinforcement learning versus swarm intelligence for autonomous multi-HAPS coordination. *SN Appl. Sci.* **2021**, *3*, 1–11. [CrossRef]
17. Hasselt, H.V.; Guez, A.; Silver, D. Deep reinforcement learning with double q-learning. In Proceedings of the Thirtieth AAAI Conference on Artificial Intelligence, Phoenix, AZ, USA, 12–17 February 2016.

18. International Telecommunications Union Radiocommunication Sector (ITU-R). *Working Document towards a Preliminary Draft New Report ITU-R M.[HIBS-CHARACTERISTICS]/Working Document Related to WRC-23 Agenda Item 1.4*; R19-WP5D Contribution 716 (Chapter 4-Annex 4.19); International Telecommunications Union: Geneva, Switzerland, 2021.
19. International Telecommunications Union Radiocommunication Sector (ITU-R). *Modelling and Simulation of IMT Networks and Systems for Use in Sharing and Compatibility Studies*; Recommendation ITU-R M.2101-0; International Telecommunications Union: Geneva, Switzerland, 2017.
20. International Telecommunications Union Radiocommunication Sector (ITU-R). *Reference Radiation Patterns of Omnidirectional, Sectoral and Other Antennas for the Fixed and Mobile Service for Use in Sharing Studies in the Frequency Range from 400 MHz to about 70 GHz*; Recommendation ITU-R F.1336-5; International Telecommunications Union: Geneva, Switzerland, 2019.
21. International Telecommunications Union Radiocommunication Sector (ITU-R). *Propagation Data Required for the Evaluation of Interference between Stations in Space and Those on the Surface of the Earth*; Recommendation ITU-R P.619-5; International Telecommunications Union: Geneva, Switzerland, 2021.
22. International Telecommunications Union Radiocommunication Sector (ITU-R). *Working Document towards Sharing and Compatibility Studies of HIBS under WRC-23 Agenda Item 1.4—Sharing and Compatibility Studies of High-Altitude Platform Stations as IMT Base Stations (HIBS) on WRC-23 Agenda Item 1.4*; R19-WP5D Contribution 716 (Chapter 4—Annex 4.20); International Telecommunications Union: Geneva, Switzerland, 2021.
23. Mnih, V.; Kavucuoglu, K.; Silver, D.; Rusu, A.A.; Veness, J.; Bellemare, M.G.; Graves, A.; Riedmiller, M.; Fidjeland, A.K.; Ostrovski, G.; et al. Human-level control through deep reinforcement learning. *Nature* **2015**, *518*, 529–533. [CrossRef] [PubMed]
24. International Telecommunications Union Radiocommunication Sector (ITU-R). *Characteristics of Terrestrial Component of IMT for Sharing and Compatibility Studies in Preparation for WRC-23*; R19-WP5D Temporary Document 422 (Revision 2); International Telecommunications Union: Geneva, Switzerland, 2021.

sensors

Article

Resource Allocation in Uplink NOMA-IoT Based UAV for URLLC Applications

Rana Karem [1,2,*], Mehaseb Ahmed [1] and Fatma Newagy [2]

1 Department of Electronics and Communications Engineering, Misr International University (MIU), Cairo 11828, Egypt; mehasb.ahmed@miuegypt.edu.eg
2 Department of Electronics and Communications Engineering, Ain Shams University (ASU), Cairo 11517, Egypt; fatma_newagy@eng.asu.edu.eg
* Correspondence: rana.kamal@miuegypt.edu.eg

Abstract: One of the main targets of future 5G cellular networks is enlarging the Internet of Things (IoT) devices' connectivity while facing the challenging requirements of the available bandwidth, power and the restricted delay limits. Unmanned aerial vehicles (UAVs) have been recently used as aerial base stations (BSs) to empower the line of sight (LoS), throughput and coverage of wireless networks. Moreover, non-orthogonal multiple access (NOMA) has become a bright multiple access technology. In this paper, NOMA is combined with UAV for establishing a high-capacity IoT uplink multi-application network, where the resource allocation problem is formulated with the objective of maximizing the system throughput while minimizing the delay of IoT applications. Moreover, power allocation was investigated to achieve fairness between users. The results show the superiority of the proposed algorithm, which achieves 31.8% delay improvement, 99.7% reliability increase and 50.8% fairness enhancement when compared to the maximum channel quality indicator (max CQI) algorithm in addition to preserving the system sum rate, spectral efficiency and complexity. Consequently, the proposed algorithm can be efficiently used in ultra-reliable low-latency communication (URLLC).

Keywords: internet of things; non-orthogonal multiple access; resource allocation; ultra reliable low latency communication; unmanned aerial vehicles; uplink transmission

Citation: Karem, R.; Ahmed, M.; Newagy, F. Resource Allocation in Uplink NOMA-IoT Based UAV for URLLC Applications. *Sensors* **2022**, *22*, 1566. https://doi.org/10.3390/s22041566

Academic Editor: Margot Deruyck

Received: 9 January 2022
Accepted: 15 February 2022
Published: 17 February 2022

1. Introduction

Due to their mobility and flexibility, unmanned aerial vehicles (UAVs)—also called drones—have become popular and have a wide range of applications. These applications include—though are not confined to—monitoring the environment, fertilizing and dusting crops, searching for mines, police monitoring, and operations of rescue. UAVs have received great attention thanks to its lower cost, high mobility, wide coverage and ease of deployment and flexibility [1]. These applications are not restricted to industrial and civilian fields only, but can also be extended to military applications [2].

In the near future, UAVs will be highly deployed in many applications in wireless networks, especially 5G networks, to serve ultra-reliable low-latency communication (URLLC) [3]. UAVs can present either in a single-UAV or multi-UAV system [4]. UAVs can be utilized as airborne base stations and flying relays to provide LoS connections to improve the coverage and power consumption for IoT devices. The recent era will witness billions of connections of Internet of Things (IoT) devices that will be served simultaneously [5]. Smart mobiles, vehicles, home appliances, and sensors are examples of these IoT devices [6].

The fifth generation (5G) network is intended to provide coverage for high-density devices with various applications of diverse quality of service (QoS) requirements [7]. Non-orthogonal multiple access (NOMA) is a promising multiple access technology which can face the challenging requirements of high throughput and low latency accompanied with the traffic of IoT devices. The basic concept of NOMA lies in the ability of NOMA to serve numerous users simultaneously over the same resource block (RB) to increase the spectrum efficiency [8]. However, due to simultaneous transmission, interference occurs between users sharing the same RB, so that successive interference cancellation (SIC) is used to detect the signal [9]. The exploitation of UAVs combined with NOMA is used to support the required massive connections of IoT devices and also provide long transmission range for IoT devices with limited transmission power capabilities. UAVs can easily overpass the IoT nodes, collect the information data and then transmit them to the data center or the other IoT devices [4]. IoT is accommodating numerous various applications with extremely precise performance requirements. Delay is one of the essential key metrics and one of the biggest challenges facing IoT applications [10]. In this paper, the UAV as a base station was exploited with NOMA to support URLLC IoT network with devices of different applications; each application has its own traffic parameters and delay requirements. A resource and power allocation scheme was proposed to manage the different QoS requirements for each application in uplink networks achieving high spectral efficiency, fairness and data rate as well as improving the delay and reliability.

The main contributions of this research are listed as follows:

- The uplink NOMA-based transmission system utilizing the UAV as a base station is modeled. Then, the resource allocation problem is formulated with the aim of maximizing the sum rate, taking into consideration the different delay requirements to serve URLLC applications.
- The proposed scheduler allocates resources jointly in both time domain and frequency domain based on the IoT devices parameters. Delay limits and priority are used by time domain; then the buffer status report (BSR) and channel quality indicator (CQI) control the frequency domain scheduler decision to allocate resources. In addition, a power allocation scheme is proposed to achieve fairness between the users allocated the same RB regardless of the different channel conditions.
- Unlike the previous works, the novelty of the presented algorithm lies in its ability to consider both the strict delay requirements of IoT devices and the system throughput while ensuring high reliability and fairness, where simulations are performed to evaluate the proposed scheduling algorithm performance. The results demonstrate the effectiveness of the proposed algorithm to serve URLLC traffic with restricted delay limits, due to the significant enhancement in delay, reliability and fairness, in addition to maximizing the sum data rate and spectral efficiency while achieving the same system complexity when compared to the maximum channel quality indicator (max CQI) algorithm.

2. Related Works

This section presents the state-of-the-art in resources scheduling in uplink 5G networks, UAV communication, and NOMA-based URLLC systems. Resource allocation and scheduling techniques in uplink 5G networks have been studied in many research papers presented in this literature. In [8], the authors proposed two resource allocation algorithms; the first one is the local rate maximization (LRM) in which the subcarrier is allocated to the user which gives the maximum rate on the chosen subcarrier. The second approach is global objective maximization (GOM) in which the allocated subcarrier is that which achieves maximum increase in the objective function.

A weighted sum rate maximization problem is modeled in [11], then a subcarrier allocation scheme which is based on iterative water-filling (IWF) algorithm is introduced. Its main idea is to initially begin with all devices allocated to all subchannels then remove the subchannel-device pair which gives the worst gain and power; repeatedly doing so until it meets the constraint that L devices are allocated to each subchannel [11]. The authors in [12] came up with a many-to-many matching model, where each subchannel forms a preference list based on the system throughput and each user creates its own preference list based on the received power. Iterative addition or substitution processes for the users to the subchannels are performed; for the purpose of enhancing the system performance until there is no more enhancement [12]. Reference [13] shows a two-sided matching and swapping technique, where firstly, each device forms its preference list based on channel gain and data rate, then each subchannel receives requests from the demanding devices and chooses the L devices with the highest energy efficiency. Then swapping operations are performed if and only if it is accompanied by an enhancement in energy efficiency [13]. UAV cellular networks were investigated with 5G technologies in [14,15] considering the channel model but the resource allocation problem was not inspected. In [16], backscatter communication technology, which is based on reflecting the incident wireless signals for the purpose of data collection, is investigated. The mobility of UAVs is exploited for maximizing the energy efficiency while optimizing the backscatter devices allocation and UAV trajectory [16]. The authors in [17] proposed NOMA scheme with index modulation to reduce the effect of contention, interference and collision in grant-free access. However, the system throughput was not considered. The authors in [18] used resource slicing and presented two user clustering mechanisms to meet the delay constraints of time stringent applications in uplink NOMA. NOMA technology is used in [19] and the resources are classified into shared and private. If the transmission and delay requirements of users cannot be achieved by the shared resources, then the private resources can be used [19]. Unlike the studies, the proposed algorithm is the first to consider the restricted delay limits of the IoT devices in addition to throughput maximization which makes the proposed algorithm suitable for URLLC applications.

The following sections in the rest of the paper are organized as follows. In Section 3, the system model is illustrated for the uplink of IoT devices served by UAV using NOMA access technique. Section 4 shows the problem formulation with the purpose of delay minimization and maximizing the network sum rate. Section 5 demonstrates the delay-rate optimization scheduling algorithm and the fairness optimization power allocation algorithm. Section 6 shows the performance evaluation of the proposed algorithm through the simulation results. Finally, the conclusion of the presented work is discussed in Section 7.

3. System Model

As shown in Figure 1, a UAV-assisted IoT network is considered where the UAV serves as an aerial base station, all IoT devices are served by a UAV that covers a cell with radius R_c. The UAV is located at 3D coordinates $(x_{UAV}, y_{UAV}, h_{UAV})$; assume that $(x_{UAV} = 0, y_{UAV} = 0, h_{UAV})$; i.e., the UAV is at the center of the cell with altitude h_{UAV}; such that $(h_{min} \leq h_{UAV} \leq h_{max})$ where h_{min} and h_{max} are the minimum and maximum allowable heights of the UAV. Assume that there are N ground IoT devices; each device is equipped with a single antenna. These devices are distributed randomly in the cell covered by the UAV; each device is located at (x_i, y_i) where $i = \{1, 2, \ldots, N\}$. Each IoT device has a packet arrival rate λ_i and transmits minimum rate R_i with delay limit D_i. Each IoT device has a transmitted power P_i $(P_{min,i} \leq P_i \leq P_{max,i})$ where $P_{min,i}$ and $P_{max,i}$ are the minimum and maximum transmitted power of IoT device i respectively. Path loss between IoT device i to the UAV expresses the large-scale fading component which is given by

$$PL_i = \frac{A}{1 + ae^{-b\ (\theta_i - a)}} + B_i, \tag{1}$$

$$A = \eta_{LOS} - \eta_{NLOS}, \tag{2}$$

$$B_i = 20\ log_{10}(d_i) + 20\ log_{10}\left(\frac{4\pi f_c}{c}\right) + \eta_{NLOS}, \tag{3}$$

where f_c is the carrier frequency, c is the light speed, $\eta_{LOS}, \eta_{NLOS}, a, b$ are constants related to the propagation environment either urban, suburban, dense urban or high-rise urban environments. Considering air to ground (A2G) communication between the UAV and the ground IoT devices, thus, each device can have line of sight (LoS) view or non-line of sight ($NLoS$) view with respect to the aerial base station with certain probability. $NLoS$ occurs when the propagation path is partially or fully obscured by physical obstacles. The *LoS* probability or *NLoS* probability are highly dependent on the device location, environment and the elevation angle between the device and the UAV, as illustrated in Figure 1. Pr_{LoS_i}, Pr_{NLoS_i} are the probabilities of IoT device i having *LoS* link or having non-line of sight (*NLoS*) link respectively between the UAV and IoT device, which could be calculated as [20]

$$Pr_{LoS_i} = \frac{1}{1 + aexp(-b[\theta_i - a])}, \tag{4}$$

$$Pr_{NLoS_i} = 1 - Pr_{LoS_i} \tag{5}$$

where θ_i is the elevation angle (measured in "degree") between IoT device i and the *UAV* as illustrated in Figure 1; it can be calculated as

$$\theta_i = \frac{180}{\pi} \sin^{-1}\left(\frac{h_{UAV}}{d_i}\right) \tag{6}$$

d_i is the distance between the IoT device i at (x_i, y_i) and the *UAV* located at (x_{UAV}, y_{UAV}, h_{UAV}) is

$$d_i = \sqrt{(x_{UAV} - x_i)^2 + (y_{UAV} - y_i)^2 + (h_{UAV})^2} \tag{7}$$

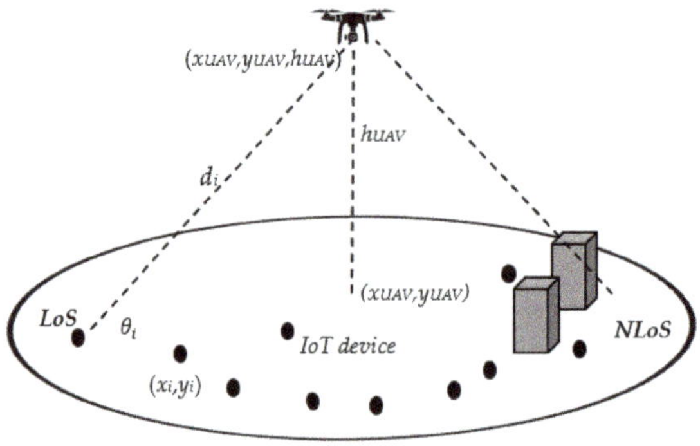

Figure 1. Illustration of the elevation angle in case of static UAV [21].

The *UAV* uses the NOMA access technique to communicate with the IoT devices. Assume that the total bandwidth allocated to the *UAV* is *B*, which is divided equally into *K* orthogonal subcarriers. Let $b_{k,i}(t)$ is the subcarriers assignment index, where $b_{k,i}(t) = 1$ means that subcarrier k is allocated to device i, otherwise $b_{k,i}(t) = 0$. There are maximum *L* IoT devices allowed to be scheduled over a single subcarrier at the same time and each IoT device gets exactly one subcarrier for simplicity. IoT devices can give preferences for the resource blocks, and their preferences are considered based on their channel quality indicator (CQI) [22] which depends on the channel quality between the *UAV* and IoT devices. The signal to noise plus interference ratio (SNIR) of IoT device i on subcarrier k is modeled as

$$\Gamma_{k,i}(t) = \frac{b_{k,i}(t) p_{k,i}(t) |g_{k,i}(t)|^2}{\sum_{f=1,\ |g_{k,f}(t)|^2 < |g_{k,i}(t)|^2}^{N} b_{k,f}(t) p_{k,f}(t) \left|g_{k,f}(t)\right|^2 + \sigma^2}, \tag{8}$$

$g_{k,i}(t)$ is the channel gain between the *UAV* and IoT device i on subcarrier k. $g_{k,i}(t)$ which is invariant in one time slot but varies over different time slots, can be modeled as:

$$g_{k,i}(t) = \sqrt{PL_{k,i}} h_{k,i}(t), \tag{9}$$

where $h_{k,i}(t)$ is the small scale fading of complex Gaussian distribution given by $h_{k,i}(t) \sim CN(0,\sigma^2)$. URLLC uses short packets to ensure low latency transmission, so Shannon capacity, which acts as the upper bound in terms of data rate, can no longer be applied. Consequently, the user data rate at finite blocklength transmission [23] is given by

$$R_{k,i}(t) = \log_2(1 + \Gamma_{k,i}(t)) - \sqrt{\frac{V_{k,i}}{m}}\, Q^{-1}(\varepsilon) \tag{10}$$

where, the decoding error probability ε [24] is given by

$$\varepsilon = Q\left(\ln 2\sqrt{\frac{m}{V}}\left(\log_2(1+\Gamma) - \frac{D}{m}\right)\right), \tag{11}$$

Such that the Q function and V are defined respectively as

$$Q(x) = \frac{1}{\sqrt{2\pi}} \int_x^{\infty} e^{\frac{-t^2}{2}}\, dt \tag{12}$$

$$V_{k,i} = 1 - (1 + \Gamma_{k,i})^{-2} \tag{13}$$

where D states the packet size, and m represents the blocklength of the channel.

The sum rate can be calculated as

$$R_{sum} = \sum_{k=1}^{K} \sum_{i=1}^{N} R_{k,i}(t), \tag{14}$$

Assume that packet arrival of the IoT device i follows a Poisson process with the average arrival rate as λ_i packets per second. The average service rate of the same IoT device is assumed to be μ_i packets per second. Both λ_i and μ_i are statistically identical and independent distributed. The queuing system model of the IoT device is shown in Figure 2. Then, the average delay $d_{av,i}$, which is based on the *M/M/1* queuing model [25], can be given by the following formula:

$$d_{av,i} = \frac{S_i}{R_{av,i} - S_i\lambda_i} \tag{15}$$

where S_i and $R_{av,i}$ are the packet size and the average rate of IoT device i respectively.

Figure 2. Queuing system model of the IoT device [26].

$R_{av,i}$ can be given as the rate of device i averaged over time as follows:

$$R_{av,i} = \frac{1}{T} \sum_{t=1}^{T_{slots}} \sum_{k=1}^{K} R_{k,i}(t) b_{k,i}(t), \tag{16}$$

where T_{slots} represents the total number of time slots.

4. Problem Formulation

The main objective is to maximize the sum data rate of all users while achieving minimum device delay through selecting the assignment index $b_{k,i}$ at every time slot.

Hence, the resource allocation problem is formulated with data rate maximization objective function as follows:

$$\underset{\{b_{k,i}(t)\}}{\text{Max}} \sum_{k=1}^{K} \sum_{i=1}^{N} R_{k,i}(t) \tag{17}$$

$$s.t.\ C1: \sum_{k=1}^{K} b_{k,i}(t) R_{k,i}(t) \geq R_{min,i} \tag{18}$$

$$C2: \sum_{k=1}^{K} b_{k,i}(t) p_{k,i}(t) \leq P_{max} \tag{19}$$

$$C3: b_{k,i}(t) \in \{0,1\} \tag{20}$$

$$C4: \sum_{k=1}^{K} b_{k,i}(t) p_{k,i}(t) \geq 0 \tag{21}$$

$$C5: \sum_{i=1}^{N} b_{k,i}(t) \leq L \tag{22}$$

$$C6: d_{av,i} < D_i \tag{23}$$

$$C7: h_{min} < h_{UAV} < h_{max} \tag{24}$$

where P_{max} is the maximum available transmission power of the IoT device. C1 is the minimum required data rate for all users to ensure QoS. C2 means that any IoT device transmit power cannot exceed P_{max}. C3 and C4 are the constraints of the assignment matrix $b_{k,i}$ and power $p_{k,i}$. C5 mentions that one subcarrier cannot be allocated to more than L users at the same time. C6 is the delay constraint, states that the average delay of the device i ($d_{av,i}$) should not exceed the delay limit requirement D_i. C7 is the UAV height constraint between the minimum and maximum allowable altitudes.

The optimization problem is non-convex due to the coupled variables $b_{k,i}$ and $p_{k,i}$ in (20) and (21). Hence, the proposed algorithm uses a linear weighted utility function to overcome the complexity of the optimization problem (17).

5. Proposed Algorithm

The optimization problem is non-convex due to the coupled variables $b_{k,i}$ and $p_{k,i}$ in (20) and (21). To find a simplified solution, the proposed scheduler in this section exploits the convexity property of the sub-problems [27] for the main non-convex problem (17). In addition, the proposed algorithm makes use of the resource element structure in NOMA which takes one subcarrier in frequency and one time symbol known as transmission time interval (TTI) [28], as demonstrated in Figure 3. Hence, the proposed algorithm operates in two domains, as illustrated in Figure 4. The first is delay minimization to guarantee the delay requirement of each IoT device which is performed by a time domain packet scheduler (TDPS). The second domain is rate maximization to increase the spectral efficiency and maximize the system sum rate which is performed by a frequency domain packet scheduler (FDPS).

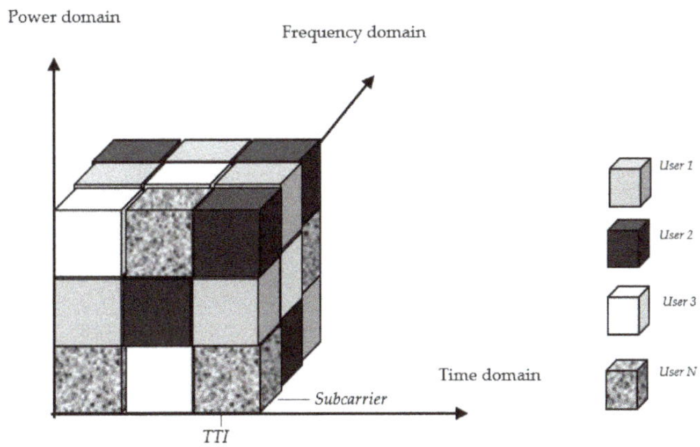

Figure 3. General NOMA resource allocation.

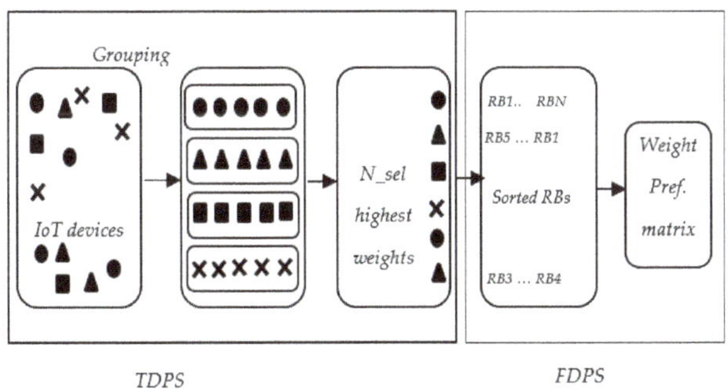

Figure 4. Illustration of the proposed scheduler.

The optimum height of the UAV should be calculated to solve constraint C7. The optimal UAV elevation angle $\theta_{optimum}$ was derived for each environment in [29] to cover a circular cell of certain radius in A2G communication independent of the multiple access technique used. Thus, according to [30]; the values of $\theta_{optimum}$ given by (25) for UAV placement corresponding to each environment either suburban, urban, dense urban or high-rise urban environments are illustrated in Table 1. Thus, the optimum UAV altitude $h_{optimum}$ to cover a cell of radius R_c can be given by

$$\theta_{optimum} = \tan^{-1}\left(\frac{h_{optimum}}{R_c}\right) \tag{25}$$

Table 1. Values of optimal elevation angle of each environment.

Environment	$\theta_{optimum}$
Suburban	20.34°
Urban	42.44°
Dense urban	54.62°
High-rise urban	75.52°

5.1. Time Domain Packet Scheduling (TDPS)

The main function of TDPS is to select a certain number of IoT devices to be scheduled in the next TTI; this process should guarantee that there are no devices exceed their maximum delay limits by minimizing the average delay. Hence, the optimization problem for time domain scheduling can be formulated using (15) as

$$\min d_{av,i} = \frac{S_i}{R_{av,i} - S_i\lambda_i}, \tag{26}$$

The device selection criteria in TDPS is based on a weighted delay metric which is essentially needed to prioritize the devices requests. The metric function is based on delay [31] to satisfy the delay constraint C6. The UAV uses the information in the BSR to compute the metric function for each requesting device. This metric aims to order the IoT devices requests to control the TDPS decision. The metric considers average delay and buffer status reports [32] as follows:

$$M_i = \alpha_j \, w_{i,j} d_{av,i} \tag{27}$$

where α_j states the priority of each traffic group, where the TDPS categorizes stations according to their delay limits into j groups, and $w_{i,j}$ is the traffic weight of user i in traffic class j, this weight can be calculated as

$$w_{i,j} = \frac{q_{i,j}}{bl}, \tag{28}$$

where $q_{i,j}$ is the queue size of user i in traffic group j, and bl is the buffer length to avoid buffer overflow of the IoT device.

After grouping and calculating metric for the IoT devices that send scheduling requests. The TDPS selects maximum N_{max} = KL devices to be scheduled in the next TTI, since there are L devices can be scheduled over the same resource block according to constraint C5 (22). Consider N_sel to be the number of chosen IoT devices to be scheduled. The rejected users send scheduling requests in the next TTI.

5.2. Frequency Domain Packet Scheduling (FDPS)

The main purpose of FDPS is allocating resources for IoT devices based on channel conditions (CQI) to increase spectral efficiency [32]. FDPS allocates resources to the IoT devices chosen by TDPS. Thus, the optimization problem of FDPS could be formulated as

$$\underset{\{b_{k,l}(t)\}}{\text{Max}} \sum_{k=1}^{K} \sum_{l=1}^{N\_sel} R_{k,l}(t) \tag{29}$$

$$s.t\ C1: \sum_{k=1}^{K} b_{k,l}(t) R_{k,l}(t) \geq R_{min,l} \tag{30}$$

$$C3: \ b_{k,l}(t) \in \{0,1\} \tag{31}$$

$$C5: \sum_{l=1}^{N\_sel} b_{k,l}(t) \leq L \tag{32}$$

In the proposed algorithm in FDPS each of the selected users prioritizes the RBs, where each IoT device sorts the RBs in descending order of CQI. The CQI value is an indication for the channel quality of each user on each subchannel. CQI value ranges from 0 to 15. The higher the value, the better the channel quality and vice versa. The CQI value is determined according to the estimated value of SINR [33]. Therefore, a preference matrix is formed containing the CQI values of all users on all RBs such that $CQI_{i,k}$ is the CQI value of user i on RB k; where the rows correspond to the IoT devices and the columns are related to the RBs. Each IoT device's preferences (CQI values) are listed in a row in this matrix; such that a higher CQI value indicates that the IoT device prefers this RB.

A weighted preference matrix is then formed which is based on the weighted metrics (27) computed in TDPS and the preferences determined in FDPS, and can be given by

$$f_{obj} = Diag(M_{1xN\_sel}) \; CQI_{N\_sel \; xK} \tag{33}$$

where M_{1xN_sel} is a row vector of the metrics of the selected devices and $CQI_{N\_sel \; xK}$ is the preference matrix which contains the CQI values of the selected devices on all RBs. Hence, the problem in (29) can be easily modified to

$$\underset{\{b_{k,l}(t)\}}{\text{Max}} \; - Diag(M_{1xN_{sel}}) \; CQI_{N_{sel}xK} b_{k,l}(t) \tag{34}$$

$$s.t \; C3: \; b_{k,l}(t) \; \in \{0,1\} \tag{35}$$

$$C5: \sum_{l=1}^{N\_sel} b_{k,l}(t) \; \leq \; L \tag{36}$$

The problem in (34) can be solved using binary integer programming which is used to solve the constrained problem to maximize f_{obj} The algorithm (Algorithm 1) to find the optimum allocation matrix is listed in detail as shown in the pseudo code.

Algorithm 1. Joint Delay-Rate Optimization Scheduler

Input: T, K, N, λ, P_{min} & P_{max}, j, L
Output: $\{s\}b_{k,i}$
Initialize: $N_{max} = L * K$
Step 1: chooses the N_{max} devices with the highest metrics in TDPS.
Arrange IoT devices in j groups according to the application type.
For $T_s = 1$ to $T_s = T$ **do**
R = IoT devices send scheduling requests
If R =< N_{max} go to Label
else, do
For i = 1 to i = R **do**
Compute the weight of device i (27):
$M_i = \alpha_i \; w_{i,j} d_{av,i}$
End for
Arrange IoT devices of each group in descending order of the weight (M_i)
Choose the N_max nodes which have the maximum weight to be scheduled in this TTI.
Label:
Choose the R nodes to be scheduled in this TTI.
End If
N_sel = the selected nodes to be scheduled
Step 2: assign each IoT device a resource block
For i = 1 to i = N_sel **do**
Form the preference matrix for each device i to all available RBs based on the CQI value.
Form the objective function which is the weighted preference matrix f_{obj}.
Initialize $\hat{k}$ = 1, correspondingly set $f_{max} = f_{obj}$.
if a solution $\grave{k}\epsilon\left\{ k\epsilon K \middle| f_{obj} \right\rangle f_{max} \}$ can be found: update $\hat{k} = \grave{k}$, and $f_{max} = f_{obj}$.
End for
Delay of the scheduled nodes is cleared, but that of waiting nodes is incremented.
Rejected nodes send scheduling requests in the next TTI.
End For

In the Joint Delay-Rate Optimization Scheduler, the UAV receives scheduling requests in every TTI. In TDPS, the UAV classifies the requests to groups according to the application. Then, the metric function is computed for all requesting devices using (27). The scheduler selects N_sel devices with the highest metrics. Then, in FDPS, the RBs allocated to each IoT device are chosen with the objective to maximize the weighted preference matrix in (33).

5.3. Power Allocation Algorithm

The Uplink power allocation scheme is used to ensure fair data rates between users sharing the same RB [34]. Consider two users share the same RB, if $g_{k,n} > g_{k,f}$ then according to (8) the SNIR of near and far users respectively are given as

$$\Gamma_{k,n}(t) = \frac{p_{k,n}(t)|g_{k,n}(t)|^2}{p_{k,f}(t)\left|g_{k,f}(t)\right|^2 + \sigma^2} \tag{37}$$

$$\Gamma_{k,f}(t) = \frac{p_{k,f}(t)\left|g_{k,f}(t)\right|^2}{\sigma^2}, \tag{38}$$

Assuming that the UAV will decode the signal of the nearest user first then decode the signal of the far user using SIC. To achieve fair data rates for both users, then,

$$\frac{p_{k,n}(t)|g_{k,n}(t)|^2}{p_{k,f}(t)\left|g_{k,f}(t)\right|^2 + \sigma^2} \quad \frac{p_{k,f}(t)\left|g_{k,f}(t)\right|^2}{\sigma^2}, \tag{39}$$

Let $|g_{k,n}(t)|^2 = g_N$ as the gain of the near user and $\left|g_{k,f}(t)\right|^2 = g_F$ as the gain of the far user. Thus, the power allocation coefficients a_N and a_F of the near and the far users respectively can be derived as

$$\frac{a_N P_{max} g_N}{a_F P g_F + \sigma^2} = \frac{a_F P_{max} g_F}{\sigma^2}, \tag{40}$$

Let $\rho = P_{max}/\sigma^2$, then divide both sides by σ^2

$$\frac{a_N \rho g_N}{a_F \rho g_F + 1} = \frac{a_F \rho g_F}{1}, \tag{41}$$

$$g_F^2 \rho^2 a_F^2 + (g_F + g_N)\rho a_F - \rho g_N = 0 \tag{42}$$

$$a_F = \frac{-(g_N + g_F) \pm \sqrt{(g_N + g_F)^2 + 4\rho g_F^2 g_N}}{2g_F^2 \rho} \tag{43}$$

where $0 < a_N < 0.5$ because $a_F > a_N$ and $a_N + a_F = 1$.

6. Simulation Results

In this section, simulation is introduced to evaluate the proposed algorithm performance. Consider a single UAV covering a cell of radius R_c = 1 km; assuming a suburban environment. An aerial UAV is centered and placed at the optimum height of the cell according to the values in Table 1. Consider IoT devices (N = 300) randomly deployed in the cell, IoT devices are supporting different applications, assuming there are 4 groups of IoT devices each has its unique traffic parameters. The maximum transmit power of each IoT device is 27 dBm and the minimum transmit power is 20 dBm. Unless stated, it is assumed that the system total available bandwidth B = 1.4 MHz consisting of 6 RBs. The noise power density $\sigma^2 = -174$ dBm/Hz. The main simulation parameters are summarized in Table 2.

To verify the efficiency of the proposed scheduler; the proposed algorithm is compared with the maximum CQI scheduling algorithm. According to [35], the best CQI algorithm shows its superiority in the achieved sum rate and spectral efficiency. The idea of the best CQI is based on scheduling the *N_sel* nodes which have the maximum CQI regardless of any other parameters.

Table 2. Simulation parameters.

Symbol	Description	Value
a	Environment Constant	4.88
b	Environment Constant	0.43
η_{LOS}	Line of sight Environment Constant	0.1
η_{NLOS}	Non-Line of sight Environment Constant	21
$\theta_{optimum}$	Optimal elevation angle	20.34°
σ^2	Noise power density	−174
U	Number of UAVs	1
P_{min}	IoT device minimum power	100 mW–20 dBm
P_{max}	IoT device maximum power	500 mW–27 dBm
R_c	Radius of the cell	1 km
TTI	Time slot	1 ms
	Simulation time	1 s
D_{limit}	Maximum delay limit	{10, 20, 30, 40} ms
λ	Arrival rate per group	{100, 250, 600, 400} (packets/s)
N	Total number of devices	300
D	Packet size	100 bits
m	Channel blocklength	100 symbols

Figure 5 shows the percentage of IoT devices that exceeds the delay limit, and as clearly seen the proposed algorithm outperforms the maximum CQI and the gap increases with the time, until it reaches a certain point at which the percentage of devices exceeding the delay limit saturates in both algorithms. This is due to its receiving large requests and scheduling *N_sel* nodes only in each TTI while rejecting the rest of nodes. The percentage of devices exceeding the delay tolerance saturates in the proposed algorithm to nearly 30% but around 44% in maximum CQI. The proposed algorithm gives priority to the nodes of the maximum buffer and least delay tolerance, resulting in 31.8% delay improvement.

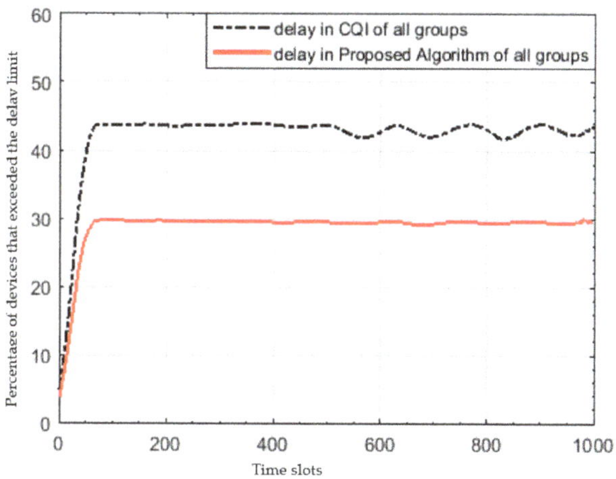

Figure 5. Percentage of devices exceeding the application delay limit at each time slot.

To evaluate the fairness of the proposed algorithm; the formula of Jain's fairness index in [8] is used, which is given by

$$Jain's\ fairness\ index = \frac{\left(\sum_{k=1}^{K} R_k\right)^2}{K \sum_{k=1}^{K} R_k^2} \tag{44}$$

where R_k represents the rate of user k and K is the total number of users. The fairness index values are confined between 0 and 1 such that the maximum value is achieved when the users have equal data rates. As obviously seen in Figure 6, the proposed algorithm is significantly fairer than the best CQI, since scheduling is performed sequentially from all groups, in addition to the fair power allocation algorithm used unlike the best CQI; resulting in 50.8% fairness enhancement at 300 users.

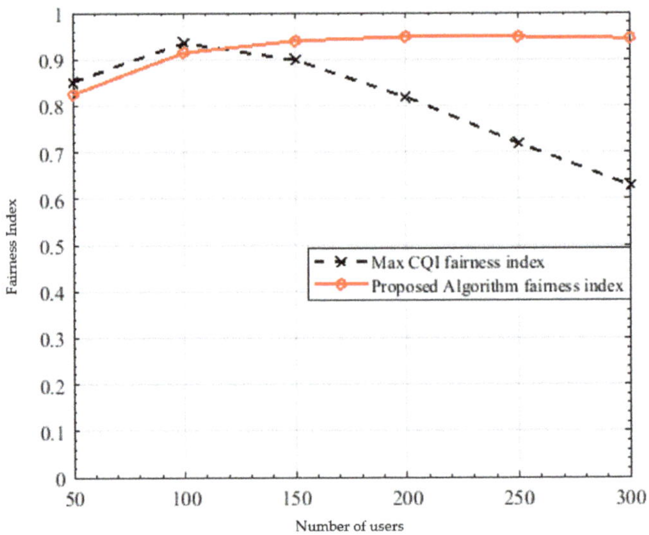

Figure 6. Fairness index versus number of users.

In Figure 7, the relationship between the transmitted power of the IoT devices and the sum rate is shown. As expected, the sum rate of the maximum CQI is greater than the proposed algorithm by only 1.6%; therefore, they still have a very close performance. By increasing the IoT device's maximum power, the sum rate is nearly the same and does not change; this is due to the slight increase in the device's maximum power that it does not affect the sum rate.

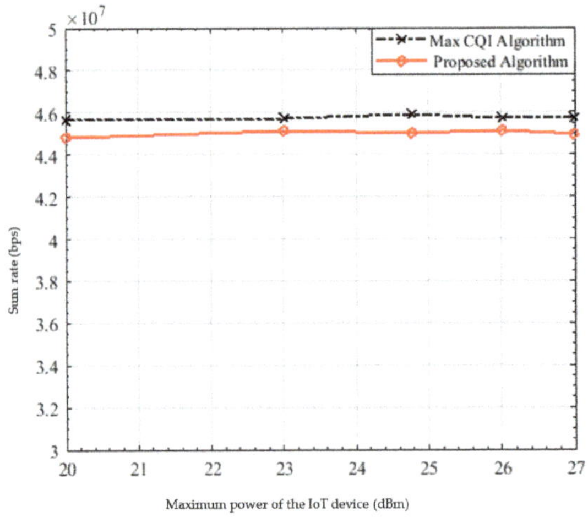

Figure 7. Sum rate versus the power of the IoT devices.

Figure 8 shows a comparison between the proposed algorithm and the best CQI in the achieved sum rate versus the number of users at different bandwidths; in case of 6, 10 and 25 resource blocks. They achieved nearly the same performance, which proves the superiority of the proposed algorithm. As the number of users increases, the sum rate increases to a certain saturation point after which the sum rate is almost constant due to scheduling the maximum number of users N_{max} in every TTI achieving almost the same sum rate. It can be noticed that the sum rate in case of 25 RBs outperforms that in case of 6 RBs and 10 RBs, while it is the least in the case of 6 RBs. It is observed that the saturation point is shifted to the right as the number of RBs increases, meaning that as the number of RBs increases, the sum rate saturates at larger number of users.

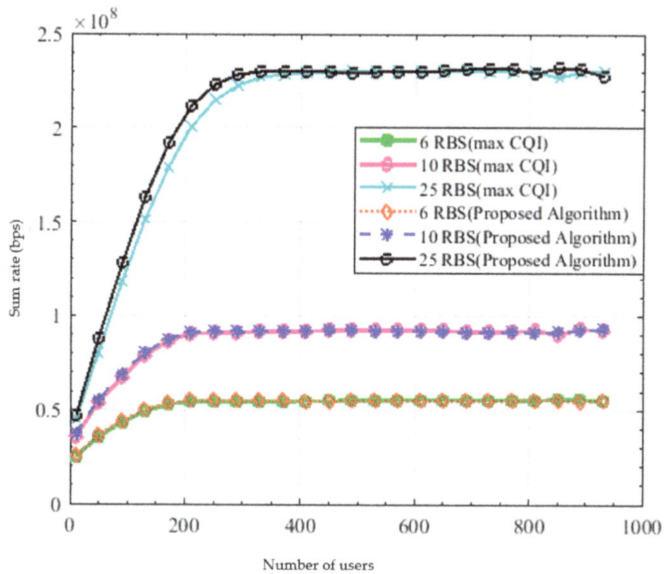

Figure 8. Sum rate versus number of users.

Figure 9 shows the achieved sum rate versus the bandwidth (number of RBs) with maximum 2 users allocated to the resource block. The spectral efficiency of the proposed algorithm is almost closer to that of the maximum CQI, which verifies the efficiency of the proposed algorithm in terms of spectral efficiency.

The results obtained in Figures 7–9, show the excellence of the proposed algorithm in terms of spectral efficiency and sum rate. This is because the proposed algorithm cannot exceed the best CQI which represents the upper bound in throughput and spectral efficiency. However, the proposed algorithm records a very close performance.

Figure 10 shows the spectral efficiency versus the maximum power of the IoT devices comparing the performance of both the proposed resource allocation algorithm and the best CQI once using the proposed power allocation and once using distance-based power allocation. As can be clearly seen, the proposed power allocation outperforms the distance=based power allocation. Moreover, as expected, the maximum CQI is superior in terms of spectral efficiency which is on average 41.2 bps/Hz followed by the proposed algorithm which achieves very close performance, nearly 40.5 bps/Hz and about 1.69% decrease only.

To evaluate the reliability of the algorithms, the relationship between the packet size and probability of decoding error is shown in Figure 11. As expected, as the packet size increases, the probability of error increases in both algorithms. However, the probability of error ε in the proposed algorithm is significantly lower than that of the max CQI, where at packet size = 1000 bits, ε is about 8.5×10^{-11} and 4.2×10^{-8} in the proposed algorithm and the max CQI respectively, resulting in nearly 99.7% performance improvement.

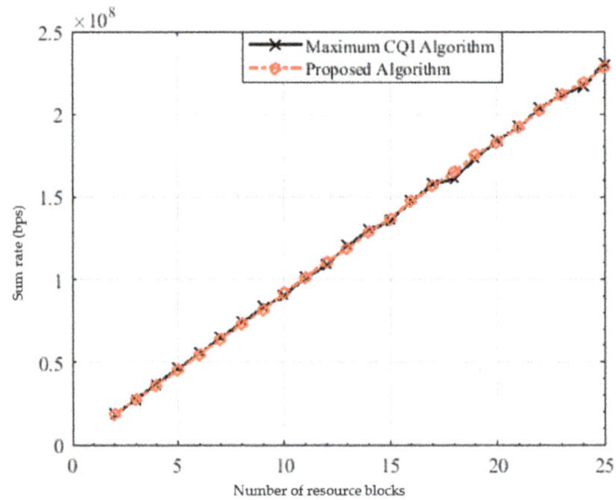

Figure 9. Sum rate versus the number of RBs.

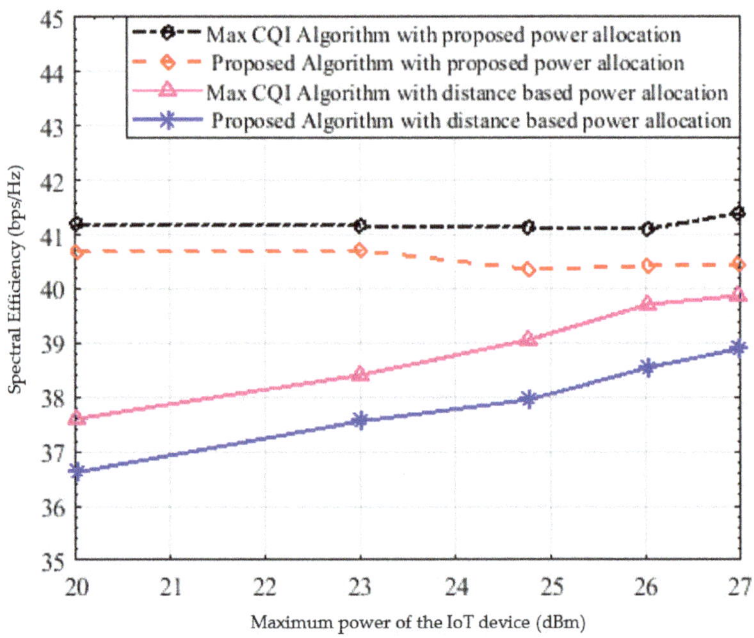

Figure 10. Spectral efficiency versus the power of the IoT devices.

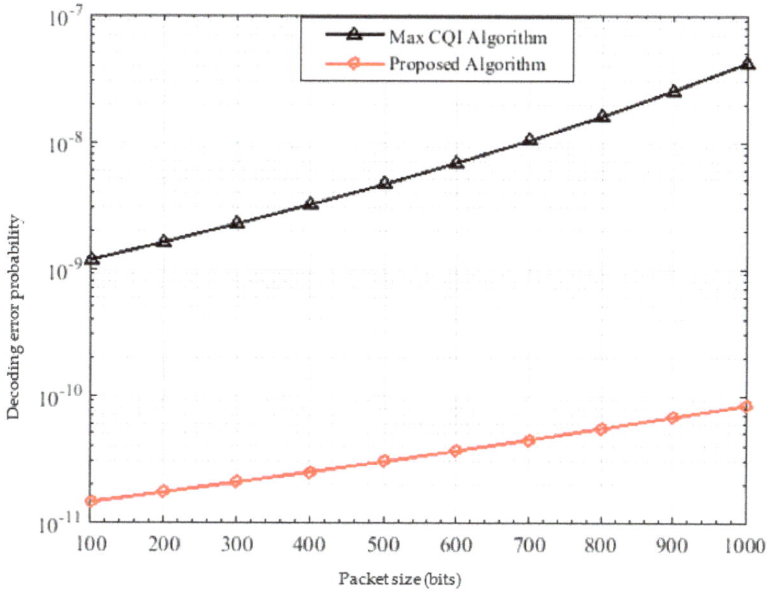

Figure 11. Decoding error probability versus the packet size.

The computational complexity of both the proposed algorithm and the best CQI is $O(KN)$ which means that they have linear complexity in the number of users and number of RBs. However, in [36], resource allocation for uplink multi carrier NOMA is developed using graph theory with complexity $O(KN^3)$. The shortest processing time (SPT) strategy is presented in [37] for uplink NOMA with complexity $O(N^2)$. Thus, the proposed algorithm has a worthy complexity improvement.

7. Conclusions

In this paper, an uplink NOMA resource allocation algorithm is proposed for a UAV-IoT-based communication network serving a large number of IoT devices of different applications. The optimization problem is formulated under constraints with the objective to maximize the data rate and minimize the delay. The scheduler works in time domain to optimize the delay, and in frequency domain to optimize the data rate. The power allocation is used to ensure fair data rates between users allocated the same RB. The simulation results show that the proposed algorithm significantly enhances the system fairness, delay and reliability, in addition to achieving a spectral efficiency and sum rate that are nearly closer to the system upper bound.

Author Contributions: Conceptualization, R.K., M.A. and F.N.; methodology, R.K. and M.A.; software, R.K.; validation, R.K., M.A. and F.N.; formal analysis, R.K.; writing—original draft preparation, R.K.; writing—review and editing, R.K. and M.A.; visualization, M.A.; supervision, M.A. and F.N. All authors have read and agreed to the published version of the manuscript.

Funding: This research received no external funding.

Institutional Review Board Statement: Not applicable.

Informed Consent Statement: Not applicable.

Data Availability Statement: Not applicable.

Conflicts of Interest: The authors declare no conflict of interest.

References

1. Abrar, M.; Ajmal, U.; Almohaimeed, Z.M.; Gui, X.; Akram, R.; Masroor, R. Energy Efficient UAV-Enabled Mobile Edge Computing for IoT Devices: A Review. *IEEE Access* **2021**, *9*, 127779–127798. [CrossRef]
2. Liu, X.; Ansari, N. Resource allocation in UAV-assisted M2M communications for disaster rescue. *IEEE Wirel. Commun. Lett.* **2018**, *8*, 580–583. [CrossRef]
3. Bennis, M.; Debbah, M.; Poor, H.V. Ultrareliable and low-latency wireless communication: Tail, risk, and scale. *Proc. IEEE* **2018**, *106*, 1834–1853. [CrossRef]
4. Duan, R.; Wang, J.; Jiang, C.; Yao, H.; Ren, Y.; Qian, Y. Resource Allocation for Multi-UAV Aided IoT NOMA Up-link Transmission Systems. *IEEE Internet Things J.* **2019**, *6*, 7025–7037. [CrossRef]
5. Muteba, K.F.; Djouani, K.; Olwal, T.O. Deep Reinforcement Learning Based Resource Allocation For Narrowband Cognitive Radio-IoT Systems. *Procedia Comput. Sci.* **2020**, *175*, 315–324. [CrossRef]
6. Zavyalova, D.; Drozdova, V. 5G Scheduling using Reinforcement Learning. In Proceedings of the 2020 International Multi-Conference on Industrial Engineering and Modern Technologies (FarEastCon), Vladivostok, Russia, 6–7 October 2020.
7. Zhang, Y.; Wang, X.; Xu, Y. Energy-Efficient Resource Allocation in Uplink NOMA Systems with Deep Reinforcement Learning. In Proceedings of the 2019 11th International Conference on Wireless Communications and Signal Processing (WCSP), Xi'an, China, 23–25 October 2019.
8. Al-Imari, M.; Xiao, P.; Imran, M.A.; Tafazolli, R. Uplink Non-Orthogonal Multiple Access for 5G Wireless Networks. In Proceedings of the 2014 11th International Symposium on Wireless Communications Systems (ISWCS), Barcelona, Spain, 26–29 August 2014.
9. Ahsan, W.; Yi, W.; Qin, Z.; Liu, Y.; Nallanathan, A. Resource Allocation in Uplink NOMA-IoT Networks: A Reinforcement-Learning Approach. *IEEE Trans. Wirel. Commun.* **2021**, *20*, 5083–5098. [CrossRef]
10. Segura, D.; Khatib, E.J.; Munilla, J.; Barco, R. 5G Numerologies Assessment for URLLC in Industrial Communications. *Sensors* **2021**, *21*, 2489. [CrossRef]
11. Al-Imari, M.; Xiao, P.; Imran, M.A. Receiver and Resource Allocation Optimization for Uplink NOMA in 5G Wireless Networks. In Proceedings of the 2015 International Symposium on Wireless Communication Systems (ISWCS), Brussels, Belgium, 25–28 August 2015.
12. Ruby, R.; Zhong, S.; Yang, H.; Wu, K. Enhanced Uplink Resource Allocation in Non-Orthogonal Multiple Access Systems. *IEEE Trans. Wirel. Commun.* **2017**, *17*, 1432–1444. [CrossRef]
13. Li, Y.; Zhang, H.; Long, K.; Choi, S.; Nallanath, A. Resource Allocation for Optimizing Energy Efficiency in NOMA-based Fog UAV Wireless Networks. *IEEE Netw.* **2019**, *34*, 158–163. [CrossRef]
14. Wang, J.Y.; Ma, Y.; Lu, R.R.; Wang, J.B.; Lin, M.; Cheng, J. Hovering UAV-based FSO communications: Channel modeling, performance analysis, and parameter optimization. *IEEE J. Sel. Areas Commun.* **2021**, *39*, 2946–2959. [CrossRef]
15. Azari, M.; Geraci, G.; Garcia-Rodriguez, A.; Pollin, S. UAV-to-UAV Communications in Cellular Networks. *IEEE Trans. Wirel. Commun.* **2020**, *19*, 6130–6144. [CrossRef]
16. Yang, G.; Dai, R.; Liang, Y. Energy-efficient UAV backscatter communication with joint trajectory and resource optimization. In Proceedings of the 2019 IEEE International Conference on Communications (ICC), Shanghai, China, 20–24 May 2019.
17. Doğan, S.; Tusha, A.; Arslan, H. NOMA with Index Modulation for Uplink URLLC Through Grant-Free Access. *IEEE J. Sel. Top. Signal Process.* **2019**, *13*, 1249–1257. [CrossRef]
18. Jaya, N.I.; Hossain, M.F. RAN Resource Slicing and Sharing with NOMA for Latency Reduction in Uplink URLLC Networks. In Proceedings of the 2020 IEEE Wireless Communications and Networking Conference Workshops (WCNCW), Seoul, Korea, 6–9 April 2020.
19. Deng, H.; Luo, F.; Li, Q. A Hybrid Resource Allocation Method for URLLC Based on NOMA. In Proceedings of the 2021 IEEE 21st International Conference on Communication Technology (ICCT), Tianjin, China, 13–16 October 2021.
20. Holis, J.; Pechac, P. Elevation dependent shadowing model for mobile communications via high altitude platforms in built-up areas. *IEEE Trans. Antennas Propag.* **2008**, *56*, 1078–1084. [CrossRef]
21. Chen, Y.; Feng, W.; Zheng, G. Optimum Placement of UAV as Relays. *IEEE Commun. Lett.* **2017**, *22*, 248–251. [CrossRef]
22. *3GPP Standard RP-020316*; Agreed CRs (Rel-5) for the WI of High Speed Downlink Packet Access—Physical Layer (Part 1). 3GPP TSG-RAN WG1 Publisher: Marco Island, FL, USA, 2002; 87–91.
23. Polyanskiy, Y.; Poor, H.V.; Verdú, S. Channel Coding Rate in the Finite Blocklength Regime. *IEEE Trans. Inf. Theory* **2010**, *56*, 2307–2359. [CrossRef]
24. Ren, H.; Pan, C.; Deng, Y.; Elkashlan, M.; Nallanathan, A. Joint Power and Blocklength Optimization for URLLC in a Factory Automation Scenario. *IEEE Trans. Wirel. Commun.* **2020**, *19*, 1786–1801. [CrossRef]
25. Bertsekas, D.P.; Gallager, R.G. Delay Models in Data Networks. In *Data Networks*, 2nd ed.; Athena Scientific: Belmont, MA, USA, 2021; pp. 162–163.
26. Garg, K.; Sharma, U. Improving Service Quality of Dhaba with the Help of Queuing Theory. *J. Manuf. Eng.* **2021**, *16*, 82–86.
27. Wang, H.; Wang, J.; Ding, G.; Wang, L.; Tsiftsis, T.A.; Sharma, P.K. Resource Allocation for Energy Harvesting-Powered D2D Communication Underlaying UAV-Assisted Networks. *IEEE Trans. Green Commun. Netw.* **2017**, *2*, 14–24. [CrossRef]

28. Lin, X.; Li, J.; Baldemair, R.; Cheng, J.F.T.; Parkvall, S.; Larsson, D.C.; Koorapaty, H.; Frenne, M.; Falahati, S.; Grovlen, A.; et al. 5G new radio: Unveiling the essentials of the next generation wireless access technology. *IEEE Commun. Stand. Mag.* **2019**, *3*, 30–37. [CrossRef]
29. Alzenad, M.; El-Keyi, A.; Lagum, F.; Yanikomeroglu, H. 3D Placement of an Unmanned Aerial Vehicle Base Station (UAV-BS) for Energy-Efficient Maximal Coverage. *IEEE Wirel. Commun. Lett.* **2017**, *6*, 434–437. [CrossRef]
30. Alzenad, M.; El-Keyi, A.; Yanikomeroglu, H. 3-D Placement of an Unmanned Aerial Vehicle Base Station for Maximum coverage of users with different QoS requirements. *IEEE Wirel. Commun. Lett.* **2017**, *7*, 38–41. [CrossRef]
31. Reyhani, A.; Song, S.; Primak, S.L.; Shami, A. Heterogeneous Delay-Power Resource Allocation in Uplink LTE. In Proceedings of the 6th Joint IFIP Wireless and Mobile Networking Conference (WMNC), Dubai, United Arab Emirates, 20–22 October 2013.
32. Mehaseb, M.A.; Gadallah, Y.; Elhamy, A.; Elhennawy, H. Classification of LTE uplink scheduling techniques: An M2M perspective. *IEEE Commun. Surv. Tutor.* **2015**, *18*, 1310–1335. [CrossRef]
33. Beschastnyi, V.; Ostrikova, D.; Melnikov, S.; Gaidamaka, Y. Modelling Multi-connectivity in 5G NR Systems with Mixed Unicast and Multicast Traffic. In Proceedings of the International Conference on Distributed Computer and Communication Networks, Moscow, Russia, 14–18 September 2020.
34. Lee, H. Power Allocation Scheme to Achieve Fair User Data Rates in NOMA Systems. In Proceedings of the 2019 34th International Technical Conference on Circuits/Systems, Computers and Communications (ITC-CSCC), Jeju, Korea, 23–26 June 2019.
35. Ashfaq, K.; Safdar, G.A.; Ur-Rehman, M. Comparative analysis of scheduling algorithms for radio resource allocation in future communication networks. *PeerJ Comput. Sci.* **2021**, *7*, e546. [CrossRef] [PubMed]
36. Zhai, D.; Zhang, R. Joint admission control and resource allocation for multi-carrier uplink NOMA networks. *IEEE Wirel. Commun. Lett.* **2018**, *7*, 922–925. [CrossRef]
37. Mohsenivatani, M.; Liu, Y.; Derakhshani, M.; Parsaeefard, S.; Lambotharan, S. Completion-Time-Driven Scheduling for Uplink NOMA-Enabled Wireless Networks. *IEEE Commun. Lett.* **2020**, *24*, 1775–1779. [CrossRef]

sensors

MDPI

Article

Cooperative Friendly Jamming Techniques for Drone-Based Mobile Secure Zone

Ga-Hye Jeon [1], Ji-Hyun Lee [2], Yeon-Su Sung [3], Hyun-Ju Park [4], You-Jin Lee [5], Sun-Woo Yun [1] and Il-Gu Lee [1,*]

[1] Department of Future Convergence Technology Engineering, Sungshin Women's University, Seoul 02844, Korea; 20171988@sungshin.ac.kr (G.-H.J.); nus0205@naver.com (S.-W.Y.)
[2] Department of Information & Technology & Data Science & Artificial Intelligence, Sogang University, Seoul 04107, Korea; easyhyun72@gmail.com
[3] Department of Convergence Security Engineering, Sungshin Women's University, Seoul 02844, Korea; yeonsu1936@gmail.com
[4] Department of Cybersecurity, Korea University, Seoul 02841, Korea; hjpark01@korea.ac.kr
[5] Department of Convergence Security, Sungshin Women's University, Seoul 02844, Korea; primarily68@gmail.com
* Correspondence: iglee@sungshin.ac.kr; Tel.: +82-2-920-7145

Abstract: Threats of eavesdropping and information leakages have increased sharply owing to advancements in wireless communication technology. In particular, the Internet of Things (IoT) has become vulnerable to sniffing or jamming attacks because broadcast communication is usually conducted in open-network environments. Although improved security protocols have been proposed to overcome the limitations of wireless-communication technology and to secure safe communication channels, they are difficult to apply to mobile communication networks and IoT because complex hardware is required. Hence, a novel security model with a lighter weight and greater mobility is needed. In this paper, we propose a security model applying cooperative friendly jamming using artificial noise and drone mobility, which are autonomous moving objects, and we demonstrate the prevention of eavesdropping and improved security through simulations and field tests. The Cooperative Friendly Jamming Techniques for Drone-based Mobile Secure Zone (CFJ-DMZ) can set a secure zone in a target area to support a safe wireless mobile communication network through friendly jamming, which can effectively reduce eavesdropping threats. According to the experimental results, the average information leakage rate of the eavesdroppers in CFJ-DMZ-applied scenarios was less than or equal to 3%, an average improvement of 92% over conventional methods.

Keywords: IoT; RF radio communication; Wi-Fi direct; D2D; drone-based mobile secure zone; friendly jamming; mobility

Citation: Jeon, G.-H.; Lee, J.-H.; Sung, Y.-S.; Park, H.-J.; Lee, Y.-J.; Yun, S.-W.; Lee, I.-G. Cooperative Friendly Jamming Techniques for Drone-Based Mobile Secure Zone. *Sensors* **2022**, *22*, 865. https://doi.org/10.3390/s22030865

Academic Editor: Margot Deruyck

Received: 17 December 2021
Accepted: 20 January 2022
Published: 24 January 2022

1. Introduction

Wireless communication network technology is evolving to meet the needs of users who want to use high-speed, high-capacity multimedia content without the limitations of location and time. However, cases of information leakage have been continually occurring owing to the fundamental limitations of wireless communication, which is vulnerable to eavesdropping [1]. To solve this problem, protocols and mechanisms have been proposed that improve the security or secure safe communication channels within the time, frequency, and space domains [2]. However, conventional wireless secure communication methods require complex hardware, which reduces the energy efficiency and data transmission performance, limiting the application to highly reliable wireless autonomous moving objects that exchange confidential information [3,4].

Among wireless network technologies, the Internet of Things (IoT) has become an essential element in all industries and everyday life, and accordingly, security vulnerabilities of wireless networks have become a larger issue [5]. Most IoT devices are light in weight with specifications that are insufficient to apply to regular personal computers or mobile devices. The existing communication systems are vulnerable to side-channel attacks, which are attacks that utilize physical information generated from the physical layer. Thus, the security protocol key can be leaked and encryption can be disabled. As they communicate in open network environments, they are easily exposed to security threats such as eavesdropping. This vulnerability can be exploited to easily collect confidential information [6].

To improve the security in IoT environments that are easily exposed to security threats, security architectures should be designed by considering the light weight and mobility of mobile IoT devices [7]. In this paper, we proposed Cooperative Friendly Jamming Techniques for Drone-based Mobile Secure Zone (CFJ-DMZ) to enhance the security in wireless communication environments for drones, which are IoT devices that can move autonomously while exchanging information with other surrounding devices. As eavesdropping is a passive attack that leaves no evidence of attacks, it is impossible to detect eavesdroppers in a wireless communication environment. Therefore, in this paper, a proactive prevention method was proposed to reduce the eavesdropping probability of unspecified potential eavesdroppers. The CFJ-DMZ uses the mobility and artificial interference of autonomous moving objects to form a secure zone, which guarantees safe data communication in the wireless communication environment of mobile IoT and flexibly controls the zone to be protected, effectively mitigating eavesdropping threats.

To evaluate the proposed CFJ-DMZ, we implemented a network simulation model and validated its performance. In the network simulation model, the transmission node moves to near the receiving node, and safe short-range device-to-device (D2D) communication is then conducted in the secure zone formed through a cooperative jammer drone. Here, three drones communicate with each other and create a locally secure zone based on the boundary of the artificial interference signal reaching range to protect the confidentiality of communication between nodes.

The main contributions of this paper are as follows:

- A cooperative friendly jamming technique is proposed to flexibly form a secure zone for confidential communication of lightweight mobile devices.
- The effect of the proposed CFJ-DMZ is analyzed through simulation and proved through field test in the implemented test bed.

The remainder of this paper is organized as follows. Section 2 introduces studies on conventional wireless communication security technologies and major studies on the wireless communication security of autonomous moving objects. Section 3 then describes the security performance criteria of friendly jamming and the friendly jamming model. Section 4 proposes a CFJ-DMZ model, and Section 5 validates the security performance of the proposed model through simulations and field experiments. Finally, Section 6 provides some concluding remarks and describes future research directions.

2. Related Work

2.1. Introduction to Friendly Jamming

Wyner defined the concept of security capacity in a study on eavesdropping channels and proved that security can be achieved according to the information theory perspective when the quality is poorer in the eavesdropper's channel than in the legitimate receiver [8]. Security channel capacity (secrecy capacity) is defined as the difference in the channel capacity between a legitimate sender and a legitimate receiver and the channel capacity between a legitimate sender and an eavesdropper. If this value is negative, the security channel capacity is zero, which means that no information can be safely sent. Here, signals and artificial interference can be generated and sent to improve the channel quality of the receiver and deteriorate the channel quality of the eavesdropper. Various studies have been conducted for this, including jamming, beamforming jamming, relay jamming, and friendly jamming techniques for improvements in wireless security. It has recently been proven that if a beam is formed using multiple antennas for jamming signals that can reduce the eavesdropping performance of malicious eavesdroppers, the communication security and reliability between legitimate communicators can be significantly improved. However, beamforming technology that uses multiple antennas or a massive antenna is complex and requires a large power consumption, making it difficult to use in IoT or mobile devices [9,10].

2.2. Friendly Jamming Security Model

Various studies are underway on friendly jamming security techniques applying jamming signals for security purposes [11–18]. A friendly jamming technique is a method of emitting friendly artificial interference signals to prevent malicious eavesdroppers from overhearing when the sender communicates with a legitimate receiver. A jammer is used to protect the wireless communication network and send messages with confidentially, allowing legitimate senders to communicate in a secure manner. In friendly jamming security technique research, security metrics are defined to demonstrate the effectiveness and validity of jamming. Friendly jamming security techniques include a method of using anti-jamming that automatically arranges multiple friendly jammers to deliver information safely between moving objects in a mobile communication network [16–18].

2.3. Friendly Jamming Security Model for Mobility Environment

Research is underway for friendly jamming security models using multiple unmanned aerial vehicles (UAVs) in mobile communication networks. The Friendly UAV Jamming (Fri-UJ) method has shown that the eavesdropping probability decreases as the number of jammers deployed near the protection zone increases because the signal to interference and noise ratio (SINR) of the eavesdropping device drops through the transmitted signals of the jammers [19]. In addition, there is a study that proved its effectiveness by applying Fri-UJ to IoT technology in the medical field [20]. However, using the Fri-UJ, as the number of UAVs increases, the number of jammers increases, thereby increasing the loss cost, and in practice, making it difficult to install an infinite number of jammers. Therefore, research is required to compare, analyze, and propose an effective friendly jamming model in a mobile communication environment based on the size of the secure zone, the number of jammers, and the security efficiency relative to cost.

In addition, research on location optimization considering the mobility of UAVs is being conducted. The location of the drone is important in order to reduce the waiting time and delay of the user or improve the quality of the service provided to users [21]. In [22], the disturbance power intensity and power trajectory path that occur when protecting legitimate nodes located on the ground using friendly jamming signals were studied. At that time, it was calculated by estimating the location of the eavesdropper, but the applicability is low because it is difficult to determine the location of the eavesdropper in the actual situations. Accordingly, in [23], randomness was modeled without identifying the location of the eavesdropper. At that time, in order to maximize the secret ratio of several legitimate receivers, the area was determined using the signal-to-noise ratio (SNR). However, as the confidentiality of the receiver in the security area is not guaranteed equally, some devices have a high potential for eavesdropping. In addition, when the optimal location is determined, only some factors change depending on the situation, so the area is not flexible and the mobility of UAVs cannot be utilized. In the case of battery usage, it is inefficient because jamming signals must be transmitted all the time.

Table 1 shows a summary of prior studies related to friendly jamming. In the table header, "Paper Title" refers to the title of the study, and "Research Topic" refers to the keywords of the study. "Number of Jammers" refers to the number of jammers used in the model: Here, "single" indicates that the model used one jammer, and "multiple" indicates that the model used two or more jammers. In addition, "Main Idea" refers to the model proposed in each paper, and "Limitation" refers to those analysis results that are a limitation of the indicated study. The models in the prior studies are inefficient and inflexible in terms of creating a secure zone for secure wireless communication because they consume large amounts of power and have no mobility. In this study, however, we proposed a model that creates an efficient and flexible secure zone by using only three drones that transmit friendly jamming signals, and we demonstrated the effectiveness of the model through simulations and field experiments.

Table 1. Comparison of prior studies on friendly jamming.

Paper Title	Research Topic	Number of Jammers	Main Idea	Limitation
UAV-enabled friendly jamming scheme to secure industrial Internet of Things [19]	Eavesdropping, Internet of Things, unmanned aerial vehicles	Multiple	Unmanned aerial vehicles (UAV)-enabled friendly jamming scheme	Need to study the optimal number of jammers
A jamming approach to enhance enterprise Wi-Fi secrecy through spatial access control [16]	Eavesdropping, Wi-Fi networks	Multiple	Defensive jamming approach in Wi-Fi networks secured by the WPA2 enterprise mode	Need to study friendly jamming techniques in a mobile environment
Achieving physical layer security with massive MIMO beamforming [17]	Antenna arrays, beamforming, cylindrical arrays, massive MIMO	Single	Beamforming with large cylindrical antenna arrays	Need to study appropriate node placement
Friendly jammer against an adaptive eavesdropper in a relay-aided network [14]	Relay-aided single-input single-output network, adaptive eavesdropping, outage probability	Single	Simulations for improvement in the secrecy capacity and SOP performances owing to the presence of friendly jamming	Need to study improved security for confident communication of nodes that want protection, and optimal number of jammers
Friendly jamming for wireless secrecy [12]	Cooperative jamming, jamming coverage, jamming efficiency, eavesdropping	Single	Cooperative/friendly jamming on the secrecy outage probability of a quasi-static wiretap fading channel	Need to study friendly jamming techniques in a mobile environment, and the optimal number of jammers

3. Friendly Jamming Technique

Equation (1) shows the information leakage rate (ILR) metrics defined for use as performance evaluation metrics of friendly jamming. The relationship between the bit error rate (BER) and friendly jamming was defined using the ILR. If the BER exceeds 0.5, then the ILR is 0, indicating that it is impossible to extract information because the BIR is high [24–26]. By contrast, if the BER is less than or equal to 0.5, then information extraction is possible, and ILR has a value greater than or equal to 0. As the ILR increases, the security decreases, indicating that more information can be extracted.

$$\mathrm{ILR} = 0,\ \text{if}\ \mathrm{BER} > 0.5\mathrm{ILR} = 1 - \frac{\mathrm{BER}}{0.5},\ \text{if}\ \mathrm{BER} \le 0.5 \tag{1}$$

Figure 1 shows the network configuration for verifying the security performance of friendly jamming. Figure 1a shows a friendly jamming technique model, which uses an open channel to send data (h^*_{sd}) to the sender node (Source, S) or receiver node (Destination, D) that uses a single antenna. Here, the eavesdropping device (Eve, E) also receives the data (h^*_{se}). A drone (Jammer, J) located near the source node transmits friendly jamming signals (h^*_{je}) to form a jamming zone. The eavesdropping quality of the eavesdropping device in the friendly jamming zone deteriorates as an effect of jamming, reducing the possibility of eavesdropping.

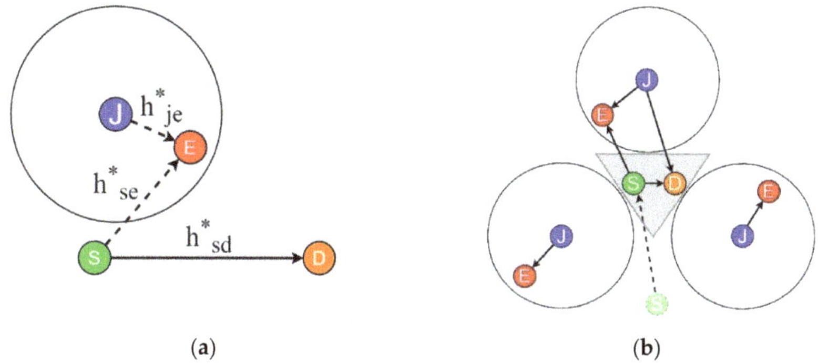

Figure 1. Network configuration for friendly jamming: (**a**) friendly jamming model and (**b**) CFJ-DMZ model.

4. CFJ-DMZ Model

This section introduces the CFJ-DMZ model that forms a secure zone by using the mobility of devices and the friendly jamming method. Three drones transmit jamming signals to the outside of the secure zone. These cooperative jamming signals reduce the eavesdropping probability of eavesdroppers and facilitate secure communication in the secure zone.

Figure 1b shows the CFJ-DMZ model proposed in this paper. The CFJ-DMZ network consists of a source node (S), destination node (D), arbitrary eavesdropper (E), and three friendly jamming drones (J). The drones transmit jamming signals after forming a secure zone, as shown in Figure 1b. The source and destination nodes communicate D2D inside the secure zone. As it is difficult to use a multi-antenna communication interface for low-power lightweight drones, we considered a communication method using a single antenna in this study.

Figure 2 shows the flowchart of a scenario for the CFJ-DMZ model, which consists of three stages.

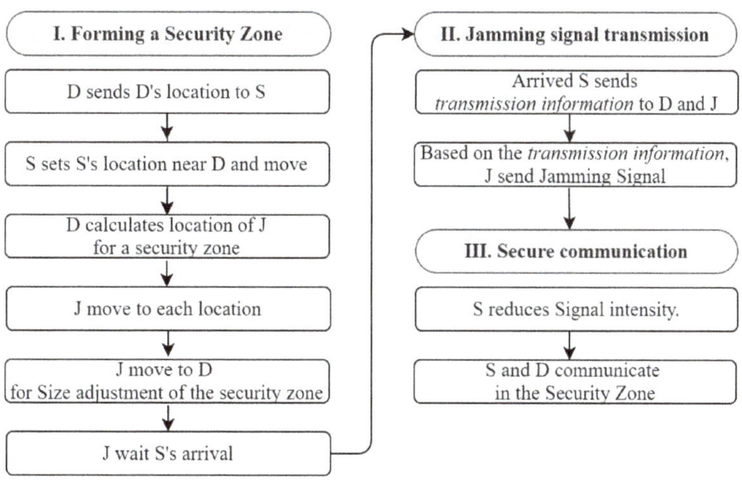

Figure 2. Simulation flow chart.

In the first stage, three drones are used to form a secure zone. When D2D communication is determined, the receiver (D) delivers its location coordinate information using GPS to the sender (D). In addition, S sends the determined location information to D and moves to the determined location, and D uses the arriving coordinate information of S and the current coordinate information of D to calculate the locations of the drones (J). Moreover, D creates a virtual circle around its position. Here, given that the maximum transmittable distance according to the strength of the transmitted jamming signals of J is x, the radius of the circle is as shown in Equation (2):

$$\text{Radius} = \frac{2}{3}\sqrt{3}x \tag{2}$$

Here, D selects three arbitrary points to make an equilateral triangle on the virtual circle created by centering on the position of D and sending the selected location to J. Each J moves to the received location. The J that arrives at each location moves to D until their respective jamming signals are no longer caught while transmitting to adjust the size of the secure zone. After the secure zone is created, J stops sending the jamming signals and waits until S arrives at the final target coordinates.

The second stage is the jamming signal transmission stage. After arriving at the final target coordinates, S sends the jamming signal transmission start time and the jamming signal transmission maintenance time to both D and J. The three units of J transmit jamming signals simultaneously based on the time information, and S and D can send secure data during the jamming signal transmission maintenance time. As such, J sends jamming signals only when communication is made according to the data transmission time. If jamming signals are continuously transmitted irrespective of the data transmission, the security inside the secure zone improves. However, continuous jamming signals interfere with the communication of other nearby transmitting and receiving objects [27]. Furthermore, it is inefficient to transmit jamming signals continuously in terms of energy. In the CFJ-DMZ, J therefore sends jamming signals only when S and D are communicating to minimize the effects of such signals on other nearby sender and receiver nodes and to use their batteries efficiently. In addition, it is assumed that the secure zone is formed only when the legitimate nodes exchange confidential information with the help of surrounding UAVs used for other purposes. Therefore, the cost of friendly jamming drones for CFJ-DMZ was not considered in this paper. On the other hand, very small control logic can be added to friendly jamming drones and the ground user's hardware to implement the proposed method, but it is assumed that the added hardware cost is trivial.

Finally, S and D communicate inside the secure zone formed by the jamming-signal transmission of J. Here, because the distance between S and D is close, the transmission signal strength of S is reduced. The three units of J transmit cooperative jamming signals. As a result, the probability of success of the eavesdroppers decreases, improving the security of D2D communication in the secure zone.

In this study, we conducted simulations and field experiments to prove the security of the CFJ-DMZ model. Octave was used for the simulations, and Raspberry Pi 3 was used in the field experiments. Section 5 discusses the security verification process and results. The experiments were focused on proving the security improvement inside the secure zone. Therefore, as a part of the process under the CFJ-DMZ scenario, we assumed that the source node has moved and that the friendly jamming drones have completed moving to form a secure zone.

5. Experiment

5.1. Simulation

5.1.1. Effect of Friendly Jamming

The CFJ-DMZ model proposed in this paper uses three drones as friendly jammers to protect the confidentiality of D2D communication. At this time, the three drones are theoretically the smallest number to make a two-dimensional space, and they protect the communication of the two legitimate nodes by forming a cost-performance effective secure zone. The jamming effect on the eavesdropper is affected by the distance between the source node and the eavesdropper (Source − Eve distance, d_{SE}) and the distance between the jamming drones and the eavesdropper (Drone − Eve distance, d_{DE}). Figure 3 shows the BER of the eavesdropper according to changes in d_{SE} and d_{DE} measured using the simulation. The BER of the eavesdropper was measured by changing each distance from 1 to 100 m. As a result, we found that when d_{DE} is shorter than d_{SE}, the BER of the eavesdropper increases, reducing the communication quality of the eavesdropper.

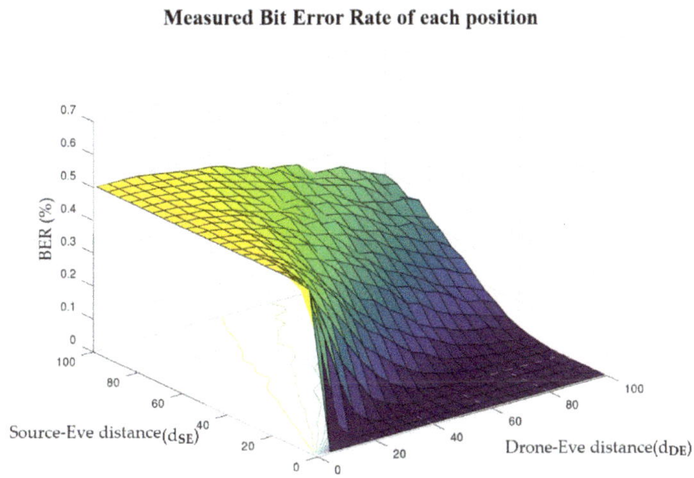

Figure 3. BER based on distance of Source-Eve and Drone-Eve.

5.1.2. Evaluation Environments

The evaluation simulator was implemented using Octave v.6.1. The simulation was conducted in a PC environment with the Windows 10 operating system, 8 GB of RAM, and an Intel i5-7200U CPU. In the simulation, a free space of 200 m × 200 m was formed, and the location of nodes was randomly arranged to prove the effectiveness of the CFJ-DMZ. The experimental environment was set as a free space that did not take into account the influence of air or other radio waves. In addition, the maximum transmission power of the transmitting node, receiving node, and friendly jamming drone in the simulation was 24 dBm each.

We configured four experimental settings to measure the leaked amount of information according to the mobility of the device and the friendly jamming technique. Figure 4 shows each experimental configuration.

After measuring the locations of the source node (S) and destination node (D), the locations of three drones (J) were calculated. The blue circles in Figure 4b,d represent the range in the cooperative jamming signals. A total of six eavesdropping nodes (eve1–6) were created, and the eavesdropping nodes (E) were located at arbitrary coordinates. The locations of the eavesdroppers were the same in all experiments for an effective comparison between the experiments. Table 2 shows a summary of the location of each node. In the simulation experiments, the distance from the source node to the destination source was set to 48.413 m in Figure 4a,b and to 2 m in Figure 4c,d.

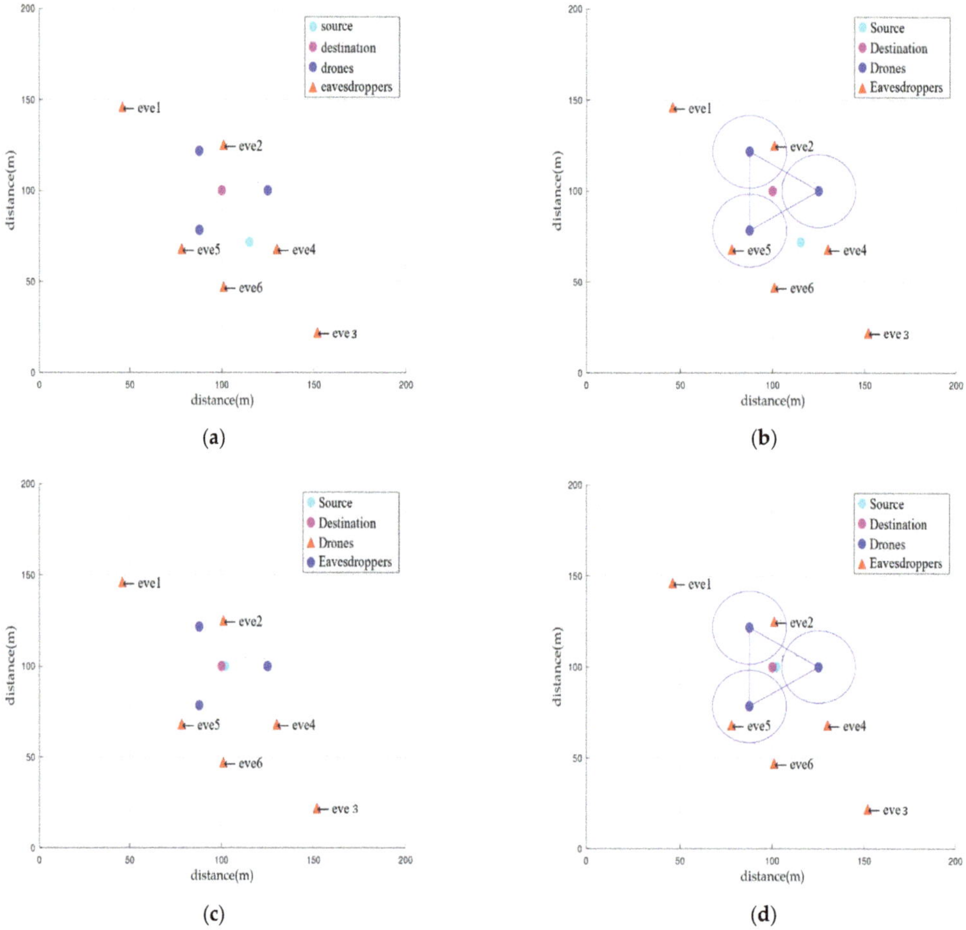

Figure 4. Network configuration for CFJ-DMZ simulation: (**a**) None mobility and None friendly jamming, (**b**) None mobility and Friendly jamming, (**c**) Mobility and None friendly jamming, and proposed scheme (**d**) Mobility and Friendly jamming.

Table 2. Location of each node in simulation.

Node	Source in Figure 4a,b	Source in Figure 4c,d	Destiantion	eve1	eve2	eve3	eve4	eve5	eve6
Location	(127.42, 60.1)	(102, 100)	(100, 100)	(23, 146)	(101, 125)	(155, 22)	(149, 89)	(73,71)	(101, 47)

The test cases of the experimental settings in Figure 4 can be summarized as shown in Table 3. In the table, "mobility" refers to with or without movement of the source node. If the mobility is O, it indicates a case in which the source node sends information after moving to near the destination node. If the mobility is X, it indicates a case in which the source node sends information without moving to near the destination node. When the source node has mobility, and the distance between the source and destination nodes is close, the transmission signal strength of the source node is reduced. "Friendly jamming" refers to whether the friendly jamming technique of the drone is used. If friendly jamming is O, it indicates a case in which the drones transmit cooperative jamming signals when sending information. If it is X, it indicates a case in which friendly jamming signals are not transmitted. "Source-Destination distance" refers to the distance between the source and destination nodes. The CFJ-DMZ model corresponds to Figure 4d in which both mobility and friendly jamming occur.

Table 3. Environment setting of Figure 4.

Case	Figure 4a	Figure 4b	Figure 4c	Figure 4d
Mobility	X	X	O	O
friendly jamming	X	O	X	O
Source-Destination distance (m)	48.413	48.413	2	2

The design process of the simulation is as follows. First, it is assumed that there is a channel h^*_{sd} in the form of a complex conjugate between the source and destination nodes. It is also assumed that P_s at the transmitter and the receiver, and the maximum transmission power of P_j of the friendly jamming drones, are both 24 (dBm). Under these assumptions, the following equations are defined to measure the BER at each object.

First, the signal that the destination node receives from the source node in Equation (3) can be expressed based on the distance between the source and destination nodes (h^*_{sd}), the distance between the jammer drones and the destination node (h^*_{jd}), the maximum transmission power of the receiver (P_s), the maximum transmission power of the jammer drones (P_j), and the noise (n_d) [28]:

$$y_e = G\sqrt{P_s}h^*_{sd}s + \sqrt{P_j}\, h^*_{jd}q + n_d \tag{3}$$

where the channel coefficient h is shown by Equation (4). In addition, d is the distance between the two communication nodes, e is a uniformly distributed random number $a + bi$, and c is the path loss exponent.

$$h = (d)^{\frac{-c}{2}} e \tag{4}$$

The amplification scale vector G can be shown through Equation (5), where N refers to the Gaussian noise.

$$G = \frac{1}{\sqrt{P_s \left|h^*_{sd}\right|^2 + N}} \tag{5}$$

Finally, the signal received by the eavesdropper in Equation (6) can be represented by the distance between the source node and the eavesdropper (h^*_{se}), the distance between the jammer drones and the eavesdropper (h^*_{je}), the maximum transmission power of the receiver (P_s), the maximum transmission power of the jammer drones (P_j), and the noise (n_d).

$$y_e = G\sqrt{P_s}h^*_{se} + \sqrt{P_j}h^*_{je}q + n_e \tag{6}$$

For the BER of the eavesdropper node in each experiment, we used an average of 1000 times, as shown in Equation (7). In Figure 4a,c, where there is no jamming signal, $G\sqrt{P_j}h^*_{je} * JamSymbols$ is calculated as zero.

$$y_e = G\sqrt{P_s}h^*_{se} * \ TrsutSymbols + G\sqrt{P_j}h^*_{je} * JamSymbols \tag{7}$$

Table 4 shows the definitions of the parameters used in the simulation pseudo-code, and Algorithm 1 shows the simulation pseudo-code itself. The detailed operating principle of Algorithm 1 is as follows.

Table 4. Defined parameters for pseudo-code.

Notation	Remark
c	Channel coefficient for free-space path loss
e	Randomized complex number
h_{se}	Channel coefficient for free-space path loss of distance between source and eve
h_{sd}	Channel coefficient for free-space path loss of distance between source and destination
h_{je}	Channel coefficient for free-space path loss of distance between jammer and eve
G	Scaling factor of amplification based on the distance source and destination

Algorithm 1. Pseudo-code for BER measurement of eve.

1: $Drone_1toEveDistatnce \leftarrow Distance\ between\ Drone_1\ and\ Eve$
2: $Drone_2toEveDistatnce \leftarrow Distance\ between\ Drone_2\ and\ Eve$
3: $Drone_3toEveDistatnce \leftarrow Distance\ between\ Drone_3\ and\ Eve$
4: $CalculateJEDistance \leftarrow Calculate\ the\ distance\ of\ the\ cloest\ drone\ from\ eve$
5: $jeDistance \leftarrow The\ distance\ between\ one\ jammer\ and\ the\ eavesdropper\ affected\ by\ the\ jammers$
6: $maxLoop \leftarrow 1000$
7: $c \leftarrow path\ loss\ exponent$
8: *procedure MeasureEveBER* (N_{bits}, c, *seDistance*, *sdDistance*, *jeDistance*, P_t, P_j)
9: $for\ 1\ to\ maxLoop$
10: $e \leftarrow randomComplexNumber(N_{bits})$
11: $h_{se} \leftarrow seDistance^{-\frac{c}{2}} * e$
12: $h_{sd} \leftarrow sdDistance^{-\frac{c}{2}} * e$
13: $h_{je} \leftarrow jeDistance^{-\frac{c}{2}}$
14: $G \leftarrow \frac{1}{\sqrt{\left(p_{t|h_{sd}|^2}\right)}}$
15: $SignalBits \leftarrow Generate\ randomly\ Signal\ Bits\ of\ N_{bits}\ at\ 0,1$
16: $SignalSymbols \leftarrow Mapping\ SignalBits\ to\ SignalSymbol\ in\ the\ form\ of\ complex\ number$
17: $JamBits \leftarrow Generate\ randomly\ JAmming\ Symbol\ of\ N_{bits}\ at\ 0,1$
18: $JamSymbols \leftarrow Mapping\ JamBits\ to\ JamSymbol\ in\ the\ form\ of\ complex\ number$
19 $eveRecieveSymbol \leftarrow G * \sqrt{P_t} * h_{se} * SignalSymbols + \sqrt{P_j} * h_{je} * JamSymbols$
20: $eveRecieveDemappedBits \leftarrow Demapping\ eveRecieve\ Symbol\ to\ Bits$
21: $MeasureBER \leftarrow \frac{Sum(SignalBits\ \neq eveRecieveDemapped\ Bits)}{N_{bits}}$, $Calculate\ BER\ with\ comparision\ between\ Signal\ Bits\ and\ eveRecieve\ Demapped\ Bits$
22: $average\ BER \leftarrow Take\ average\ for\ the\ BER\ results\ of\ each\ loop\ so\ the\ BER\ has\ minimized\ the\ bias$
23: $end\ for$
24: *end procedure*

In line 1–5, the distance between the jamming drones and eves is measured, and the measured distance is used for the influence of jammer on eve. Line 6 refers to the number of simulation repetitions, and the average BER is evaluated by repeating 1000 times. The c of line 7 is the path loss index and is used to evaluate the channel coefficient. Line 8–24 is a function of calculating the average BER of the eavesdropper and is repeatedly performed by a value specified in line 6. This function receives the size of data to be transmitted, the path loss index, the maximum source power of jammer, and the distance between source node and eavesdropper, source node and destination node, and jammer and receiver as parameters. The e of line 10 is a random number in the form of a complex number and is used to evaluate the channel coefficient. In line 11–13, channel coefficients are calculated using c and e. The G of line 14 is a scaling coefficient amplified according to the distance between the source node and the destination node. In line 15–16, 0 and 1 are randomly generated as the size of the data to be transmitted and then mapped in a complex number form. In line 17–18, data to be used as jamming signals are randomly generated and then mapped in complex number form. Line 19 refers to a signal received by the eavesdropper, and line 20 refers to the bit formed as signals. In line 21, the BER is calculated using the bits sent by the source node and the bits received by eve. In line 22, the average BER is evaluated.

5.1.3. Results of Simulation

The source node transmitted 100,000 SignalSymbol data, and the BER was measured for the destination node and the eavesdroppers. The average BER was obtained by repeating the simulation 1000 times. In every case, the average BER of the destination node was 0, which means that all data sent by the source node were received. Figure 5 shows a graph that applies the average BER results of the eavesdroppers to the ILR metrics defined in Section 3.

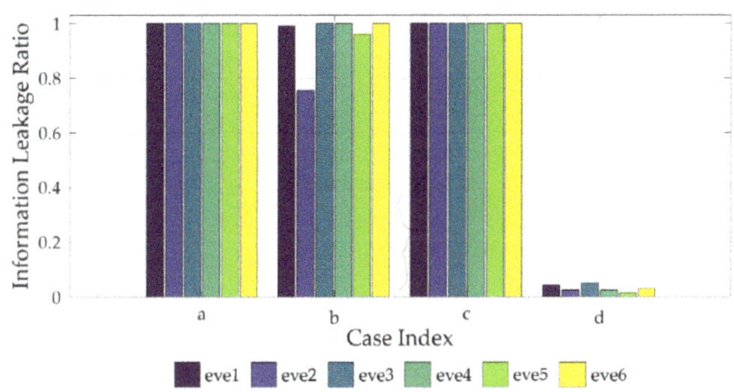

Figure 5. Simulation results: conventional schemes (**a**–**c**) and proposed scheme (**d**).

In Figure 4a, the ILR of every eavesdropper is 1. Therefore, if the devices have no mobility and the friendly-jamming technique is not used, the eavesdropping probability of the eavesdroppers is high.

In Figure 4b, the characteristics of Wang's Friendly UAV Jamming model (Fri-UJ) [19] are included. Fri-UJ reduces the eavesdropping probability by using multiple UAVs as jammers to transmit friendly jamming signals in a mobile communication network environment. However, unlike CFJ-DMZ, the mobility of the source node is not considered. At this time, the ILR of eve2 and eve5, which are within the cooperative jamming signal range of the drones, is 0.754 and 0.960, respectively. The ILR of the eavesdroppers outside the cooperative jamming signal range of the drones is 1. Therefore, if the friendly jamming technique is used without the mobility of the devices, the eavesdropping probability of eavesdroppers is high.

In Figure 4c, the ILR of every eavesdropper is 1. As the distance between the source and destination nodes is close, the signal strength of the source node is reduced. However, it does not have a significant impact on the communication quality of the eavesdroppers, because the simulation environment is a free space. Therefore, if the devices have mobility but the friendly jamming technique is not used, the eavesdropping probability of the eavesdroppers is high.

In Figure 4d, which is the case of the CFJ-DMZ model, the ILR of eve2 and eve5, which are within the cooperative jamming signal range of the drones, is 0.024 and 0.014, respectively. Furthermore, the ILR of the eavesdroppers outside the cooperative jamming signal range of the drones is close to zero, and the average ILR of all eavesdroppers is 0.03. Therefore, when the mobility of the devices and the friendly-jamming technique are used together, the communication quality of the eavesdropper decreases. Compared to Figure 4b that includes the characteristics of Fri-UJ, the average ILR is reduced by 92%, and compared to Figure 4a,c, the average ILR is reduced by 97%, facilitating secure communication in the secure zone.

5.2. Field Experiment

5.2.1. Effect of Friendly Jamming

In this section, we conducted experiments using Raspberry Pi. The friendly jamming drones and the eavesdroppers were also implemented using Raspberry Pi. In the communication between Raspberry Pi devices, packets were sent using the D2D communication method. The source node became the AP using the host mode, and the destination node was connected to the AP of the source node. The maximum transmission power of each node was 24 dBm. Jamming signals were generated using the Ping of Death method.

For the data, a string "1" consisting of a total of 256 bits was used, and as the preamble bits for synchronization, a string "a" consisting of 128 bits was used. Figure 6 shows how the preamble bits were processed. If "a" with less than 64 bits is received, it corresponds to a case in which the ILR is less than 0.5. Therefore, the data received through the corresponding packet are all processed as an unanalyzable state. If "a" with 64 or more bits is received, the number of error bits is obtained after removing the preamble data. The number of error bits is calculated by adding the number of lost bits and the number of unmatched bits. The number of lost bits is calculated by subtracting the number of receiving bits from the number of sending bits, and the number of unmatched bits is calculated by counting 1 after the sequential Xor. The BER of the destination node and the eavesdroppers is obtained using the error bits.

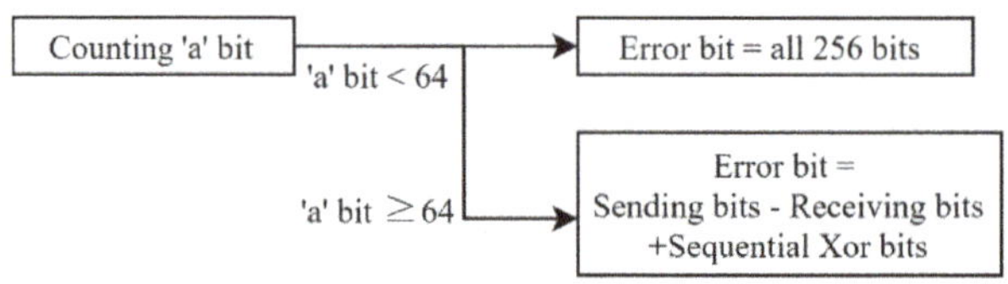

Figure 6. Synchronization with preamble bit.

First, the friendly jamming model of Figure 1a was implemented to verify the effectiveness of the jamming signals. One jamming drone node was deployed based on the locations of the source and destination nodes.

The transmitted data of the source node, and the BER of both the destination node and the eavesdroppers, were measured. Table 5 shows the average BER obtained by repeatedly applying the model of Figure 1a 1000 times. The average BER of the destination node is zero. In other words, the destination node can communicate normally with the source node because it is unaffected by the jamming signals. The average BER of the eavesdroppers located around the drone nodes (eve3, eve4, and eve5) is 0.975, 0.992, and 0.938, respectively, showing a result close to 1. This experimental result shows that the legitimate destination node can safely deliver information while reducing the amount of information leaked to the surrounding eavesdroppers by friendly jamming. Furthermore, the average BER of the eavesdroppers that are not located around the drones (eve1, eve2, and eve6) is 0.389, 0.012, and 0.562, respectively, demonstrating relatively low values. Therefore, if the friendly jamming technique is used, the eavesdropping probability of the eavesdroppers decreases, improving the communication security.

Table 5. Results of friendly jamming (field experiment).

Node	Destination Node	eve1	eve2	eve3	eve4	eve5	eve6
BER metric	0	0.389	0.012	0.975	0.992	0.938	0.562

5.2.2. Experimental Settings

The field experiment was executed using Raspberry Pi 3 Model B+. The experiment was conducted in an empty lot of 50 m $\times$ 50 m with a Quad-core 64-bit ARMv8 CPU and

1 GB of RAM. At this time, the empty lot was used to minimize the influence of other radio waves. The maximum transmission power of Raspberry Pi is 24 dBm, so the maximum transmission power of the source node, receiving node, and friendly jamming drone used in the experiment is 24 dBm each. The field experiments were conducted under the same environmental configuration as used in the simulations. That is, the devices were placed according to the experimental settings of Figure 4. Similar to the simulations, to achieve an effective comparison between the experiments, the locations of the eavesdropper were the same in every experiment. Furthermore, considering the signal ranges of the source node and the jamming nodes, we placed the eavesdroppers at the locations where the effects of the signals received according to the experimental environment were the same as those received by the eavesdroppers during the simulations.

5.2.3. Experiment Results

The BERs of the destination node and the eavesdroppers, respectively, were measured when the source node sent the data. The average BER was obtained by measuring the BER 1000 times repeatedly. In every case, the average BER of the destination node was 0, which means that the data sent by the source node were all received. Figure 7 shows a graph applying the average BER results of the eavesdroppers to the ILR metrics defined in Section 3.

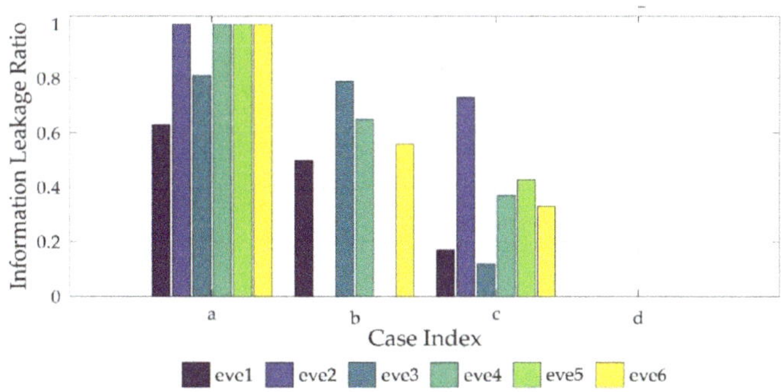

Figure 7. Field experiment results; conventional schemes (**a**–**c**) and proposed scheme (**d**).

In Figure 4a, the ILR of the eavesdroppers, except for eve1 and eve3, is 1. The ILR of eve1 and eve3 is 0.639 and 0.819, respectively. Therefore, if the devices have no mobility and the friendly jamming technique is not used, the eavesdropping probability of the eavesdroppers is high. Furthermore, it was found that, as the distance between the eavesdropper and the source node increases, the eavesdropping probability of the eavesdropper decreases.

In Figure 4b, the ILR of every eavesdropper decreases because the cooperative jamming signals affect the communication quality of the eavesdroppers. The ILR of both eve2 and eve5, which are within the cooperative jamming signal range of the drones, is zero. However, the ILR of the eavesdroppers outside the cooperative jamming signal range of the drones is greater than or equal to 0.5. Therefore, if the friendly jamming technique is used without the mobility of the devices, the eavesdropping probability of the eavesdroppers outside the cooperative jamming signal range is high.

In Figure 4c, the transmission signal strength is reduced because the distance between the source and destination nodes is close. As a result, the ILR of eve1 and eve3 is 0.180 and 0.121, respectively. However, the ILR of eve2, eve4, eve5, and eve6, which are relatively close to the source node, is 0.732, 0.380, 0.433, and 0.338, respectively. Although the ILR of every eavesdropper decreases compared to that of Figure 4a, the ILR of some eavesdroppers is high. Therefore, if the devices have mobility but the friendly jamming technique is not used, the eavesdropping probability of the eavesdroppers is high.

In Figure 4d, which is the case of the CFJ-DMZ model, the ILR of every eavesdropper is zero. In other words, when the mobility of the devices and the friendly jamming technique are used together, the communication quality of the eavesdroppers decreases, reducing the eavesdropping probability. Therefore, secure communication is facilitated in the secure zone.

5.3. Evaluation

The results obtained from the simulations and the field experiments are as follows. Based on the ILR of eve1 and eve3 in Field Experiment Figure 4a, it was found that as the distance between the source node and the eavesdropper increases, the communication quality of the eavesdropper deteriorates.

Furthermore, based on the ILR of eve5 and eve6 in Figure 4b, it was found that as the eavesdropper reaches closer to the friendly jamming drones, the communication quality of the eavesdropper deteriorates.

A comparison of the average ILR of the eavesdroppers between Figure 4a–d shows the effect of the device mobility. It can be seen that when the source node moves to the receiving node to reduce the distance between the two devices and decreases the transmission signal strength, the communication quality of the eavesdroppers deteriorates.

In Figure 4d, where the mobility of the devices and the friendly jamming technique are used together, the average ILR of the eavesdroppers is 0.03 in the simulation results and zero in the field experiment results. This means that the CFJ-DMZ model can reduce the communication quality of eavesdroppers to reduce the eavesdropping probability in the zone where the possibility of eavesdropping is high. Therefore, the communication security in the secure zone improves.

6. Conclusions

In this paper, we proposed the CFJ-DMZ method, which improves the security of wireless communication by using the mobility of mobile IoT devices and jamming signals, and we verified its effectiveness in drone communication environments. The drones conducting the cooperative friendly jamming move to locations where a secure D2D communication will be performed and transmit jamming signals to form a secure zone. The formed secure zone can effectively prevent eavesdropping and is flexible because the location and size can be easily changed. Furthermore, because jamming signals are transmitted only when the data transmitter and receiver communicate, the effect of the jamming signals on other source and destination nodes is minimized, and the batteries are efficiently used. Through the CFJ-DMZ model-applied simulations and fields tests, the BER of the eavesdropping devices was measured, which confirmed that the receiving performance of the eavesdroppers deteriorated, reducing the normal packet reception rate. Furthermore, we defined the ILR as a metric for a security performance evaluation and confirmed experimentally that the information leakage decreased with the proposed scheme.

By integrating it with IoT networking environments across future social systems, including logistics, delivery, and unmanned moving objects, the proposed CFJ-DMZ method can be used not only for military drone communications but also as a model that can actually be commercialized. As a limitation of this study, the effects of jamming were examined in two-dimensional planes. To consider the intrinsic emission characteristics of RF, the effectiveness of the CFJ-DMZ method should also be verified in three dimensions. Furthermore, low latency is important for the proposed method to be applied to real-time systems, not only for delay-tolerant applications. However, in this study, time complexity could not be analyzed, and the study was conducted from the perspective of an information leakage ratio to verify whether confidential communication is possible using the proposed method. Therefore, experiments and verifications are required for various environments, including a case in which an eavesdropper enters the secure zone formed by the drones. In follow-up studies, we will make improvements on this limitation by applying actual environmental parameters to mathematical models and simulation models for assessment of time and resource complexity. Based on these studies, we expect to design a more advanced security architecture and increase the level of security in confidential and complex zones.

Author Contributions: Conceptualization, Y.-J.L., J.-H.L. and I.-G.L.; methodology, H.-J.P., Y.-J.L. and J.-H.L.; software, Y.-S.S. and G.-H.J.; validation, Y.-S.S., H.-J.P. and Y.-J.L.; formal analysis, G.-H.J., Y.-S.S. and H.-J.P.; investigation, H.-J.P., Y.-J.L. and S.-W.Y.; resources, J.-H.L. and H.-J.P.; data curation, G.-H.J. and Y.-S.S.; writing—original draft preparation, Y.-J.L., H.-J.P. and Y.-S.S.; writing—review and editing, G.-H.J., J.-H.L., S.-W.Y. and I.-G.L.; visualization, Y.-S.S. and G.-H.J.; supervision, I.-G.L.; project administration, I.-G.L.; funding acquisition, I.-G.L. All authors have read and agreed to the published version of the manuscript.

Funding: This work was partly supported by a National Research Foundation of Korea (NRF) grant funded by the Korea government (MSIT) (No. 2020R1F1A1061107) and a Korea Institute for Advancement of Technology (KIAT) grant funded by the Korea Government (MOTIE) (P0008703, The Competency Development Program for Industry Specialist).

Institutional Review Board Statement: Not applicable.

Informed Consent Statement: Not applicable.

Data Availability Statement: Not applicable.

Conflicts of Interest: The authors declare no conflict of interest.

Glossary/Nomenclature/Abbreviations

The table summarizes the notation used in the paper.

Nation	Meaning
CFJ-DMZ	Cooperative Friendly Jamming Techniques for Drone-based Mobile Secure Zone
IoT	Internet of Things
D2D	Device to Device
UAV	Unmanned Aerial Vehicle
Fri-UJ	Friendly UAV Jamming
SINR	Signal to Interference and Noise Ratio
ILR	Information Leakage Rate
BER	Bit Error Rate
S	Source
D	Destination
E	Eve
J	Jammer
d_{SE}	Source-Eve distance
d_{DE}	Drone-Eve distance

References

1. Waliullah, M.D.; Gan, D. Wireless LAN security threats & vulnerabilities. *Int. J. Adv. Comput. Sci. Appl.* **2014**, *5*, 176–183.
2. Chen, J.; Jiang, M.; Liu, Y. Wireless LAN security and IEEE 802.11i. *IEEE Wirel. Commun.* **2005**, *12*, 27–36. [CrossRef]
3. Kim, Y.S.; Tague, P.; Lee, H.; Kim, H. Carving Secure Wi-Fi Zones with Defensive Jamming. In Proceedings of the 7th ACM Symposium on Information, Computer and Communications Security, Seoul, Korea, 2–4 May 2012.
4. Allouche, Y.; Arkin, E.M.; Cassuto, Y.; Efrat, A.; Grebla, G.; Mitchell, J.S.; Sankararaman, S.; Segal, M. Secure communication through jammers jointly optimized in geography and time. *Pervasive Mob. Comput.* **2017**, *41*, 83–105. [CrossRef]
5. Yang, Y.; Wu, L.; Yin, G.; Li, L.; Zhao, H. A survey on security and privacy issues in Internet-of-Things. *IEEE Internet Thing* **2017**, *4*, 1250–1258. [CrossRef]
6. Bose, T.; Bandyopadhyay, S.; Ukil, A.; Bhattacharyya, A.; Pal, A. Why Not Keep Your Personal Data Secure yet Private in IoT?: Our Lightweight Approach. In Proceedings of the 2015 IEEE Tenth International Conference on Intelligent Sensors, Sensor Networks and Information Processing (ISSNIP), Singapore, 7–9 April 2015.
7. Zhang, X.; He, J.; Wei, Q. Security Considerations on Node Mobility in Wireless Sensor Networks. In Proceedings of the 2009 Fourth International Conference on Computer Sciences and Convergence Information Technology, Seoul, Korea, 24–26 November 2009.
8. Wyner, A.D.; Aaron, D. The wire-tap channel. *Bell Syst. Tech. J.* **1975**, *54*, 1355–1505. [CrossRef]
9. Bakr, O.; Mudumbai, R. A New Jamming Technique for Secrecy in Multi-Antenna Wireless Networks. In Proceedings of the 2010 IEEE International Symposium on Information Theory, Austin, TX, USA, 13–18 June 2010; pp. 2513–2517.

10. Cumanan, K.; Xing, H.; Xu, P.; Zheng, G.; Dai, X.; Nallanathan, A.; Ding, Z.; Karagiannidis, G.K. Physical layer security jamming: Theoretical limits and practical designs in wireless networks. *IEEE Access* **2016**, *5*, 3603–3611. [CrossRef]
11. Vilela, J.P.; Bloch, M.; Barros, J.; McLaughlin, S.W. Wireless secrecy regions with friendly jamming. *IEEE Trans. Inf. Forensics Secur.* **2011**, *6*, 256–266. [CrossRef]
12. Vilela, J.P.; Bloch, M.; Barros, J.; McLaughlin, S.W. Friendly Jamming For Wireless Secrecy. In Proceedings of the 2010 IEEE International Conference on Communications, Cape Town, South Africa, 23–27 May 2010.
13. He, B.; She, Y.; Lau, V.K.N. Artificial noise injection for securing single-antenna system. *IEEE Trans. Veh. Technol.* **2017**, *66*, 9577–9581. [CrossRef]
14. Giti, J.E.; Sakzad, A.; Srinivasan, B.; Kamruzzaman, J.; Gaire, R. Friendly Jammer Against an Adaptive Eavesdropper in a Relay-Aided Network. In Proceedings of the 2020 International Wireless Communications and Mobile Computing (IWCMC), Limassol, Cyprus, 15–19 June 2020.
15. Kishk, M.A.; Dhillon, H.S. Stochastic geometry-based comparison of secrecy enhancement techniques in D2D networks. *IEEE Wirel. Commun. Lett.* **2017**, *6*, 394–397. [CrossRef]
16. Kim, Y.S.; Tague, P.; Lee, H.; Kim, H. A jamming approach to enhance enterprise Wi-Fi secrecy through spatial access control. *Wirel. Netw.* **2015**, *21*, 2631–2647. [CrossRef]
17. Yaacoub, E.; Husseini, M.A. Achieving Physical Layer Security with Massive MIMO Beamforming. In Proceedings of the 2017 11th European Conference on Antennas and Propagation (EUCAP), Paris, France, 19–24 March 2017.
18. Sharma, V.; You, I.S.; Andersson, K.; Palmieri, F.; Rehmani, M.H.; Lim, J.D. Security, privacy and trust for smart mobile-Internet of Things (M-IoT). *IEEE Access* **2020**, *8*, 167123–167163. [CrossRef]
19. Wang, Q.; Dai, H.; Wang, H.; Xu, G.; Sangaiahs, A.K. UAV-enabled friendly jamming scheme to secure industrial Internet of Things. *J. Commun. Netw.* **2019**, *21*, 481–490. [CrossRef]
20. Li, X.; Dai, H.N.; Wang, Q.; Imran, M.; Li, D.; Imran, M.A. Securing Internet of Medical Things with Friendly-jamming schemes. *ELSEVIER Comput. Commun.* **2020**, *160*, 431–442. [CrossRef] [PubMed]
21. Zhang, S.; Ansari, N. Latency Aware 3D Placement and User Association in Drone-Assisted Heterogeneous Networks with FSO-Based Backhaul. *IEEE Trans. Veh. Technol.* **2021**, *70*, 11991–12000. [CrossRef]
22. Duo, B.; Hua, H.; Li, Y.; Hua, Y.; Zhu, X. Robust 3D trajectory and power design in probabilistic LoS channel for UAV-enabled cooperative jamming. *ELSEVIER Veh. Commun.* **2021**, *32*, 100387. [CrossRef]
23. Lohanan, P.; Mishrab, D.; Nguyen, J.; Gupta, V. Secrecy-aware UAV position-aided jamming for practical eavesdropper localization models. *ELSEVIER Veh. Commun.* **2021**, *33*, 100430.
24. Hamamreh, J.M.; Furqan, H.M.; Arslan, H. Classifications and applications of physical layer security techniques for confidentiality: A comprehensive survey. *IEEE Commun. Surv. Tutor.* **2019**, *21*, 1173–1828. [CrossRef]
25. Leung-Yan-Cheong, S.K.; Hellman, M.E. The Gaussian wiretap channel. *IEEE Trans. Inf. Theory* **1978**, *IT-24*, 451–456. [CrossRef]
26. Demijan, K.; Jeongseok, H.; Steven, W.M.; Joao, B.; Byungjae, K. LDPC Codes for the Gaussian Wiretap Channel. *IEEE Trans. Inf. Forensics Secur.* **2011**, *6*, 532–540.
27. D'Oro, S.; Ekici, E.; Palazzo, S. Optimal power allocation and scheduling under jamming attacks. *IEEE/ACM Trans. Netw.* **2017**, *25*, 1310–1323. [CrossRef]
28. Abdullah, Z.; Chen, G.; Abdullah, M.A.M.; Chambers, J.A. Enhanced secrecy performance of multihop IoT networks with cooperative hybrid-duplex jamming. *IEEE Trans. Inf. Forensics Secur.* **2020**, *16*, 161–172. [CrossRef]

sensors

MDPI

Article

Dynamic Selection Techniques for Detecting GPS Spoofing Attacks on UAVs

Tala Talaei Khoei *, Shereen Ismail and Naima Kaabouch

School of Electrical Engineering and Computer Science, University of North Dakota, Grand Forks, ND 58202, USA; shereen.ismail@ndus.edu (S.I.); naima.kaabouch@ndus.edu (N.K.)
* Correspondence: tala.talaeikhoei@ndus.edu

Abstract: Unmanned aerial vehicles are prone to several cyber-attacks, including Global Positioning System spoofing. Several techniques have been proposed for detecting such attacks. However, the recurrence and frequent Global Positioning System spoofing incidents show a need for effective security solutions to protect unmanned aerial vehicles. In this paper, we propose two dynamic selection techniques, Metric Optimized Dynamic selector and Weighted Metric Optimized Dynamic selector, which identify the most effective classifier for the detection of such attacks. We develop a one-stage ensemble feature selection method to identify and discard the correlated and low importance features from the dataset. We implement the proposed techniques using ten machine-learning models and compare their performance in terms of four evaluation metrics: accuracy, probability of detection, probability of false alarm, probability of misdetection, and processing time. The proposed techniques dynamically choose the classifier with the best results for detecting attacks. The results indicate that the proposed dynamic techniques outperform the existing ensemble models with an accuracy of 99.6%, a probability of detection of 98.9%, a probability of false alarm of 1.56%, a probability of misdetection of 1.09%, and a processing time of 1.24 s.

Keywords: unmanned aerial vehicles; global positioning system; GPS spoofing attacks; detection techniques; machine learning; dynamic selection; hyperparameter tuning

Citation: Talaei Khoei, T.; Ismail, S.; Kaabouch, N. Dynamic Selection Techniques for Detecting GPS Spoofing Attacks on UAVs. *Sensors* **2022**, *22*, 662. https://doi.org/10.3390/s22020662

Academic Editor: Omprakash Kaiwartya

Received: 21 November 2021
Accepted: 12 January 2022
Published: 15 January 2022

1. Introduction

The use of unmanned aerial vehicles (UAVs) in military and civilian applications has exponentially increased over the last decade. Military applications include inspection and patrol, surveillance, reconnaissance, area mapping, and strike and rescue missions. Civilian applications include multimedia shooting, agricultural monitoring, meteorological monitoring, disaster detection, traffic control, cargo transportation, delivery services, and emergency rescue. Middle and long-distance applications rely heavily on Global Positioning Systems (GPSs) for navigation and precise positioning tasks [1].

Huge technical advances in the design, control, and automation have been made over the last two decades; however, the security aspect of UAVs has been largely overlooked [2]. UAVs can be subject to several cyber-attacks, such as GPS spoofing and jamming, which can impact the safety of civilians and airspace. Several UAV security incidents were reported during warfare and conflicts in Iran, Ukraine, and Iraq. During these attacks, malicious users transmitted fake GPS signals with incorrect positional and timing data that could be easily detected, resulting in erroneous navigation. These signals are similar to those from by satellites and are indistinguishable from authentic GPS signals.

A number of techniques have been proposed to detect GPS spoofing attacks. These methods can be classified into three categories [3]: cryptography-based, signal processing methods, and external UAV characteristics. Cryptography techniques encrypt the GPS signals, which require a key to decrypt [4]. Techniques under the second category extract spatial and geometrical characteristics, or physical layer characteristics, such as angle-of-arrival, signal strength, signal phase, and discontinuities from legitime GPS signals. The third category is based on external UAV characteristics, such as speed and acceleration, that that can be measured by the sensors of the UAV flight control system, such as barometer, inertial measurement unit (IMU), and compass.

These GPS spoofing detection methods have some drawbacks that limit their application. For instance, cryptographic methods are not practical for civil applications as they require encryption/decryption keys (GPS signals have to be unencrypted for these applications). Methods based on signal processing or external characteristics of UAVs may require additional hardware (sensors or antenna) or auxiliary equipment, apply changes to the interface specifications, or need extensive signal processing capabilities, which adversely affect real-time system performance or require additional communication overhead.

Inertial navigation systems (INS) techniques require continuous inertial sensor calibration as the error in position estimates and their covariance continuously grow without bounds. For the detection of GPS spoofing attacks, these techniques can be only used when the quality of sensors with respect of size and cost is high. Therefore, using such detection techniques for small drones is not possible [5]. In addition, using sensors, such as gyroscope and accelerator, also involves some limitations in detecting GPS spoofing attacks. For example, the accelerator can only measure changes in the velocity [6]. Furthermore, this sensor cannot measure the rotation around its own axis of movement. Therefore, it has to be used with a gyroscope to measure angular velocities. The accelerator is also sensitive to temperature, which makes it difficult for it to perform properly in different environmental situations of UAVs. Moreover, Gyroscope is a sensor that does not measure linear motion in any direction or any static angle of orientation [7]. Therefore, as these two sensors are the two main components of INS and IMU-based techniques, and based on their drawbacks, they cannot be used in detecting GPS spoofing attacks on UAVs [8].

Several studies based on traditional machine learning (ML) techniques have been proposed to classify and detect GPS spoofing attacks on UAVs. Examples of such techniques include artificial neural networks [2] and tree models [9]. Such models provide effective solutions for detecting these in the detection of GPS spoofing. Such models provide effective solutions in the detection of GPS spoofing [10].

Ensemble learning techniques are considered as one of the main developments in machine learning in the past decade as they perform better than traditional machine learning methods [11]. Examples of ensemble models that have been proposed for detecting cyber-attacks are bagging, boosting, and stacking [12]. The stacking classifiers apply meta-learning algorithms to select the best combinations of the base machine learning algorithms. The bagging methods use a combination of repetitive techniques to generate several sets from the original data and evaluate the performance simultaneously. In Boosting algorithms, the weight of observation is adjusted based on the last classification. Therefore, these three ensemble techniques can provide a better performance than a single conventional machine learning (ML) model. However, these models deal with some limitations, such as the difficulty of interpreting outputs and low or high rates of bias, which may lead to under or over-fitting issues.

Therefore, a holistic solution that can easily interpret and perform better than any single conventional ML model in detecting GPS spoofing attacks is known as multiple classifier systems (MCS). In this technique, a pool of classifiers is competing to provide the best prediction for a data sample, and the final result belongs to the most efficient base classifier. One example of MCS approach is the dynamic classifier selection (DCS) [13], which focuses on learning methods that automatically choose a subset of techniques in the prediction process. DCS focuses on fitting several ML classifiers on a training dataset and choosing the model that provides the best result in the prediction process based on specific proposed factors.

In this work, we propose two dynamic-based selection methods that detect GPS spoofing attacks on UAVs: Metric-Optimized Dynamic (MOD) selection and Weighted-Metric-Optimized Dynamic (WMOD) selection. We implement ten well-known supervised machine learning classifiers in both the proposed methods. These models are Support

Vector Machine, Naive Bayes, Decision Tree, K Nearest Neighbor, Linear Discriminative Analysis, Random Forest, Artificial Neural Network, Logistic Regression, Elastic Net, and AdaBoost. The two proposed classifier selection methods are trained and tested using a dataset with 13 GPS signals features built from real-time experiments and MATLAB attack simulations. The evaluation is conducted in terms of probability of detection (P_d), probability of false alarm (P_{fa}), probability of misdetection (P_{md}), accuracy(ACC), and processing time.

The contributions of this paper are:

- A one-stage ensemble feature selection technique to identify correlated and low importance features simultaneously.
- Two dynamic-based selection methods, MOD and WMOD, for efficient detection of spoofing signals.
- Performance comparison of the MOD and WMOD dynamic methods with bagging, boosting, and stacking-based ensemble models for validating the proposed techniques.

The remainder of this paper is organized as follows: Section 2 reviews the related works, while Section 3 illustrates the proposed architecture. Section 4 describes the materials applied in this work and highlights the methodology of the study. Section 5 discusses the simulation results. A conclusion is presented in Section 6.

2. Related Work

Several studies have been performed on GPS spoofing detection and mitigation methods. For instance, the authors of [7] proposed a GPS spoofing detection method that depends on the acceleration error calculated by estimating the acceleration from the GPS receiver and the acceleration measured from the IMU. In [9], the authors used IMU measurements (angle, velocity, and acceleration) and GPS data (longitude and latitude) in a two-step method that applies the XGBoost model and a Genetic Algorithm, to detect GPS-spoofing attacks. XGBoost was applied to learn the relationship between the IMU and GPS data, while the Genetic Algorithm was applied to tune the training parameters. An approach based on an artificial neural network was proposed in [2] to detect GPS spoofing signals. Several features, such as pseudo-range, doppler shift, and signal-to-noise ratio (SNR), were used to perform the GPS signal classification. Different neural network configurations were analyzed and tested. The proposed method revealed an acceptable efficiency in terms of probability of detection and probability of false alarm.

In [14], the authors proposed an anti-spoofing model that used linear regression to predict and model the optimal UAV route to its destination and used Long Short-Term Memory in the trajectory prediction. The model provides more than one detection scheme for GPS spoofing signals to improve UAV flight security and sensitivity to deception signal detection. Simulation experiments have determined that this method could enhance the ability to resist GPS spoofing without increasing hardware costs. Another GPS spoofing detection method was proposed in [15], based on the vision sensor combined with a UAV's sensors, monocular camera, and IMU. This method used vision sensors combined with IMU data to detect GPS spoofing. Another vision-based UAV spoofing detection method that utilized Visual Odometry was presented in [16], which uses the UAV camera since fake GPS signals would not alter its images. The UAV relative trajectory can be extracted from images using Visual Odometry. This extracted trajectory is compared with flight trajectory information obtained from GPS positions, to detect the spoofed signals.

In [17], the authors proposed a GPS spoofing-detection framework that needs minimal prior configuration and applies information fusion. The real-time detection scheme derives the current UAV location from IMU and compares it to the location information received by the GPS receiver to determine if the UAV system was experiencing a GPS spoofing attack. In [18], the authors proposed a new algorithm to handle GPS spoofing attacks that caused unknown sudden system state variable changes. The compensation of the GPS spoofing effect was manipulated using a prediction discrepancy based on a particle filter algorithm. The proposed algorithm decreases the effects of GPS spoofing errors and estimates the true position of the UAV in the presence of GPS spoofing attacks. In [19], the authors proposed a spoofing detection and classification algorithm based on Least Absolute Shrinkage and Selection Operator. They used some signal processing techniques to observe the decomposition of two code-phase values for authentic and spoofed signals using a certain threshold to mitigate false alarms. The proposed method achieves a promising detection error rate for a spoofer attack in nominal signal-to-noise ratio conditions.

In [20], the authors proposed a methodology that consists of several ML models with a set of values for K-folds where voting techniques are integrated to choose the learning model that achieves the highest accuracy. In [21], a hardware-based solution was proposed to detect GPS spoofing attacks. The authors demonstrated a simple method to detect hijacking based on gyroscopes measurements and GPS data. A switching mode resilient detection and estimation framework for GPS spoofing attacks has been studied in [22]. The authors tried to address the sensor drift issue by keeping the estimation errors to remain in a tolerable region with high probability.

Machine learning methods do not require additional hardware, which may be attractive for small civilian UAVs. For instance, in [23], the authors proposed an approach to detect UAV GPS spoofing attacks based on the analysis of state estimation using Support Vector Machine. The proposed method detects GPS spoofing attacks to some extent; however, the system experienced performance degradation during long attacks due to the interaction with the GPS sensor, especially with the Micro-Electro-Mechanical Systems sensors. In [24], a GPS spoofing detection method was proposed that leverages the uplink received signal strength measurements collected from base stations to identify the adaptive trustable residence area, which represents the trust region within which the UAV GPS position should be located to be classified as authentic or non-spoofed. In [3], the authors proposed a method for GPS spoofing attack detection based on a machine learning algorithm, Long Short-Term Memory, and compared the results to a method based on specifically designed UAV flight paths. This method can detect attacks well when the flight trajectory is not complicated. Table 1 provides a summary of existing studies in literature with their advantages and limitations.

Table 1. Existing Literature on Detecting GPS Spoofing on UAVs.

Category	Approach	Advantages	Limitations
External UAV characteristics	Acceleration error [7]	• Uses magnitude acceleration error to provide better performance	• Depends on accelerator error. • Pre-defined probability of false alarm.
	IMU-based [9]	• Provides a detection rate of 96.3% and 100% in hijacked and non-hijacked cases.	• Only detects attacks with similar behaviors during training.
	IMU-based [25]	• In best cases, detection rate of 98.6%,within 8 s when the system is under attack. • Provides a precision of 97%, a recall of 97%, and F1-score of 97%.	• In worse cases, the detection of GPS spoofing attacks can table over 28 s after the UAV started its mission.

Table 1. *Cont.*

Category	Approach	Advantages	Limitations
	Gyroscopes measurement- based [21]	• Easy to be implemented in any drone.	• Needs motion sensors (gyro-scopes and accelerators for detection. These are power hungry.
Artificial Intelligence Method	Artificial Neural Network-based [2]	• Provides an accuracy of 98.3%, a probability of detection of 99.2%, a probability of misdetection of 2.6%, and probability of false alarm of 0.8%.	• Uses a dataset with only 5 features and very limited samples.
	Linear regression-based and long short term memory [14]	• Works effectively in a case of UAV flying along the specified rout	• Lack of optimization methods from the perspective of UAV sensor integrated navigation and UAV attitude control.
	Prediction-discrepancy based [18]	• Reduces the effects of GPS spoofing errors and estimates the true position of the UAV in the presence of GPS spoofing attacks.	• Evaluated only based on accuracy and redundancy.
	Least Absolute Shrinkage and Selection Operator [19]	• Provides a 0.3% detection error rate for a spoofing attack in nominal signal-to-noise ratio conditions and an authentic-over-spoofer power of 3 dB.	• Uses a public old dataset, namely, Texas spoofing test battery as benchmark, • Lack of using common evaluation metrics, such as the probability of misdetection.
	K-learning based [20]	• Provides an accuracy of 99%, a precision of 98%, a recall of 99%, and F-score of 98%.	• Uses only Shimmer and Jitter as features in the dataset.
	Resilient State Estimation [22]	• Addresses the sensor drift problem.	• Evaluated only based on estimated error, and statistics of attacks.
	Support Vector machine [23]	• Improves the performance in case of using magnetometer sensors.	• Performance degradation during long attacks.
	5G-assisted position monitoring [24]	• A detection rate of 95%, and F1-score of 88%.	• Lacks of several evaluation metrics.

Table 1. *Cont.*

Category	Approach	Advantages	Limitations
	Long-Short Term Memory [3]	• A comprehensive comparison with encryption-based detection techniques in terms of detection rate and time cost.	• Detection rate of 78% and a time cost of 3s. • Detection rate is high when the flight trajectory is not complicated.
Signal Processing	Vision-based [15]	• Detects spoofing attacks with an average of 5s based on several parameters.	• Only applied when the attacker is visible.
	Vision-based [16]	• Detects spoofing in the long-range UAV flights when the changes in UAV flight direction is larger than 3°and in the incremental UAV spoofing with the redirection rate of 1°.	• Only applied when the attacker is visible.

Although many spoofing detection techniques have been proposed in the literature, spoofers are continually evolving to produce new GPS spoofing attacks that are hard to detect, which increases the necessity to develop new mechanisms to prevent this kind of attack. Ensemble learning techniques can be a practical solution to address the limitations of the existing methods. In literature, there are no studies to investigate the performance of such approaches in detecting GPS spoofing attacks targeting UAVs; however, ensemble approaches, namely, bagging, boosting, and stacking, have been frequently utilized in detecting cyber-attacks in wireless communication systems. For instance, in [26], the authors proposed a stacked-based ensemble model to classify and detect attacks on wireless networks. The proposed approach consists of several base learning methods, namely, Support Vector Machine, Decision Tree, Random Forest, and Artificial Neural Network. The stacking approach outperforms the base learners. In [27], the authors compared different ensemble models, namely, bagging, boosting, and stacking, for predicting received signal power on UAVs. Their results demonstrate that the stacking model, including Support Vector Machine, Artificial Neural Network, and Gaussian Process, outperformed other base classifiers.

Dynamic classifier selection methods have been recently proposed as ensemble approaches that select the best performance ML model among all base models. To the best of our knowledge, no studies proposed such a technique for classifying and detecting GPS spoofing attacks on UAVs. Therefore, to fill the existing gap, two dynamic-based selection methods are proposed that use ten machine learning models. These methods select the ML method that provides best results to detect the presence or absence of an attack. To validate our proposed techniques and demonstrate that they provide optimal results, we compared our proposed methods with the three most known ensemble models, namely, bagging, boosting, and stacking, with our proposed techniques.

3. Proposed Architecture

The proposed system architecture is shown in Figure 1. This system consists of three phases: dataset building, data pre-processing and feature selection, and training and classification. For the dataset building, real-time experiments were conducted to collect real GPS signals, while attacks were generated through simulations. Features were identified and extracted from the real GPS signals, and the attack simulated signals [10]. The features for all samples are included in a dataset to be pre-processed. The second phase is the data pre-processing and feature selection, which focuses on missing value imputation, categorical data encoding, feature scaling, identifying correlated features, and discarding low importance features. In this study, we use feature scaling and transfer categorical feature values to numerical values to avoid any bias in the corresponding dataset.

Two feature selection techniques are applied: Spearman Correlation and Information Gain. The ensemble feature selection can simultaneously identify the correlated and low importance features and discard them from the corresponding dataset [28]. The primary aim of using ensemble feature selection is to decrease the dimensionality of the dataset and identify the most important features [29] that can enhance the performance of the proposed model.

For the training, testing, and classification phases, we implement ten traditional ML techniques: Support Vector Machine, Naive Bayes, Decision Tree, K Nearest Neighbor, Linear Discriminative Analysis, Random Forest, Artificial Neural Network, Logistic Regression, Elastic Net, and AdaBoost. To get the optimal results of each model, a hyperparameter tuning technique, Bayesian optimization, is used.

Two dynamic methods are implemented for detecting GPS spoofing attacks targeting UAVs. The proposed methods dynamically choose the classifier that achieves the best results for the considered performance metrics. Incoming signals are classified as authentic or spoofed in the prediction phase, and their probabilities is evaluated.

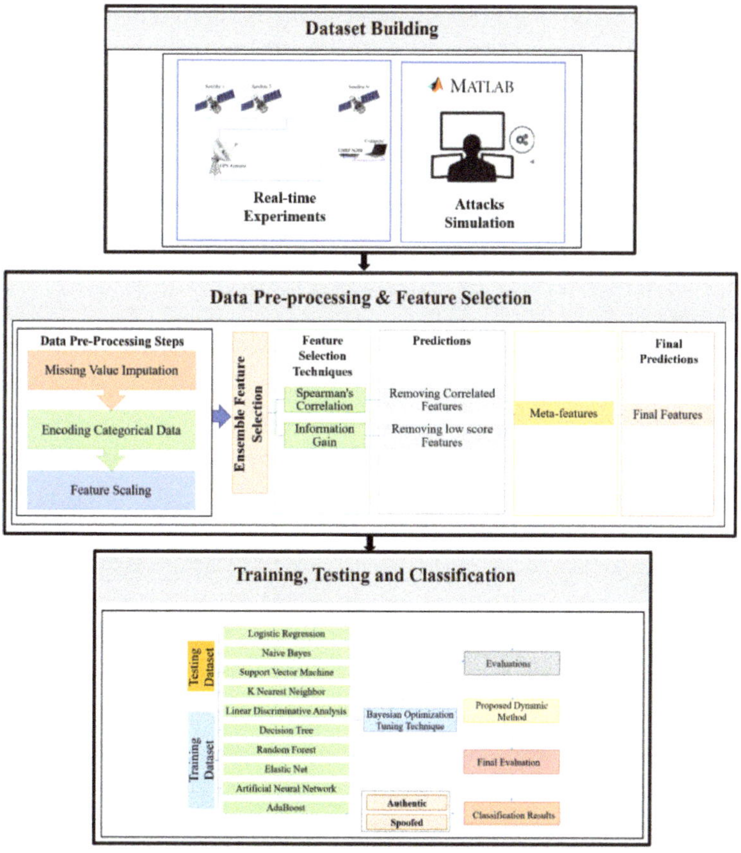

Figure 1. Overview of the Proposed Architecture.

4. Methodology

In the following, we discuss the dataset, data pre-processing, feature selection, description of the proposed models, and hyperparameter tuning that are used in this study, as follows:

4.1. Dataset

In this study, the used dataset was built in the work described was implemented in [10] . Real-time experiments and simulations were conducted to collect a dataset of authentic and spoofed signals at different dates in several locations. The hardware used in the implementation consisted of a universal software radio peripheral (USRP), a front-end active GPS antenna, and an I5-4300U laptop with 8 G RAM running with Ubuntu 16.04.7 LTS version. GPS attacks were simulated using MATLAB by considering three types of spoofing attacks with different complexity levels: simplistic, intermediate, and sophisticated. Each of these attacks impacts specific features of the GPS signals, such as Doppler Shift Measurement, Receiver Time, and Pseudo Range. In simplistic spoofing attacks, a fake GPS signals, which was unsynchronized with the authentic signals, was generated. In this case, higher Doppler Shift measurements were out of the normal range of ±20 Hz, leading to a signal drift. In this type of attacs, GPS spoofing signals are also transmitted at a higher power level, compared to that of authentic GPS signals, resulting in a higher Signal-to-Noise Ratio value.

In intermediate spoofing attacks, the attacker has a knowledge of UAV position. The intermediate attacker is able to control of the generated GPS signals. In this type of attack, the Doppler Shift Measurements and Pseudo Range values are kept within the normal ranges. In sophisticated attacks, the spoofer gains control over several channels of multiple synchronized antennas. This type of attack is the most threatening spoofing attack, due to the effect of multipath signals and the motion of the satellites and receiver.

Thirteen features were extracted from various receiver stages, starting from the tracking loop to the observable block. The extracted features from the received GPS signals with their short descriptions are listed in Table 2 . The corresponding dataset is balanced and contains 10,055 samples, of which 5028 are authentic signals and the remaining are equally divided between the three types of GPS spoofing attack signals. A sample of dataset is presented in Figure 2.

PRN	DO	PD	RX	TOW	CP	EC	LC	PC	PIP	PQP	TCD	CN0	Output
6	1767.341	25848800.1	490421	490421	-104336	139807.6	141797.4	169440.4	-167659	-24503.4	1833.427	47.76715	Authentic
11	4288.248	27044270	490421	490420.9	-163777	133518.9	142117.5	151754.5	-151718	3305.043	3945.384	44.15442	Authentic
2	3248.301	27396904.1	490421.1	490421	-136937	159911.6	126664.1	154882.1	-153995	-16556	3423.554	45.07013	Authentic
6	1767.21	25848793.3	490421.1	490421	-104371	180763.1	161273.1	201418	-201404	2342.733	1830.553	47.64579	Authentic
11	4286.317	27044255.5	490421.1	490421	-163863	103654.7	128726.2	122319.6	-122231	-4654.91	3945.384	44.22315	Authentic
19	-260.117	26153548.2	490517.8	490517.7	11988.13	100229.6	98970.64	107500.6	104006.2	-27186.4	-380.939	40.00464	Spoofed
24	-1248.21	28744890.7	490627.6	490627.5	173506.1	123478.4	118979.6	88462.14	201274.5	155462.2	-1037.09	41.59442	Spoofed
12	3732.485	27248858.7	490627.6	490627.5	-610456	129473.2	129473.2	-155462	148543.3	-8138	3932.341	44.65298	Spoofed
6	1598.385	25779592.2	490627.6	490627.5	-252809	145475.3	134349.6	158606.3	-157522	-18515.2	1793.507	45.97645	Spoofed
11	3710.505	26843081.9	490701.6	490701.5	-279101	149870.1	145513.3	160052.3	-159972	5060.913	3656.549	44.0468	Spoofed

Figure 2. Sample of Dataset.

4.2. Data Pre-Processing

The dataset was previously pre-processed by identifying and removing any null, unknown, and noisy values during the missing value identification step [30]. The next step is to encode any categorical values to numerical values. There are only two categorical data values in our dataset, which represent the signals as attack or normal. For this purpose, we encode normal signals as 0 and spoofed signals as 1. Afterward, feature scaling is performed by applying normalization and standardization methods. Normalization can re-scale the values into ranges between 0 and 1. In this study, we use the power transformer technique based on the Yeo-Johnson transformer. Unlike other techniques, this method can handle positive, negative, and zero data values. We also applied a simple standardization technique, which re-scaled the values to a mean of 0 and a standard deviation of 1.

Table 2. List of Features.

Feature	Abbreviation	Description
Satellite Vehicle Number	PRN	Identifying uniquely each satellite in orbit.
Doppler Shift Measurement	DO	Difference in the frequency of a GPS receiver moving relatively to its source. Difference in the frequency of a GPS receiver moving relatively to its source.
Pseudo Range	PD	Difference between the transmission and the reception time.
Receiver Time	RX	Time of transmission of the navigation messages.
Decoded Time Information	TOW	Information regarding the reception time of a subframe.
Carrier Phase Shift	CP	Beat frequency difference between the received carrier and a receiver-generated carrier replica.
Prompt Correlator	PC	Happens when the replica signal generated from the receiver is compatible with the incoming signals.
Late Correlator Output	LC	Occurrs at the 1/2 chip spacing after the prompt correlator.
Early Correlator Output	EC	Happens at the 1/2 chip spacing before the prompt correlator.
Prompt In-phase Prompt	PIP	In-phase component of the Prompt correlator amplitude.
Prompt Quadrature Prompt	PQP	Quadrature component of the prompt correlator amplitude.
Carrier Loop Doppler Measurements	TCD	Doppler shift that is measured during the correlation stage.
Signal to Noise Ratio	CN0	Doppler shift that is measured during the correlation stage. Ratio of the power signal to noise.

4.3. Feature Selection

Ensemble feature selection techniques are widely used to enhance the robustness of feature selection techniques. These techniques are classified into two categories, namely, homogeneous and heterogeneous. In homogeneous ensemble feature selection, the same method is used with different sizes of training data, while heterogeneous ensemble feature selection mostly focuses on different feature selection methods with similar training datasets. This study employs a heterogeneous ensemble feature selection technique using two traditional feature selection techniques, namely, Spearman's Correlation and Information Gain. The goal of selecting these two feature-selection techniques is to remove correlated and unimportant features from the given dataset.

Spearman Correlation [31] primarily calculates the association and direction between each two features by calculating the score τ given by:

$$\tau = 1 - \frac{6\sum_{i=1}^{n}(d_i)^2}{n(n^2-1)} \quad (1)$$

where d_i is the difference between the two ranks of each observation, i is the index of the observation, and n is the number of observations. A feature is correlated if it attains a coefficient over 0.9. We consequently removed a feature from each pair of correlated features.

We also used the information-gain feature-selection technique, called mutual information [32], for feature importance to estimate the gain of each variable in terms of the target variable. The information gain, also known as entropy, is calculated for every feature; features with high entropy are selected as important features, and those with low entropy values are considered of low importance. Any feature that achieves an entropy less than 0.1 is discarded from the dataset in this work.

4.4. Hyperparameter Tuning

Several types of tuning techniques have been proposed in the literature; however, Bayesian optimization has emerged as an effective approach, outperforming other techniques such as random search and grid search since grid search suffers from the curse of dimensionality and random search is not suitable for training complex models [33–35]. Bayesian optimization can provide a practical solution to optimize functions using a computationally cheap surrogate model [36]. This approach can offer robust solutions for optimizing the black-box functions, applying a non-parametric Gaussian process to simulate unknown functions. A surrogate utility function, also known as the acquisition function, is another main component of Bayesian optimization, which is defined as a way to improve the optimality of the underlying function [37]. In this study, considering the benefits of Bayesian Optimization and shortcomings of other techniques, we employ this technique for optimization tuning.

4.5. Description of the Proposed Methods

Dynamic classifier selection techniques consist of a pool of homogenous or heterogeneous base classifiers. Homogenous classifiers are defined as using a set of classifiers that are of the same type built upon various data. In contrast, heterogeneous classifiers are designed using a group of classifiers belonging to various types built upon same data. In this work, we employed a set of heterogeneous base classifiers: Support Vector Machine (SVM), Naive Bayes (NB), Decision Tree (DT), K Nearest Neighbor (KNN), Linear Discriminative Analysis (LDA), Random Forest (RF), Artificial Neural Network (ANN), Logistic Regression (LR), Elastic Net (EN), and AdaBoost. The primary reason behind selecting heterogeneous classifiers is to increase the final model diversity without changing any model parameters [38].

We propose two dynamic selection methods for detecting GPS spoofing attacks on UAVs: MOD and WMOD classifiers, as shown in Figure 3. These methods focus on evaluating the ML models in terms of the probability of detection P_d, probability of false alarm P_{fa}, probability of misdetection P_{md}, and accuracy ACC. Figure 3a depicts the M base models and K performance metrics, i represents the base model index, and j represents the performance metric index. We initially calculate the performance metrics $(K_{i,j})$ for every base model M_i to find the optimal results; then,we determine the count $(K_{i,j})$ where M_i achieves the best results for every base model M_i. As a model achieves higher $P_d$and ACC values and lower P_{fa} and P_{md} values, the model is considered better at detecting GPS spoofing attacks; therefore, the model with higher P_d and ACC, and lower P_{md} and P_{fa}, will be selected for the final incoming GPS spoofing signal detection. This concept is implemented in the proposed MOD classifier approach since the algorithm will identify the model with the highest number of best metrics, using $\text{Max}(K_{i,j})$ for final detection.

MOD classifier is simpler to implement compared to other approaches, does not need extensive processing, does not require additional hardware, and has low time complexity. This classifier heavily depends on the selected base algorithm, which achieves the best metrics. As a result, no additional cost of processing or computational complexity was added to the overall algorithm. However, if two base models achieve the same number of best metrics , the MOD classifier will select one of the two classifiers trivially as the best model, in some cases when two base models achieve the same number of best metrics. To address this issue, we propose another approach: WMOD classifier. WMOD classifier, shown in Figure 3b, assigns a weight, w, for each performance metric to calculate a score using sum $(w_{i,j})$ for each base model. In this model, a weight is assigned to each of the selected metrics based on their importance. We consider the importance of accuracy higher than the importance of P_d, P_{fa}, and P_{md}. Therefore, we assign a weight of 0.4 for accuracy, while other metrics, such as P_d, P_{fa}, and P_{md}, have each a weight of 0.2. We have to determine the count $(K_{i,j})$ for every model M_i, where M_i obtains the highest weights according to the defined weights. The model that achieves the best score is used for the final detection of any incoming signal.

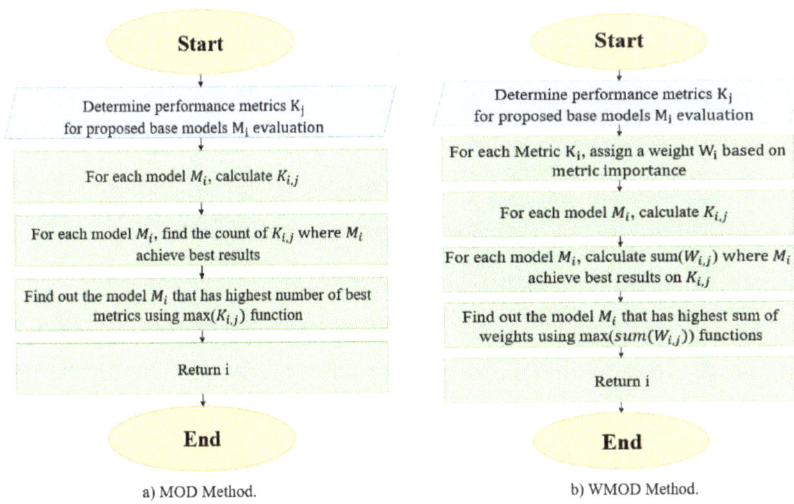

Figure 3. Flowcharts of the Proposed Dynamic Selection Methods.

5. Results

We ran our simulation on intel core i7-10750H, CPU of 2.60 GHz, and 16.0 GB memory. We used four evaluation metrics to assess the proposed model's efficiency: the probability of detection (P_d), probability of false alarm (P_{fa}), probability of misdetection (P_{md}), accuracy *(ACC)*, and processing time. These metrics were calculated using the following equations:

$$P_d = \frac{T_P}{T_P + F_N} \tag{2}$$

$$P_{fa} = \frac{F_P}{T_F + F_N} \tag{3}$$

$$P_{md} = \frac{F_N}{T_N + F_P} \tag{4}$$

$$ACC = \frac{T_P + T_N}{T_P + T_N + F_P + F_N} \tag{5}$$

where T_P is the number of corrected predicted malicious flows, T_N is the number of predicted normal flows, F_P is the number of incorrectly predicted malicious flows, and F_N is the number of incorrectly predicted normal flows.The processing time is defined as a time to train and test the classifiers. This metric highly depends on the ML model and dataset size. A three-fold cross-validation technique was applied to train 80% of the data and test 20% of the remaining dataset. The simulation analysis for the proposed dynamic methods was compared with the ten base selected classifiers in terms of the selected evaluation metrics.

Figure 4a,b show the results of the ensemble feature selection techniques, Spearman's Correlation and Information Gain, respectively. We discarded one feature of each pair of correlated features with a mutual coefficient > 0.9 as well as features of low importance with scores < 0.1 from the corresponding dataset. As one can see in Figure 4a, two pairs of features have a high correlation; the first pair, *DO* and *TCD*, has a correlation of 95%, and the second pair, TOW and RX, has a correlation of 94%. In addition, *DO* has a higher importance than *TCD*, and TOW has a higher importance than *RX*. Therefore, *TCD* and *RX* were discarded from the dataset. The remaining features are the relevant features selected to classify GPS signals.

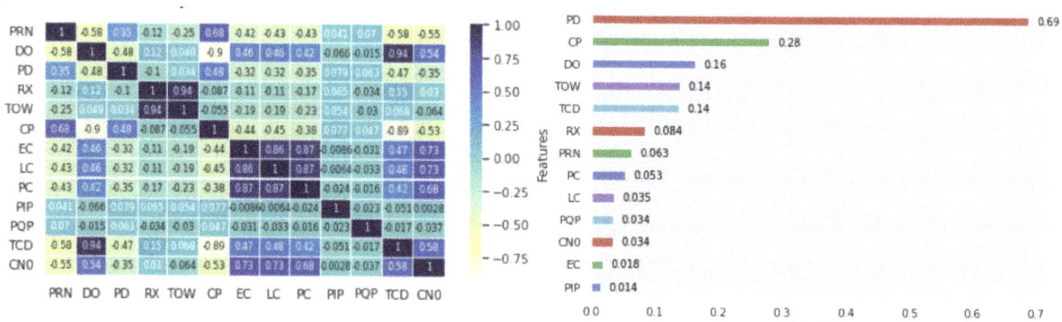

(a) Spearman's Correlation Coefficient. (b) Information Gain.

Figure 4. Importance of Features based on Ensemble Feature Selection: Spearman's Correlation Coefficient and Information Gain.

Table 3 provides the parameter setting and best parameter results obtained after applying the Bayesian Optimization algorithm. As can be seen, we specify several parameter settings with multiple values to check the optimality of every ML model. The parameter setting for every ML model is selected based on the provided values in Scikit-learn. Scikit-learn tool is a simple and efficient library that provides the suitable implementation for training, testing , and validating ML models, along with parameter settings for every ML model. The lists of setting parameters are provided in the table. These values are applied to achieve the best performance for each of the individual models. For instance, the activation function in the NN model is set to identity, logistic, tanh, or relu, and our selected tuning technique identifies Tanh as the activation function that guarantees the highest performance for the NN model. The NN model also has other parameters, including solver and alpha, that are required to be provided by tuning technique to achieve optimal results. In addition, the NB model consists of a parameter, namely, var_smoothing. This parameter is set to several values, provided in Scikit-learn tool. The Bayesian optimization technique identifies 1×10^{-3} as a hyperparameter, among other values, that ensures the best possible performance for the NB model. The same observations can be seen for the other selected models. To this end, these best parameters are used in training the selected models to ensure the optimality of the results.

Table 3. Parameter Setting Results.

Model	Parameter Setting	Best Parameters
SVM	C = [0.1, 1, 10, 100], degree = [1, 2, 3, 4, 5], gamma = [1, 0.1, 0.01, 0.001, 0.0001].	C = 10, degree = 5, gamma = 0.1.
NB	var_smoothing = [1e-2, 1e-3, 1e-4, 1e-5, 1e-6, 1e-7, 1e-8, 1e-9, 1e-10, 1e-11, 1e-12, 1e-13, 1e-14, 1e-15].	var_smoothing = 1e-3.
DT	Criterion = ['gini', 'entropy'], Splitter = ['best', 'random'], max_features = ['auto', 'sqrt', 'log2'], max_depth = range (1, 32).	criterion = 'entropy', splitter = 'best', max_features = 'auto', max_depth = 26.0.
RF	n_estimators = [10, 100, 1000, 10,000], max_depth = range (10, 200), min_samples_split = range (2, 10).	n_estimators = 1000, max depth = 110, min_samples_split = 2.
KNN	n_neighbors = range (1, 20), p = range (1, 10).	n_neighbors = 6, p = 1.0.
LDA	Solver = ['svd','lsqr'].	solver = 'lsqr'.
NN	Activation = ['identity', 'logistic', 'tanh', 'relu'], Solver = ['lbfgs', 'sgd', 'adam'], Alpha = linspace(0.0001, 0.5, num = 50).	activation = 'tanh', solver = 'lbfgs' alpha = 0.0409,
LR	l1_ratio = linspace(0.0001, 1, num = 50), C = [0.1, 1, 10, 100], Solver = ['newton-cg', 'sag', 'lbfgs'].	l1_ratio = 0.0001, C = 100.0, solver = 'lbfgs'.
EN	l1_ratio = linspace(0.0001, 1, num = 50), alpha = linspace(0.0001, 2, num = 50), selection = ["random", "cyclic"].	l1_ratio = 0.190, alpha = 0.1409, selection = 'cyclic'.
AD	n_estimators = [10, 100, 1000, 10,000].	n_estimators = 100.

Ten ML models are used with their best parameters' values in implementing the proposed dynamic classifiers. WMOD is proposed to handle a limitation of MOD. Such limitation occurs when two ML classifiers have the same number of metrics with the best results; therefore, two classifiers are selected as optimal. To address this issue, WMOD is proposed to return only the model with the best metric results.

Figure 5a provides the results of the proposed methods and the three ensemble models in terms of accuracy. As one can observe, the proposed MOD and WMOD dynamic methods provide the best results in terms of accuracy in comparison with bagging, boosting, and stacking-based ensemble models. As shown, the MOD and WMOD classifiers both have an accuracy of 99.8%. The stacking classifier has an accuracy of 99.7%, followed by bagging and boosting classifiers. Bagging model has an accuracy of 99.6%, while the boosting-based ensemble has the lowest accuracy of 99.56% compared to the other classifiers.

Figure 5b presents the results of the five models in terms of probability of detection. As can be seen, the proposed dynamic methods outperform the three ensemble models in terms of probability of detection with a slight difference. MOD and WMOD have the highest probability of detection of 99.9%, followed by the stacking, bagging, and boosting models. The stacking model has a probability of detection of 99.8%, bagging model has a probability of detection of 99.6%, and the boosting model has the lowest probability of detection of 99.35%.

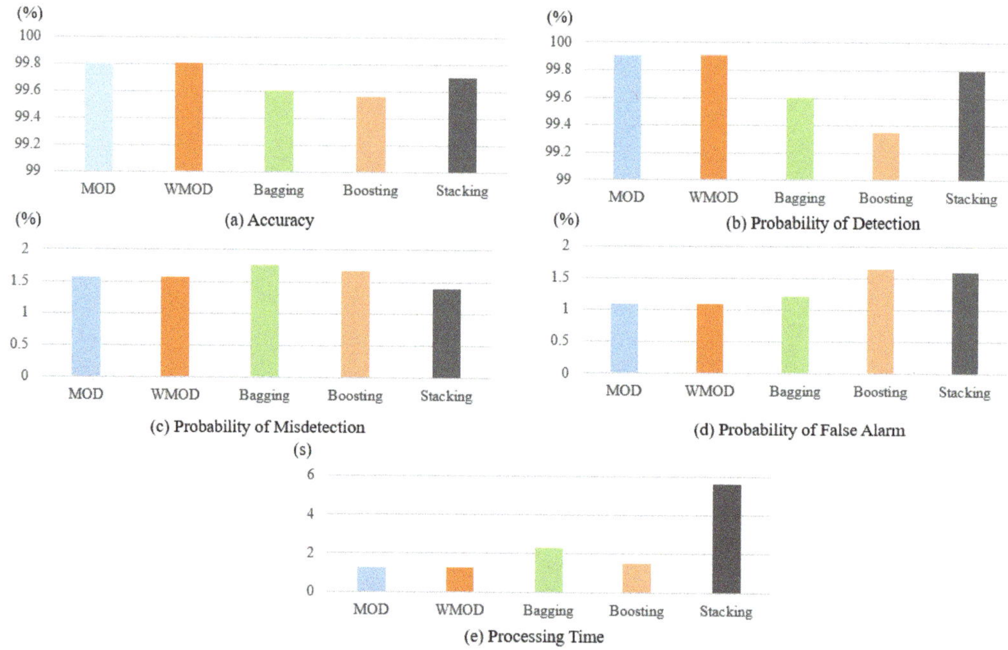

Figure 5. Evaluation results of the selected methods in terms of accuracy, probability of detection, probability of misdetection, probability of false alarm, and processing time.

Figure 5c illustrates the results of the proposed dynamic selection methods and the three ensemble models in terms of the probability of misdetection. As one can observe, MOD and WMOD have an acceptable probability of misdetection; however, the lowest probability of misdetection belongs to the stacking model. The proposed dynamic selection methods have a probability of misdetection of 1.56%, while the stacking model has a probability of misdetection of 1.4%. The other two ensemble models, bagging and boosting, also provide acceptable probability of misdetection results of 1.67% and 1.76%, respectively.

Figure 5d provides the results of the 5 models in terms of the probability of false alarm. As can be seen, MOD and WMOD have the lowest and best probability of false alarm compared to the other ensemble models. These methods have a probability of false alarm of 1.09%, the bagging classifier has a probability of false alarm of 1.2%, and the stacking and boosting have a probability of false alarm of 1.6% and 1.64%, respectively.

Figure 5e provides the results of the selected models in terms of their processing time. As one can see, the MOD and WMOD classifiers require a processing time of 1.24 s, which is considered much lower in comparison with other techniques, such as bagging, boosting, and stacking. The bagging classifier has a processing time of 1.321 s, while the boosting classifier achieves a processing time of 1.987 s. The stacking classifier has a processing time of 5.432 s, which is significantly higher than MOD and WMOD.

The number of false positives (f_p) is another important factor in evaluating models that compares the number of false positively predicted samples to total number of samples that are negatively predicted. Figure 6 provides the number of false positive for the highlighted methods. As one can observe, the MOD and WMOD provides the best number of false positives, followed by bagging, stacking, and boosting. The proposed dynamic selection methods have a number of false positives of 10.9 per second. In contrast, the bagging classifier has a number of false positives of 12 per secondm and the stacking and boosting classifiers have a number of false positives of 16 and 16.4 per second.

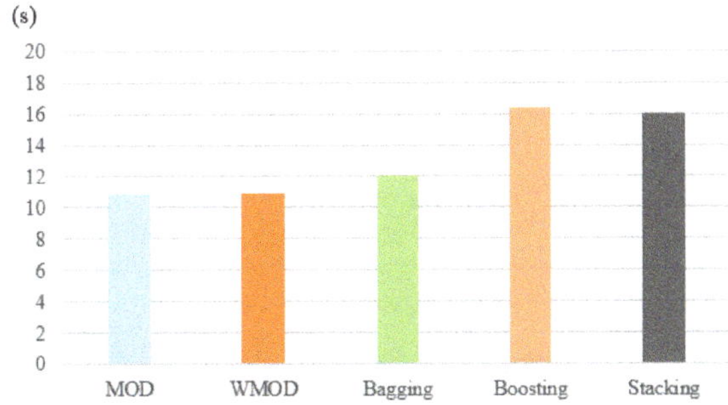

Figure 6. Number of False Positives for the Proposed Dynamic Selection Methods: MOD and WMOD Against the Classical Ensemble Techniques.

Table 4 provides the results of the proposed dynamic selection methods and the ensemble techniques, namely, bagging, boosting, and stacking. This table shows that MOD and WMOD have the best results in terms of accuracy, probability of detection, and probability of misdetection. In contrast, the stacking model provides the best result in terms of the probability of misdetection. It can be noticed that the proposed methods, MOD and WMOD, provide a probability of misdetection of 1.56%, which is higher than the stacking model by 0.16% considered as an insignificant difference. In contrast, the stacking model has a probability of false alarm of 1.6%, which is 0.51% higher than the probability of false alarm of these proposed methods. In addition, this stacking model has an accuracy of 99.7% and a probability of detection of 99.8%, which are 0.1% lower than the accuracy and probability of detection of the proposed dynamic selection methods.

As one can observe, the processing time of the proposed classifiers is 1.24 s, which is significantly lower than that of the other ensemble approaches. The bagging classifier has a processing time of 2.321 s, which is 1.081 s higher than that of MOD and WMOD. The boosting classifier has a processing time of 1.511 s, which is 0.271 s higher than that of the proposed classifiers. The stacking classifier has the worst processing time, which is 4.41 s higher than the processing time of MOD and WMOD.

To shed more light on the effectiveness of the proposed methods, we calculate the number of false positives (f_p) for MOD and WMOD, followed by bagging, stacking, and boosting. The proposed methods provide a number of false positives of 10.9 per second, which is 1.1 lower than the bagging classifier, 5.5 lower than the boosting classifier, and 5.1 lower than the stacking classifier. To conclude, our proposed classifiers provide higher accuracy and probability of detection, and lower probability of misdetection, false alarm processing time, and the number of false alarms compared to the other classical ensemble techniques.

Table 4. Evaluation Results of the Proposed Dynamic Selection Methods and Ensemble Models.

Methods	**Metrics**					
	ACC **(%)**	P_d **(%)**	P_{md} **(%)**	P_{fa} **(%)**	**Processing Time (s)**	f_p **(s)**
MOD	99.8	99.9	1.56	1.09	1.24	10.9
WMOD	99.8	99.9	1.56	1.09	1.24	10.9
Bagging	99.6	99.6	1.76	1.2	2.321	12
Boosting	99.56	99.35	1.67	1.64	1.511	16.4
Stacking	99.7	99.8	1.4	1.6	5.65	16

In short, the key insights can be summarized as follows:

- Ensemble feature selection removes the correlated and low importance features and decreases computational power and time.
- MOD and WMOD methods dynamically select one best classifier between the implemented ML models.
- The proposed dynamic methods can choose the best metric among the implemented models based on the considered metrics, which means such methods can be easily extended to include additional metrics that can significantly enhance the selection criteria.
- Comparison of the ensemble models with the proposed dynamic methods shows that the two dynamic methods can achieve good results in detecting GPS spoofing attacks on UAVs.

6. Conclusions

Interest in detecting GPS spoofing attacks on UAVs has increased significantly in the last decade, leading to considerable progress in different technologies. Several techniques have been proposed to identify and detect these vulnerabilities; however, this field of study still needs to address several challenges and limitations, such as high misdetection and false alarm rates. In this work, we used a one-stage heterogeneous ensemble feature selection to discard correlated and low importance features from the considered dataset using Spearman Correlation and Information Gain. As a result, two features, RX and TCD, were discarded from the given dataset. We implemented two dynamic selection methods, MOD and WMOD, which dynamically selected the best ML model among the ten implemented. However, MOD has a limitation when two ML classifiers have the same number of metrics with the best results. WMOD addresses this limitation and perfectly optimizes the selection criteria. The results show that MOD and WMOD have an accuracy of 99.6%, a probability of detection of 98.9%, a probability of false alarm of 1.56%, a probability of misdetection of 1.09%, and a processing time of 1.24%. These results outperform those of the existing ensemble learning models.

Author Contributions: Investigation, T.T.K. and S.I.; Methodology, T.T.K. and S.I.; Supervision, N.K.; Writing—original draft, T.T.K. and S.I.; Writing—review & editing, N.K. All authors have read and agreed to the published version of the manuscript.

Funding: This research was funded by of the National Science Foundation (NSF), Award Number 2006674.

Informed Consent Statement: Not applicable.

Data Availability Statement: Dataset will be available in Github and other sites.

Acknowledgments: The authors acknowledge the support of the National Science Foundation (NSF), Award Number 2006674.

Conflicts of Interest: The authors declare no conflict of interest.

References

1. Kumar, K.; Kumar, S.; Kaiwartya, O.; Sikandar, A.; Kharel, R.; Mauri, J.L. Internet of unmanned aerial vehicles: QoS provisioning in aerial ad-hoc networks. *Sensors* **2020**, *20*, 3160. [CrossRef]
2. Manesh, M.R.; Kenney, J.; Hu, W.C.; Devabhaktuni, V.K.; Kaabouch, N. Detection of GPS Spoofing Attacks on Unmanned Aerial Systems. In Proceedings of the 2019 16th IEEE Annual Consumer Communications and Networking Conference, CCNC 2019, Las Vegas, NV, USA, 11–14 January 2019. [CrossRef]
3. Wang, S.; Wang, J.; Su, C.; Ma, X. Intelligent detection algorithm against Uavs' GPS spoofing attack. In Proceedings of the International Conference on Parallel and Distributed Systems—ICPADS, Hong Kong, China, 2–4 December 2020; pp. 382–389. [CrossRef]
4. Sheet, D.K.; Kaiwartya, O.; Abdullah, A.H.; Cao, Y.; Hassan, A.N.; Kumar, S. Location information verification using transferable belief model for geographic routing in vehicular ad hoc networks. *IET Intell. Transp. Syst.* **2017**, *11*, 53–60. [CrossRef]
5. Manickam, S. GPS Signal Authentication Using INS-A Comparative Study and Analysis. Master's Thesis, University of Calgary, Calgary, AB, Canada 2016.
6. Jiang, P.; Wu, H.; Xin, C. DeepPOSE: Detecting GPS spoofing attack via deep recurrent neural network. *Digit. Commun. Netw.* **2021**, in press.

7. Kwon, K.C.; Shim, D.S. Performance analysis of direct GPS spoofing detection method with AHRS/Accelerometer. *Sensors* **2020**, *20*, 954. [CrossRef] [PubMed]
8. Narain, S.; Ranganathan, A.; Noubir, G. Security of GPS/INS based on-road location tracking systems. In Proceedings of the 2019 IEEE Symposium on Security and Privacy (SP), San Francisco, CA, USA, 19–22 May 2019; pp. 587–601.
9. Feng, Z.; Guan, N.; Lv, M.; Liu, W.; Deng, Q.; Liu, X.; Yi, W. Efficient drone hijacking detection using two-step GA-XGBoost. *J. Syst. Archit.* **2020**, *103*, 101694. [CrossRef]
10. Aissou, G.; Ould Slimane, H.; Benouadah, S.; Kaabouch, N. Tree-Based Supervised Machine Learning Models for Detecting GPS Spoofing Attacks on UAS. In Proceddings of the 2021 IEEE 12th Annual Ubiquitous Computing, Electronics & Mobile Communication Conference (UEMCON), New York, NY, USA, 1–4 December 2021
11. Khoei, T.T.; Ismail, S.; Kaabouch, N. Boosting Models with Tree-Structured Parzen Estimator Optimization to Detect Intrusion Attacks on Smart Grid. In Proceedings of the 2021 IEEE 12th Annual Ubiquitous Computing, Electronics & Mobile Communication Conference (UEMCON), New York, NY, USA, 1–4 December 2021.
12. Khoei, T.T.; Aissou, G.; Hu, W.C.; Kaabouch, N. Ensemble learning methods for anomaly intrusion detection system in smart grid. In Proceedings of the 2021 IEEE International Conference on Electro Information Technology (EIT), Mt. Pleasant, MI, USA, 13–15 May 2021.
13. Cruz, R.M.; Sabourin, R.; Cavalcanti, G.D. Dynamic classifier selection: Recent advances and perspectives. *Inf. Fusion* **2018**, *41*, 195–216. [CrossRef]
14. Meng, L.; Yang, L.; Ren, S.; Tang, G.; Zhang, L.; Yang, F.; Yang, W. An Approach of Linear Regression-Based UAV GPS Spoofing Detection. *Wirel. Commun. Mob. Comput.* **2021**, *2021*, 5517500. [CrossRef]
15. Qiao, Y.; Zhang, Y.; Du, X. A Vision-Based GPS-Spoofing Detection Method for Small UAVs. In Proceedings of the 13th International Conference on Computational Intelligence and Security, CIS 2017, Hong Kong, China, 15–18 December 2017; pp. 312–316. [CrossRef]
16. Varshosaz, M.; Afary, A.; Mojaradi, B.; Saadatseresht, M.; Parmehr, E.G. Spoofing detection of civilian UAVs using visual odometry. *ISPRS Int. J. Geo-Inf.* **2019**, *9*, 6. [CrossRef]
17. Liang, C.; Miao, M.; Ma, J.; Yan, H.; Zhang, Q.; Li, X.; Li, T. Detection of GPS spoofing attack on unmanned aerial vehicle system. In Proceedings of the International Conference on Machine Learning for Cyber Security, Xi'an, China, 19–22 September 2019; pp. 123–139.
18. Majidi, M.; Erfanian, A.; Khaloozadeh, H. Prediction-discrepancy based on innovative particle filter for estimating UAV true position in the presence of the GPS spoofing attacks. *IET Radar Sonar Navig.* **2020**, *14*, 887–897. [CrossRef]
19. Schmidt, E.; Gatsis, N.; Akopian, D. A GPS spoofing detection and classification correlator-based technique using the LASSO. *IEEE Trans. Aerosp. Electron. Syst.* **2020**, *56*, 4224–4237. [CrossRef]
20. Shafique, A.; Mehmood, A.; Elhadef, M. Detecting Signal Spoofing Attack in UAVs Using Machine Learning Models. *IEEE Access* **2021**, *9*, 93803–93815. [CrossRef]
21. Feng, Z.; Guan, N.; Lv, M.; Liu, W.; Deng, Q.; Liu, X.; Yi, W. Efficient drone hijacking detection using onboard motion sensors. In Proceedings of the Design, Automation Test in Europe Conference Exhibition (DATE) 2017, Lausanne, Switzerland, 27–31 March 2017; pp. 1414–1419. [CrossRef]
22. Yoon, H.J.; Wan, W.; Kim, H.; Hovakimyan, N.; Sha, L.; Voulgaris, P.G. Towards Resilient UAV: Escape Time in GPS Denied Environment with Sensor Drift. *IFAC-PapersOnLine* **2019**, *52*, 423–428. [CrossRef]
23. Panice, G.; Luongo, S.; Gigante, G.; Pascarella, D.; Di Benedetto, C.; Vozella, A.; Pescapè, A. A SVM-based detection approach for GPS spoofing attacks to UAV. In Proceedings of the 2017 23rd International Conference on Automation and Computing (ICAC), Huddersfield, UK, 7–8 September 2017; pp. 1–11.
24. Dang, Y.; Benzaïd, C.; Shen, Y.; Taleb, T. GPS Spoofing Detector with Adaptive Trustable Residence Area for Cellular based-UAVs. In Proceedings of the GLOBECOM 2020–2020 IEEE Global Communications Conference, Taipei, Taiwan, 7–11 December 2020; pp. 1–6.
25. Liu, R.; Liu, E.; Yang, J.; Li, M.; Wang, F. Optimizing the Hyper-Parameters for SVM. *Intell. Control. Autom.* **2006**, *344*, 712–721.
26. Rajadurai, H.; Gandhi, U.D. A stacked ensemble learning model for intrusion detection in wireless network. *Neural Comput. Appl.* **2020**, 1–9 . [CrossRef]
27. Goudos, S.K.; Athanasiadou, G. Application of an ensemble method to UAV power modeling for cellular communications. *IEEE Antennas Wirel. Propag. Lett.* **2019**, *18*, 2340–2344. [CrossRef]
28. Ben Brahim, A.; Limam, M. Ensemble feature selection for high dimensional data: a new method and a comparative study. *Adv. Data Anal. Classif.* **2018**, *12*, 937–952. [CrossRef]
29. Hoque, N.; Singh, M.; Bhattacharyya, D.K. EFS-MI: An ensemble feature selection method for classification. *Complex Intell. Syst.* **2018**, *4*, 105–118. [CrossRef]
30. Ismail, S.; Khoei, T.T.; Marsh, R.; Kaabouch, N. A Comparative Study of Machine Learning Models for Cyber-attacks Detection in Wireless Sensor Networks. In Proceedings of the 2021 IEEE 12th Annual Ubiquitous Computing, Electronics & Mobile Communication Conference (UEMCON), New York, NY, USA, 1–4 December 2021
31. Reading, F.; Aspects, M. Spearman Rank Correlation Coefficient. *Concise Encycl. Stat* **2008**, 502–505 .
32. Kent, J.T. Information gain and a general measure of correlation. *Biometrika* **1983**, *70*, 163–173. [CrossRef]

33. Snoek, J.; Larochelle, H.; Adams, R.P. Practical Bayesian optimization of machine learning algorithms. *Adv. Neural Inf. Process. Syst.* **2012**, *4*, 2951–2959.
34. Joy, T.T.; Rana, S.; Gupta, S.; Venkatesh, S. Fast hyperparameter tuning using Bayesian optimization with directional derivatives. *Knowl. Based Syst.* **2020**, *205*, 106247. [CrossRef]
35. Shahriari, B.; Swersky, K.; Wang, Z.; Adams, R.P.; De Freitas, N. Taking the human out of the loop: A review of Bayesian optimization. *Proc. IEEE* **2016**, *104*, 148–175. [CrossRef]
36. Nguyen, V. Bayesian optimization for accelerating hyper-parameter tuning. In Proceedings of the IEEE 2nd International Conference on Artificial Intelligence and Knowledge Engineering, AIKE 2019, Sardinia, Italy, 3–5 June 2019; pp. 302–305. [CrossRef]
37. Shende, S.; Gillman, A.; Yoo, D.; Buskohl, P.; Vemaganti, K. Bayesian topology optimization for efficient design of origami folding structures. *Struct. Multidiscip. Optim.* **2021**, *63*, 1907–1926. [CrossRef]
38. Britto, A.S.; Sabourin, R.; Oliveira, L.E. Dynamic selection of classifiers—A comprehensive review. *Pattern Recognit.* **2014**, *47*, 3665–3680. [CrossRef]

sensors

MDPI

Article

Enhanced Dynamic Spectrum Access in UAV Wireless Networks for Post-Disaster Area Surveillance System: A Multi-Player Multi-Armed Bandit Approach

Amr Amrallah [1,*], Ehab Mahmoud Mohamed [2,3], Gia Khanh Tran [1] and Kei Sakaguchi [1]

[1] Department of Electrical and Electronic Engineering, School of Engineering, Tokyo Institute of Technology, Meguro, Tokyo 152-8550, Japan; khanhtg@mobile.ee.titech.ac.jp (G.K.T.); sakaguchi@mobile.ee.titech.ac.jp (K.S.)
[2] Electrical Engineering Department, College of Engineering, Prince Sattam Bin Abdulaziz University, Wadi Addwasir 11991, Saudi Arabia; ehab_mahmoud@aswu.edu.eg
[3] Electrical Engineering Department, Faculty of Engineering, Aswan University, Aswan 81542, Egypt
* Correspondence: amrallah@mobile.ee.titech.ac.jp

Abstract: Modern wireless networks are notorious for being very dense, uncoordinated, and selfish, especially with greedy user needs. This leads to a critical scarcity problem in spectrum resources. The Dynamic Spectrum Access system (DSA) is considered a promising solution for this scarcity problem. With the aid of Unmanned Aerial Vehicles (UAVs), a post-disaster surveillance system is implemented using Cognitive Radio Network (CRN). UAVs are distributed in the disaster area to capture live images of the damaged area and send them to the disaster management center. CRN enables UAVs to utilize a portion of the spectrum of the Electronic Toll Collection (ETC) gates operating in the same area. In this paper, a joint transmission power selection, data-rate maximization, and interference mitigation problem is addressed. Considering all these conflicting parameters, this problem is investigated as a budget-constrained multi-player multi-armed bandit (MAB) problem. The whole process is done in a decentralized manner, where no information is exchanged between UAVs. To achieve this, two power-budget-aware PBA-MAB) algorithms, namely upper confidence bound (PBA-UCB (MAB) algorithm and Thompson sampling (PBA-TS) algorithm, were proposed to realize the selection of the transmission power value efficiently. The proposed PBA-MAB algorithms show outstanding performance over random power value selection in terms of achievable data rate.

Keywords: unmanned aerial vehicles; dynamic spectrum access; quality of service; reinforcement learning; multi-armed bandit

Citation: Amrallah, A.; Mohamed, E.M.; Tran, G.K.; Sakaguchi, K. Enhanced Dynamic Spectrum Access in UAV Wireless Networks for Post-Disaster Area Surveillance System. *Sensors* **2021**, *21*, 7855. https://doi.org/10.3390/s21237855

Academic Editor: Margot Deruyck

Received: 5 November 2021
Accepted: 23 November 2021
Published: 25 November 2021

1. Introduction

The fast development of UAVs, which are commonly known as drones, has received much attention in various domains [1,2]. Recently, UAVs have been leveraged for future civil applications although their usage was restricted to military applications only during the last few years. This is considered a promising direction since UAVs have unique properties that can support this goal. UAVs are capable of various functions as they are able to fly, are maneuverable, and are easy to deploy. Hence, UAVs can handle different tasks as delivery services, traffic monitoring, aerial photography, disaster management, rescue operations, and wireless communications [1,2]. In recent years, major disasters have occurred around the world such as the great Tohoku earthquake and tsunami, which hit Japan in 2011; Hurricane Sandy on the northeastern coast of the USA in 2012; the Nepal earthquake in 2015, the massive explosion in the port of Beirut, Lebanon, in 2020; and the global wildfires in North America and Europe in 2021. All these natural disasters caused terrible damage to infrastructure and loss of human lives. The first few hours after the disaster are considered the golden relief time to provide support and emergency aid to save these precious lives. Therefore, this paper focuses on wireless communications applications for UAVs to support a post-disaster area surveillance system. Specifically, UAVs can fly over the post-disaster areas to collect live photos of the current situation and send this collected information to a disaster management center to be analyzed. This will enable rescue teams to get information promptly about the actual situation in the affected area, which will enhance their response time [3].

On the other hand, and due to the persistent increase in demand for mobile services, spectrum resources are becoming more and more scarce [4]. Therefore, it is expected that future mobile networks will host a modern communications technology that supports unsurpassed networking architecture and energy-efficient devices. To realize these novel concepts, new fundamental challenges have appeared on the surface. Unlike wired communications systems, due to the national spectrum regulations and the hardware limitation, the wireless world has limited links to distribute. Consequently, it will be mandatory for the traditional regulation of the spectrum to have a fundamental reform so that it can allow more efficient use of spectrum resources. Spectrum inefficiency has become a major concern; hence it is imperative to search for an effective solution to deal with the resource allocation problems from the spectrum and power-efficiency points of view. This solution should achieve three main goals. Firstly, it should be amenable to the distributed implementation. Secondly, it should be capable of dealing with the uncertainty caused by the lack of information. Thirdly, it should deal with users' selfishness. One of the most promising solutions is the DSA system [5], which can be implemented as a CRN [6]. A DSA system has the ability to enhance the spectrum utilization efficiency [7]. Hence, CRN allows unlicensed Secondary Users (SUs) to coexist with the licensed Primary Users (PUs) in the licensed band without causing any harmful impact on PUs in terms of different Quality of Services (QoS) aspects. In other words, SUs can utilize a portion of the licensed PUs spectrum under certain QoS constraints [6]. Therefore, to enhance the network efficiency, SUs' spectrum utilization should be maximized while keeping an eye on the QoS level of the high-priority traffic, i.e., the PUs traffic, to avoid any services interruption to the highly prioritized data transmission.

The concept of this resource allocation issue is considered a challenging problem for two reasons. First, the resource allocation process can be made with a large number of orthogonal communication dimensions such as time, frequency, code, space, and antenna direction [8]. Second, in order to enhance the spectrum utilization, QoS for both PUs and SUs should be maximized. To achieve this, there are different conflicting parameters that need to be jointly optimized as transmitted power, channel occupation, total throughput, and mutual interference level between simultaneous users. Therefore, for a certain number of PUs and SUs, there are indispensable targets for the optimization algorithm such as the interference threshold for each PU, the channel state information, and the geographical location for both of PUs and SUs. Moreover, this optimization scenario can be decentralized,; in other words, there is no need to deploy a fusion center to collect enough information from the environment and complete the optimization process to the end. Since energy levels are not observed in general, and both PUs and SUs form a distributed network, it can benefit from that distribution to sense the available energy at each node. From this point of view, the design of an efficient future wireless network needs to deal with the uncertainty of information besides different users' competition and selfishness. Hence, it becomes mandatory to search for a powerful mathematical tool that can deal with such unprecedented network problems.

Machine learning (ML) algorithms, more precisely reinforcement learning (RL) algorithms, are leveraged to deal with these kinds of optimization problems [9]. The reason behind selecting RL algorithms is their capability to achieve tremendous results in generalization and efficiency, leading to their capability to tackle real-life problems, and especially in field of wireless communications [9]. Furthermore, RL algorithms are able to deal with conflicting optimization parameters of the resource allocation problem for the DSA system [10]. Without prior information about the environment, an agent can learn to enhance its future actions based on its past experience. MAB algorithms are considered one of such

RL algorithms. MAB algorithms can be described as a set of actions (arms) of a bandit machine that each arm leads to a certain reward [11]. A player needs to maximize their accumulated reward over the playing epoch by choosing one arm to pull in each playing round. Moreover, this player has no idea about the reward behind each arm. So, this instantaneous reward behind each arm is revealed once the player decides to select this arm. Therefore, for this hidden setting, the player may lose some reward in each trial due to not selecting the arm that leads to the highest reward value instead of the chosen arm. This loss is denoted by regret [12]. Thus, each player should select a sequence of arms to pull to maximize their total reward over horizon, in other words, to minimize their total regret over horizon. This is a common dilemma faces MAB algorithms and it is called the exploration–exploitation trade off [13–15].

Over the last decade, with the rapid increase in the number of natural disasters occurring throughout the world, there has become an urgent need to develop a smart post-disaster surveillance system. This smart system should operate in a fully decentralized manner, i.e., without having a controlling center, to speed up collection and analysis of data for a post-disaster area to enhance the performance and reduce the response time of the rescue operations. DSA systems are considered a rich topic that was deeply investigated in the early 2000s for some quite old applications such as analog TV white spaces, especially in the Very High Frequency (VHF) and the Ultra High Frequency (UHF) bands [16]. Hence, we aimed to refurbish the well-known DSA system by exploiting the benefit of using ML algorithms as a modern optimization tool. Furthermore, UAVs, which are capable of flying and capturing high-resolution videos using attached cameras, were leveraged recently to support various applications in the civilian life. All these ideas motivated us to develop a smart and cheap post-disaster surveillance system by combining the advantages of DSA system, UAVs, and ML algorithms. In addition, this system is presented as unconventional method to solve the spectrum scarcity problem. In this way, DSA-system-aided ML algorithms can open the gate to unprecedented applications in the field of UAVs wireless communication networks.

In this paper, we aim to design and evaluate a spectrum allocation for a DSA system using MAB algorithms to support a post-disaster surveillance system. From a MAB perspective, UAVs, which are considered SU transmitters, will act as the player who aims to maximize their long-term reward, i.e., data rate. Furthermore, this player is constrained by a limited power budget. On the other hand, different transmitting power levels will act as arms of the bandit machine. The MAB algorithm is considered the most suitable algorithm for our optimization problem as it can deal with online optimization problems without any prior information about the environment except the player's observations of the achieved reward while playing. Our paper adapts two different MAB algorithms, the Upper Confidence Bound (UCB) [15] and Thompson Sampling (TS) [17], to address such an optimization problem. In this paper, a modified version of MAB algorithms is proposed to treat our optimization problem. This is called the Power-Budget-Aware PBA-MAB (MAB) algorithm. The key idea behind the PBA-MAB algorithm is to include the available power budget for each UAV in the decision-making process when choosing the most appropriate transmitting power value.

From the point of view of the DSA system, the SU network, which consists of UAVs and temporary base stations, shared the spectrum resources as a CRN with the PU network, which consists of highway Electronic Toll Collection (ETC) gates and cars passing these ETC gates, under certain QoS constrains. Hence, SU transmitters are allowed to send their data without causing a harmful interference to the most precious data of the PU network. It should be mentioned that our design allows both the PU network and the SU network to coexist at the same time under a certain signal-to-interference-plus-noise ratio (SINR) threshold. Furthermore, we need to utilize the multi-objective formulation. Given the location of each PU and the power budget of each SU, we seek to design for a joint optimization problem considering different conflicting objects such as interference coordination, sum-rate maximization, and total number of active SUs in the network, subject to QoS constraints for both PUs and SUs. Despite the adversarial problem definition and the selfish behavior of each UAV toward achieving its maximum data rate, modified MAB algorithms learn how to select the most suitable action over time to enhance the overall system performance as discussed in [18–20] and illustrated in our paper. The main contributions of this paper can be summarized in the following points:

- The selection of the transmitted power value for UAVs aiding a post-disaster area surveillance system is formulated as an optimization problem aiming to maximize the achievable data rate while considering the limited available power budget for each UAV. This is done in a decentralized manner as there is no exchange of information among UAVs.
- Integrating the post-disaster surveillance system as a CRN is considered an unconventional solution for the spectrum scarcity problem. Furthermore, it can reduce the overhead cost of renting dedicated frequency channels for post-disaster surveillance operations, while they are rarely used just when a disaster occurs.
- Despite the nature of original MAB algorithms to maximize the long-term reward, i.e., the achieved data rate, MAB algorithms are modified to take into account the limited power budget for transmission. Therefore, the selection of the transmitted power not only aims to maximize the data rate for the current channel but also considers the remaining power budget to maximize the data rate for the next available channel.

The rest of the paper is organized as follows. Section 2 overviews the related work. Section 3 introduces the system model and the power value selection optimization problem. Section 4 introduces proposed PBA-MAB algorithms and how these algorithms can deal with this kind of optimization problem. Section 5 gives simulation and analysis of the proposed optimization scenario. Finally, we summarize the result and point out the future research in Section 6.

2. Related Works

Since the early 21th century, the idea of DSA gained increasing attention, especially in the US and Europe, due to the spectrum congestion [21]. An overview of the major technical and regularity issues of DSA systems was presented in [21]. The authors of [22] introduced the concept of multi-dimensional spectrum sensing and discussed the challenges associated with it. They developed prediction algorithms based on the past multi-dimensional spectrum utilization information to predict the future usage of the spectrum. With the aid of the DSA system, CRN can be established to support different applications as public safety, smart grid, broadband cellular, and medical applications. Ref. [23] discussed some challenges that faced the practical application toward this idea. An overview of CRN design layers, such as the physical layer (PHY), the medium-access control layer (MAC), and the network layer, is presented in [24]. Furthermore, the authors showed how these layers can interact with each other. The authors of [25] investigated the throughput improvements in a CRN using different channel selection techniques such as frequency hopping, frequency tracking, and frequency coding. Ref. [26] investigated the CRN formed by the incorporating radio capabilities of a Wireless Sensor Network (WSN). It addressed both advantages and limitations of CRN for WSN in conjunction with the existing applications and techniques. A continuous-time Markov chain model is implemented in [27] for a DSA system in an open spectrum wireless network. The authors of [28] examined how CRN devices can find an available spectrum channel under different system capabilities, spectrum policies, and environmental conditions. They defined this problem as a "rendezvous" problem. With the aid of RL algorithms, the authors of [29] proposed a framework for Internet of Things (IoT) devices to capture and model the traffic behavior of short-time spectrum occupancy in order to determine the existing interference in the shared bands. In [30], a novel information and energy cooperation method were introduced for cognitive Heterogeneous Networks (HetNets). This method aimed to enhance energy efficiency by solving an energy efficiency maximization problem with respect to joint time allocation and power control. The authors of [31] proposed an enhanced fusion center rule for soft decision cooperative spectrum sensing using energy detection to mitigate the noise uncertainty effect and to enhance the sensing performance of CRNs.

In recent years, there have been research efforts for using UAVs to support post-disaster area applications. In [32], the authors used UAVs with conjunction with cellular network and WSN to aid disaster management applications. A genetic algorithm was used in [33] for UAV location optimization to enhance the overall coverage and data rate of the wireless network. The authors of [34] proposed an effective method to support rescue operations in locating victims of a natural disaster. This was done with the aid of lidar and infrared depth cameras attached to UAVs to build a detecting system independent of the illumination intensity. A video recorder and a geolocation module attached to UAV were used in [35] to search for survivors in a post-disaster area. In [36], the authors examined flying communication services using Wi-Fi, video camera, and web servers attached to UAVs. They aimed to enable affected users after a disaster to use their smartphones for texting and video communication in real-time. The authors of [37] proposed a mobility model based on self-deployment of an aerial ad hoc network based on the Jaccard dissimilarity metric for a post-disaster area. The software simulation integrates the mobility of victims and generate a corresponding UAVs mobility model to trace those victims. In [38], authors proposed an energy efficient task scheduling for the collected data by UAVs from ground IoT network to support a disaster management system.

In [39], UAVs were used as on-demand airborne relays to connect remote users with a cellular BS when they were separated by vast obstacles. Furthermore, UAVs can be used in WSNs to distribute and collect information in both of Control Plane (CP) and Data Plane (DP) from wireless sensors deployed on the ground level [40,41]. UAVs are being used to assist the management and control of Vehicle Ad hoc NETworks (VANETs) and extend its coverage [42]. All the above existing research works assume a full awareness of the network parameters, which is not the case of our paper, where there is no information change among UAVs while trying to maximize the achievable data rate, as the network is fully decentralized.

On the other hand, RL algorithms have become a promising optimization technique for solving chronic UAV problems that have occurred as a result of integrating UAVs in wireless communication applications. RL algorithms are well known for their capability to achieve near optimal results in generalization and efficiency. Therefore, they are used to tackle real-time problems in the field of wireless communications. Detailed discussion about different MAB algorithms can be found in [43,44]. It has been shown in several works that MAB algorithms can be adapted to tackle such problems related to DSA systems. The authors of [45] proposed MAB learning algorithms for CRN, and particularly for spectrum sensing in a DSA system in licensed bands [7]. Different MAB algorithms, such as UCB and TS, have been used to improve the spectrum access in unlicensed Wi-Fi networks [45,46]. The authors of [47] considered a set of policies for multiple-user-independent and identical distributed (iid) and rested MAB problems with the assumption that each SU declares its action to others, e.g., the selected channel, which is considered a strong constraint. A disputed learning and spectrum access policy for iid rewards is discussed in [48], and it was proven that this policy has a logarithmic order regret. In [49], the decentralized learning for DSA system with multiple SUs spectrum access has been studied. The authors of [50] proposed a modified MAB algorithm to solve the gateway selection in UAV wireless network for post-disaster area applications. These algorithms are considering the battery life while searching for the most suitable gateway UAV to maximize the total system throughput. A dynamic wireless channel selection based on the MAB algorithm with laser chaos time sequence is proposed in [51]. The adaptive channel selection achieved a higher throughput using four channels Wireless Local Area Network (WLAN) based on IEEE802.11a system. The authors of [52] proposed a simple and powerful tug-of-war MAB algorithm. Since this algorithm is very simple, it can be applied in wireless network selection for devices with small processing capabilities as IoT devices and smartphones. Ref. [53] studied the millimeter-wave (mmWave) two-hop relaying as a single-player MAB problem in order to enable one relay probing while maximizing the achievable spectral efficiency. This was done by using modified versions of MAB algorithms. The authors of [54] studied the problem of joint neighbor discovery and selection in mmWave device to device (D2D) networks using a stochastic budget-constraint MAB algorithm.

3. System Model

This section discusses the network architecture of the post-disaster area surveillance system using UAVs and the used channel model for transmitting the collected data.

3.1. Post-Disaster Area Surveillance System Architecture

Figure 1 shows a simplified version of the system architecture of the UAV wireless network in a metropolitan post-disaster area. Since the first few hours after the occurrence of the natural disaster (such as flood or earthquake) are considered the golden relief time to save human lives, as discussed in the introduction section, UAVs should collect pivotal information about victims in the damaged area using an attached high-definition camera. The collected data can be further analyzed by the disaster management center to identify victim's exact location, number, age, gender, and injury status. On the other hand, temporary base stations are deployed in the disaster area to collect this information from surveillance UAVs and send them to the disaster management center to aid rescue teams. These temporary base stations are used as charging stations for UAVs. Furthermore, they are considered the starting flying points. UAVs fly over the disaster area to capture live photos of certain points at the damaged area. The way in which these temporary base stations transmit the collected data to the disaster management center, and the method for selecting surveillance points, are outside the scope of this paper. Moreover, we assumed in this paper that the different locations in the affected area have the same weight of importance, so these points were chosen on random bases.

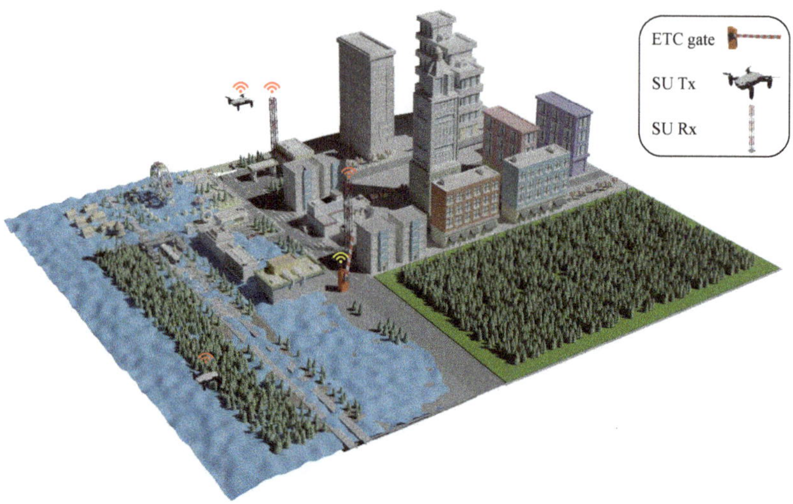

Figure 1. UAV surveillance-system-assisted DSA for a metropolitan post-disaster area.

On the other hand, our system aims to build this surveillance system using CRN. Therefore, the SU network, which is represented by UAVs and temporary base stations, will utilize the same frequency band of the PU network. The PU network is represented by ETC gates and bypassing vehicles in a nearby highway. In this way, we aimed to reduce the cost of reserving dedicated channels for surveillance system while it is being used during the time of natural disasters only. Each UAV collects and sends data to its corresponding temporary base station. Furthermore, each UAV should not deal harmful interference to the transmitted data between ETC gates and vehicles on the nearby highway. It should be mentioned that our optimization problem design is considered a soft-spectrum allocation. The difference between conventional spectrum allocation that have been studied in [55,56] and our optimization problem is that the conventional optimization problem treats the spectrum allocation as a hard allocation problem; i.e., no two users (PU and SU) can share same channels at the same time. However, our design introduces other orthogonal dimensions of the threshold to enable more than one user to coexist at the same frequency band if their QoS constraints are not violated. Furthermore, for the sake of generalization, we supposed that all PU channels, which connect every ETC gate and nearby vehicles which are passing this ETC gate, are always active and occupied with the PU network traffic. In this way, we considered the worst-case design scenario in which the QoS constraints should be carefully verified during the optimization process.

3.2. Problem Formulation

In the following, our design employs the physical model proposed in [57], which provides a path-loss model to realize the communication environment. It is assumed UAVs can communicate to temporary base station via air-to-air wireless communication link. Basically, this type of link can be called a Line of Sight (LoS) wireless communication link. Since the design is built using CRN, which shares the spectrum between PU network and SU network, this shared frequency band is split into Q independent sub-bands, and each sub-band has a bandwidth W in Hertz. Each primary and secondary transmitter receiver pair, referred to as primary and secondary users, is numbered by indices $\psi \in \Psi = \{PU_1, \ldots, PU_\Psi\}$ and $\omega \in \Omega = \{SU_1, \ldots, SU_\Omega\}$. Hence, at any time r, the general path-loss formula between any transmitter α and any receiver β can be expressed by:

$$L_{\alpha\beta,q}(r) = \frac{G_{Tx,\alpha} G_{Rx,\beta}}{d_{\alpha\beta}^{\xi}} \left(\frac{c}{4\pi f_q(r)} \right)^2 \tag{1}$$

where $G_{Tx,\alpha}$ and $G_{Rx,\beta}$ are the transmit and receive antenna gains, respectively, $d_{\alpha\beta}$ is the distance between α transmitter and β receiver, c is the speed of light, $f_q(r)$ is the carrier frequency of sub-band q, and ξ is the attenuation constant for the LoS wireless communication link. For the current design, it is assumed that the pass loss is the dominant loss factor for the received power. Hence, the effect of multi-path fading and shadowing is ignored. Furthermore, we assumed the transmitted signal is affected by an Additive White Gaussian Noise (AWGN) channel with zero mean and N_0 variance. Therefore, the SINR of SU ω in carrier q at time r can be given by:

$$\gamma_{\omega,q}(r) = \frac{p_{\omega,q}(r) L_{\omega\omega,q}(r)}{N_0 + \sum_{\lambda \in \Psi \cup \Phi, \lambda \neq \omega} p_{\lambda,q}(r) L_{\lambda\omega,q}(r)} \tag{2}$$

where $p_{\omega,q}(r)$ and $p_{\lambda,q}(r)$ denote the transmitted power of the ω-th SU and the λ-th PU or SU, respectively. For a successful established communication link, the SINR should satisfy a condition that the achievable SINR must be greater than the threshold SINR, which is given by $\gamma_{\omega,q}(r) > \gamma_{\omega\mathrm{TH},q}(r)$. Under these assumptions, the achievable data rate can be calculated by:

$$R_{\omega,q}(r) = \begin{cases} W \sum_{q=1}^{Q} \log_2(1 + \gamma_{\omega,q}(r)), & \text{if } \gamma_{\omega,q}(r) > \gamma_{\omega\mathrm{TH},q}(r) \\ 0, & \text{otherwise} \end{cases} \tag{3}$$

where W is the bandwidth of the communication channel.

Since the data rate is measured from the receiver side, we assumed this value is reported to the SU transmitter through a feedback channel. The concept behind this assumption comes from how modern communication systems are supposed to offer high flexibility in different ways. One of these ways is to split user and control planes to support software defined networking applications to allow flexible placement of processing function between different network nodes [58]. For PUs, it is assumed that they operate in a narrowband network, which means a pre-determined power value is assigned to each PU. This design criterion is suitable when licensed users have to operate on narrowband channels. On the other hand, for a wideband PU network, straightforward extension can be done without affecting this methodology. Since SUs need to utilize these multiband channels, where each sub-band is previously assigned to a certain PU, each SU has a power budget denoted by $P_{\max}$. Whereas it is assumed that our PUs and SUs networks use omnidirectional antennas, the communication channel can be established according to (1) with considering antennas gain $G_{Tx,\alpha} = G_{Rx,\beta} = 1, \forall \alpha, \beta$. Furthermore, it is assumed that each PU transmits using only single sub-band, and PUs operate in disjoint sub-bands. As a result, we have the number of PUs equal to number of channels and hence $\Psi = Q$. The main target of the optimization algorithm is to maximize the sum-rate, the total throughput, for the SUs network. This can be achieved by optimizing the power levels allocated for each SU within each shared traffic channel. The power allocation vector can be defined as $\mathbf{p}_\omega = [p_{\omega,1}, \ldots, p_{\omega,Q}]^{\mathrm{T}}$, where each element represents the power value for SU ω for each sub band q. In case that a SU has a power vector equal to zero, it means that this SU in inactive. On the other hand, for PUs, it is allowed for a single PU to transmit only on a single sub-band so that they are operating in disjoint sub-bands. Moreover, during data transmission of SUs, they should avoid causing any harmful interference to the high priority traffic that belongs to PUs network. It is mandatory for each SU to satisfy this condition and not exceed its allowed power budget during transmission as well. Considering all these power budget limitations and interference constraints, the sum-rate maximization problem can be formulated as:

$$\begin{aligned} &\max \frac{1}{\mathcal{R}} \sum_r \sum_\omega \sum_q R_{\omega,q}(r) \\ &\text{s.t. } \gamma_{\psi,q}(r) > \gamma_{\psi\mathrm{TH},q}(r) \\ &\qquad \gamma_{\omega,q}(r) > \gamma_{\omega\mathrm{TH},q}(r) \end{aligned} \tag{4}$$

where $\mathcal{R}$ is the total time spent for data transmission , $r = 1, \ldots, \mathcal{R}$, and $\gamma_{\psi,q}(r) > \gamma_{\psi\mathrm{TH},q}(r)$, $\gamma_{\omega,q}(r) > \gamma_{\omega\mathrm{TH},q}(r)$ are the SINR constraint conditions for all PUs and all SUs, respectively. Thus, for SUs, it is mandatory to satisfy both SINR conditions to utilize a sub-band channel from PUs channels.

Since our network is designed in a decentralized way with no information exchange between different network elements, the only information available to UAVs are the location, the channel frequency and the transmission power of each ETC gate system. Therefore, we have developed a method to let UAVs estimate the interference caused by self-transmission and calculate the corresponding SINR value for each PU's receiver. With the aid of Equations (1) and (2), each UAV will calculate the expected SINR value at each ETC gate under the interference effect of its own data transmission. Then, each UAV can check individually for the satisfaction of SINR conditions for both the PU network and the SU network. In such a way, there is no need to deploy a fusion center to share the SINR information between different SU network nodes, and therefore the network can be implemented in a decentralized way.

4. Proposed Power Budget Aware MAB Algorithm

This section discusses two proposed algorithms to tackle this rate maximization problem. These algorithms are called Power Budget Aware Upper Confidence Bound (PBA-UCB) and the Power Budget Aware Thomson Sampling (PBA-TS).

4.1. Proposed PBA-UCB Algorithm

UCB is considered one of the efficient MAB algorithms that can achieve balancing for the exploration-exploitation dilemma of the MAB algorithm. UCB enhances the confidence of the arm selection by decreasing the uncertainty behind the reward that will be revealed. Algorithm 1 illustrates a modified version of the UCB algorithm, which is called the PBA-UCB algorithm. This algorithm is applied to each UAV to select the most suitable transmission power in a selfish way to maximize the system rate. It is assumed that each UAV has information about the location of surrounding ETC gates operating in the surveillance area. Furthermore, they know the transmitting frequency for each ETC gate. The method of how UAVs can detect the location and the operating frequency of each ETC gate is behind the scope of this paper. Hence, each UAV tries to maximize its own data rate while competing with other UAVs to increase its transmission power while keeping an eye on the SINR threshold. At the beginning, i.e., the first $\mathcal{N}$ rounds, PBA-UCB algorithm, which is enabled on each UAV, tests the data rate that can be achieved by transmitting on all available channels with random transmission power and observes the achievable data rate. Afterwards, for the remaining rounds, $\mathcal{N}+1 \leq r \leq \mathcal{R}$, the PBA-UCB algorithm picks a power value in a way that satisfies:

$$p^{*}_{\omega,q}(r) = \underset{p_{\omega,q} \in \mathbf{p}_{\omega}}{\arg\max} \left(\hat{\mu}_{\omega,q}(r-1) + \sqrt{\frac{\eta \ln(r)}{T^{(p)}_{\omega,q}(r-1)}} - \frac{p_{\omega,q}}{p_{\omega,M}} \right) \tag{5}$$

where $p_{\omega,q} \in \mathbf{p}_{\omega}$ is the average reward obtained for transmission power value p in channel q up to the last previous round $(r-1)$, $\hat{\mu}_{\omega,q}(r-1)$ is the average achievable data rate to the last previous round $(r-1)$ using transmission power value p in channel q, and it can be calculated as:

$$\hat{\mu}_{\omega,q}(r-1) = \frac{1}{T^{(p)}_{\omega,q}(r-1)} \sum_{m=1}^{T^{(p)}_{\omega,q}(r-1)} R_{\omega,q}(m) \tag{6}$$

where $R_{\omega,q}(m)$ is the achievable data rate, which can be obtained from Equation (3). $T^{(p)}_{\omega,q}(r-1)$ is a count of the number of selections of this transmitting power value until the last previous round $(r-1)$. $p_{\omega,q}$ is the selected power value for transmission and $p_{\omega,M}$ is the total available power budget for UAV that can be used. This equation illustrates how PBA-UCB works. If a transmission power value is selected many times, which makes $T^{(p)}_{\omega,q}(r-1)$ become large, the confidence bond term $\sqrt{\frac{\eta \ln(r)}{T^{(p)}_{\omega,q}(r-1)}}$ decreases, and that causes the UAV to seek to explore other power values that are less selected in the previous rounds. On the other hand, when a transmission power value achieved a high reward, i.e., high data rate, during the past rounds, which means $\hat{\mu}_{\omega,q}(r-1)$ becomes large, the UAV seeks to exploit this high-gain arm in order to achieve the maximum achievable reward during this round. Originally, the PBA-UCB algorithm sets parameter η to a positive value of 2 in most cases [13], but empirically, when it is set to $\eta = 0.5$, the performance is improved [12]. In that way, the PBA-UCB algorithm can solve the exploration–exploitation trade-off in an efficient way. Furthermore, the term $\frac{p_{\omega,q}}{p_{\omega,M}}$ shows how a UAV can balance between selecting a power value to achieve a high data rate and consider for the remaining power budget to be used in transmission on next available channels. It should be mentioned that this last term defines the contribution behind our proposed PBA-UCB algorithm. Since the original UCB algorithm could achieve only balancing between exploration and exploitation, our proposed PBA-UCB algorithm enables a novel way to keep an eye on the remaining power budget while balancing between exploration and exploitation. Furthermore, when selecting a transmission power, the PBA-UCB algorithm checks for the satisfaction of both PU and SU SINR conditions. Once it is satisfied, the algorithm confirms the use of this transmission power value, starts to transmit data, and calculates the corresponding rate. Otherwise, it sets the transmission power to zero and also sets the corresponding data rate to zero. In this way, the PBA-UCB algorithm can make sure there is no harmful interference that affects the PU data transmission. On the other hand, it also counts for the interference threshold on other SUs data transmission. Since the SINR condition is considered a critical design issue, this operation is done in both of the initialization phase and the rate maximization phase to ensure the feasibility of the proposed PBA-UCB algorithm. Algorithm 1 illustrates the proposed PBA-UCB algorithm.

Algorithm 1 PBA-UCB transmission power selection

---

for $\omega \leftarrow 1$ to Ω **do**
 for $1 \leq r \leq \mathcal{N}$ **do**   ▷ initialization phase
  for $q \leftarrow 1$ to Q **do**
   Select a random value for $p_{\omega,q}(r)$
   if $\gamma_{\psi,q}(r) > \gamma_{\psi\text{TH},q}(r)$ **then**
    if $\gamma_{\omega,q}(r) > \gamma_{\omega\text{TH},q}(r)$ **then**
     Obtain $R_{\omega,q}(r)$
     $T^{(p)}_{\omega,q}(r) \leftarrow 1$
    else
     $p_{\omega,q}(r) \leftarrow 0$
    end if
   else
    $p_{\omega,q}(r) \leftarrow 0$
   end if
  end for
 end for
 for $r \leftarrow \mathcal{N}+1$ to $\mathcal{R}$ **do**   ▷ rate maximization phase
  Set $p_{\omega,M}$ max SU Tx power
  for $q \leftarrow 1$ to Q **do**
   $p^{*}_{\omega,q}(r) = \underset{p_{\omega,q} \in \mathbf{P}_{\omega}}{\arg\max} \left(\hat{\mu}_{\omega,q}(r-1) + \sqrt{\frac{\eta \ln(r)}{T^{(p)}_{\omega,q}(r-1)}} - \frac{p_{\omega,q}}{p_{\omega,M}} \right)$
   if $\gamma_{\psi,q}(r) > \gamma_{\psi\text{TH},q}(r)$ **then**
    if $\gamma_{\omega,q}(r) > \gamma_{\omega\text{TH},q}(r)$ **then**
     Obtain $R_{\omega,q}(r)$ using $p^{*}_{\omega,q}(r)$
     $T^{(p^*)}_{\omega,q}(r) \leftarrow T^{(p^*)}_{\omega,q}(r-1)+1$
     $\hat{\mu}_{\omega,q}(r) \leftarrow \frac{1}{T^{(p^*)}_{\omega,q}(r)} \sum_{m=1}^{T^{(p^*)}_{\omega,q}(r)} R_{\omega,q}(m)$
     $p_{\omega,M} \leftarrow p_{\omega,M} - p^{*}_{\omega,q}$
    else
     $p^{*}_{\omega,q}(r) \leftarrow 0, R_{\omega,q}(r) \leftarrow 0$
    end if
   else
    $p^{*}_{\omega,q}(r) \leftarrow 0, R_{\omega,q}(r) \leftarrow 0$
   end if
  end for
 end for
end for

4.2. *Proposed PBA-TS Algorithm*

TS algorithm copes with the exploration–exploitation dilemma using a different method than the previously discussed UCB algorithm. Basically, the reward gained by laying with different arms using the TS algorithm is drawn from a pure Bayesian probabilistic model [59]. In the beginning, TS uses a prior distribution for the reward based on the initialization of parameters of the probabilistic model. Afterward, it tries to keep tracking of the reward posterior distribution using the observation from the environment during the learning process. Thus, it can randomly choose a suitable arm that is matched to be optimal according to the probability model. Thus, at each round, random samples are drawn from the constructed reward's posterior distribution. TS selects an arm to play that can maximize the selected sampled value. Then, the arm's posterior distribution is updated by modifying its model parameters. This updated distribution will be used for the arm selection of the upcoming rounds. It is known that TS has a superb empirical performance and even better than the achieved performance of the UCB algorithm.

In our proposed PBA-TS algorithm, it is assumed that the reward, i.e., the achieved data rate, is affected by AWGN noise and mutual interference from other PUs and SUs occupying the same channel. Hence, the assumption of the Gaussian distribution is compatible with our problem formulation. The selection of the most suitable power value for transmission, which can maximize the achieved data rate, can be expressed as:

$$p^*_{\omega,q}(r) = \underset{p_{\omega,q} \in \mathbf{p}_\omega}{\arg\max} \left(\varphi_{\omega,q}(r-1) - \frac{p_{\omega,q}}{p_{\omega,M}} \right) \tag{7}$$

where $\varphi_{\omega,q}(r-1)$ is a sample for the previously constructed posterior distribution from the achieved data rate by a UAV ω at channel q with transmission power $p_{\omega,q}$. The posterior distribution is constructed from the Gaussian distribution $\mathcal{N}(\hat{\mu}_{\omega,q}(r), \sigma^2(r))$, where $\hat{\mu}_{\omega,q}(r)$ and $\sigma^2(r)$ are the mean and the variance of the distribution according to the model in [20], and they can be calculated as:

$$\hat{\mu}_{\omega,q}(r) = \frac{1}{T^{(p)}_{\omega,q}(r)} \sum_{m=1}^{T^{(p)}_{\omega,q}(r)} R_{\omega,q}(m) \tag{8}$$

$$\sigma^2(r) = \frac{1}{T^{(p)}_{\omega,q}(r) + 1} \tag{9}$$

where $R_{\omega,q}(m)$ is the achievable data rate and can be obtained from Equation (3), $T^{(p)}_{\omega,q}(r)$ is the counted number of selections of this transmitting power value until the last previous round $(r-1)$, and $R_{\omega,q}(m)$ is the achieved data rate. The term $\frac{p_{\omega,q}}{p_{\omega,M}}$ is deduced form the distribution to balance between the rate maximization process and the remaining power budget that should be used to transmit data over the next channels. At each round r, a sample $\varphi_{\omega,q}(r-1)$ is taken from the previously constructed Gaussian distribution. Then, the optimum power value $p^*_{\omega,q}$ that maximizes Equation (7) will be selected for transmission. After that, UAV ω starts to transmit over a channel q using $p^*_{\omega,q}$, its corresponding number of selections $T^{(p^*)}_{\omega,q}(r)$ is updated, and the achievable data rate $R_{\omega,q}(r)$ is observed to construct the Gaussian distribution for the next round $r+1$. This process is conducted till the last round $\mathcal{R}$. Furthermore, along with the PBA-UCB algorithm, the SINR conditions of both of PU and SU networks are examined at each time when choosing a certain power value for data transmission. If both SINR conditions are satisfied, the PBA-TS algorithm starts to use this transmission power value and counts the corresponding data rate. Otherwise, the PBA-TS algorithm sets the transmission power to zero, which leads to zero achievable data rate. The whole process of the proposed PBA-TS algorithm is summarized in Algorithm 2.

4.3. Complexity Analysis of the Proposed Algorithms

In this paper, we spotlight the task of UAVs to build a post-disaster surveillance system as a CRN by finding the optimal policy for each UAV. In Algorithms 1 and 2, learning processes can find the optimal transmission power value for both PBA-UCB and PBA-TS by examining various transmission power values over every channel for all UAVs using different policies. On the other hand, it tries to keep the interference level under certain thresholds. Let Ξ represent the total number of available arms, i.e., total elements of the power vector $\mathbf{p}$. It is assumed that the action space is deterministic; i.e., all actions are well known to each UAV. Therefore, the number of iterations of PBA-UCB is at most of the order of $\mathcal{O}(\Omega \cdot Q \cdot \Xi)$ steps. In particular, the complexity of PBA-UCB can be expressed as $\mathcal{O}(\Omega \cdot Q^2)$, if the total number of the available power levels Ξ in the power vector $\mathbf{p}$ is equal to the total number of channels Q. This means the complexity of the PBA-UCB algorithm is a polynomial in Ω and Q. Moreover, the PBA-TS has the same computational complexity $\mathcal{O}(\Omega \cdot Q^2)$ as the PBA-UCB algorithm. However, the update strategy in the PBA-TS algorithm is based on sampling from the Gaussian distribution $\mathcal{N}(\hat{\mu}_{\omega,q}(r), \sigma^2(r))$; hence it may impose a slightly higher complexity depending on the sampling process.

Algorithm 2 PBA-TS transmission power selection

for $\omega \leftarrow 1$ to Ω **do**
  Set $\hat{\mu}_{\omega,q} \leftarrow 0, \sigma^2 \leftarrow 1$
  for $r \leftarrow 1$ to $\mathcal{R}$ **do**
    Set $p_{\omega,M}$= max SU Tx power
    for $q \leftarrow 1$ to Q **do**
      Draw a sample $\varphi_{\omega,q}(r-1)$ from the distribution $\mathcal{N}(\hat{\mu}_{\omega,q}(r), \sigma^2(r))$
      $p^*_{\omega,q}(r) = \underset{p_{\omega,q} \in \mathbf{P}_\omega}{\arg\max}\left(\varphi_{\omega,q}(r-1) - \frac{p_{\omega,q}}{p_{\omega,M}}\right)$
      if $\gamma_{\psi,q}(r) > \gamma_{\psi \mathrm{TH},q}(r)$ **then**
        if $\gamma_{\omega,q}(r) > \gamma_{\omega \mathrm{TH},q}(r)$ **then**
          Obtain $R_{\omega,q}(r)$ using $p^*_{\omega,q}(r)$
          $T^{(p^*)}_{\omega,q}(r) \leftarrow T^{(p^*)}_{\omega,q}(r-1) + 1$
          $\hat{\mu}_{\omega,q}(r) \leftarrow \frac{1}{T^{(p^*)}_{\omega,q}(r)} \sum_{m=1}^{T^{(p^*)}_{\omega,q}(r)} R_{\omega,q}(m)$
          $\sigma^2(r) \leftarrow \frac{1}{T^{(p^*)}_{\omega,q}(r)+1}$
          $p_{\omega,M} \leftarrow p_{\omega,M} - p^*_{\omega,q}$
        else
          $p^*_{\omega,q}(r) \leftarrow 0, R_{\omega,q}(r) \leftarrow 0$
        end if
      else
        $p^*_{\omega,q}(r) \leftarrow 0, R_{\omega,q}(r) \leftarrow 0$
      end if
    end for
  end for
end for

5. Simulation Results

In this section, the simulation results of our proposed algorithms are evaluated in terms of solution performance. We distributed each PU and SU transmitter randomly in a 5 km $\times$ 5 km area, while PUs and SUs receivers are deployed in a certain area from PUs and SUs transmitters to comply with the SINR constraint. The SINR threshold is chosen to be 30 dB for the PUs network, which is relatively high to ensure that the accumulated data transmission from SUs will not cause any harmful interference to the most valuable traffic. On the other hand, the SINR value for SUs network is set to 5 dB to ensure a successful data transmission. The transmission powers for PUs and SUs networks are set to 24 dBm and 30 dBm, respectively. We deployed 10 armed bandits to represent 10 different levels of UAVs' transmission power. These power levels are uniformly distributed with separation equal to the maximum transmission power divided by number or armed bandits. Both PU and SU networks operate at 5.8 GHz band with a bandwidth equal to 10 MHz. Since both PUs and SUs networks operate in an open area, the attenuation constant parameter is set to 3 for a free-space communication in a metropolitan area. Table 1 summarizes the system's parameters which are used for simulation.

Table 1. Simulation parameters.

Notation	Value
No. of armed bandits	10
Simulation area	5 km × 5 km
PU Tx power	24 dBm
P_{max}	30 dBm
W	10 MHz
f_q	5.8 GHz
c	3×10^8 m/s
ξ	3
$\gamma_{\psi \text{TH},q}$	30 dB
$\gamma_{\omega \text{TH},q}$	5 dB
N_0	−100 dBm
η	0.5

Figure 2 shows an example of PUs and SUs transmitter/receiver pairs deployment. The deployment of PU receivers, i.e., cars, in the simulation area was done in a random way within δ distance from their corresponding transmitters, while δ is chosen to achieve 30 dB at the boundary of their deployment region. The number of sub-bands is set to be equal to the number of PUs, and hence $\Psi = Q$, as described previously in Section 3.

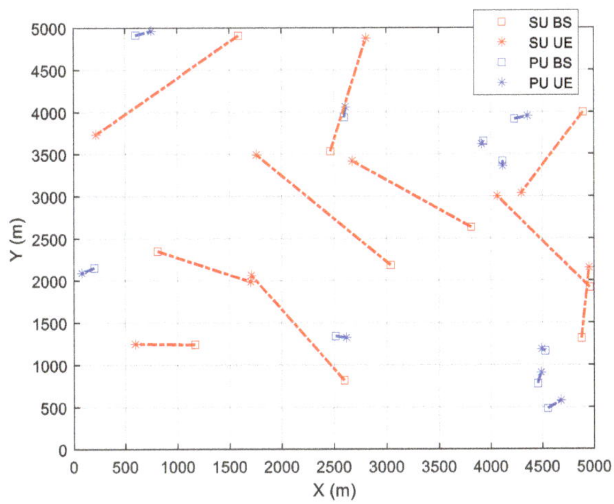

Figure 2. Distribution of PUs and SUs Tx/Rx pairs.

5.1. Average Total System Rate

This section shows the performance of the total average system rate in bps/Hz against different values of UAVs and ETC gates.

Figure 3 shows the total average system rate using 10 UAVs while increasing the number of ETC gates. It is shown in this figure that the PBA-TS algorithm achieved the highest data rate performance compared to both the PBA-UCB algorithm and transmission using a random power value. The reason behind this is that PBA-TS algorithm is constructed using posterior distributions for the obtained data rates through the integrated Bayesian strategy. On the other hand, transmission using a random power value has the worst performance due to the randomness in the selection of this power value for transmission in each round. Thus, each UAV experiences random interference from not only ETC gates but also other UAVs that share these channels. Furthermore, when the number of ETC gates increases and each ETC gate has its own separate channel, the number of available spectrum resources increases as well. This leads to each UAV becoming able to transmit data over a wider band of channels and causes the total achievable average system rate to increase for both the PBA-TS algorithm and the PBA-UCB algorithm. On the other hand, and due to the randomness illustrated in this section, the increase in the achievable total average system rate using a random power value data transmission is not as high as the achievable data rate using either the PBA-TS algorithm or the PBA-UCB algorithm.

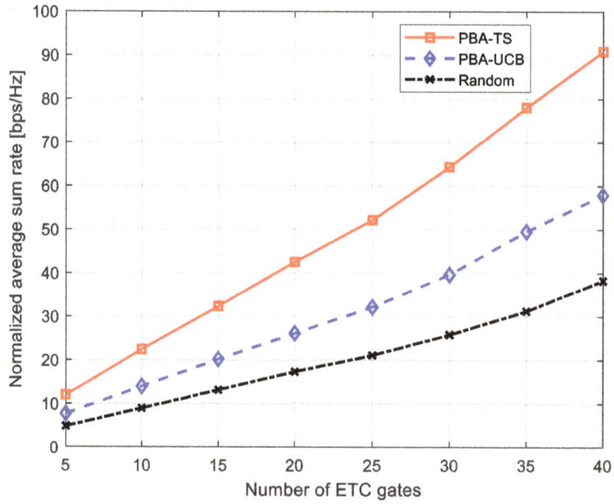

Figure 3. Normalized average sum rate against number of ETC gates using 10 UAVs.

Figure 4 shows the performance of the achievable total average system rate against an increasing number of UAVs while keeping the number of ETC gates equal to 10. It is interesting that at the beginning with a few increments of the number of UAVs, the achievable data rate, using our proposed PBA-MAB algorithms, is increased till a certain point. Then, the achievable data rate begins to decrease with any increment in the number of deployed UAVs. The reason behind that is that while increasing the number of UAVs, the mutual interference between UAVs increases as well. Our proposed PBA-MAB algorithms succeeded in mitigating the interference effect, which is reflected in the achievable data rate reduction. Furthermore, the proposed PBA-TS algorithm can still achieve the highest data rate performance compared to the proposed PBA-UCB algorithm and the transmission using a random power value.

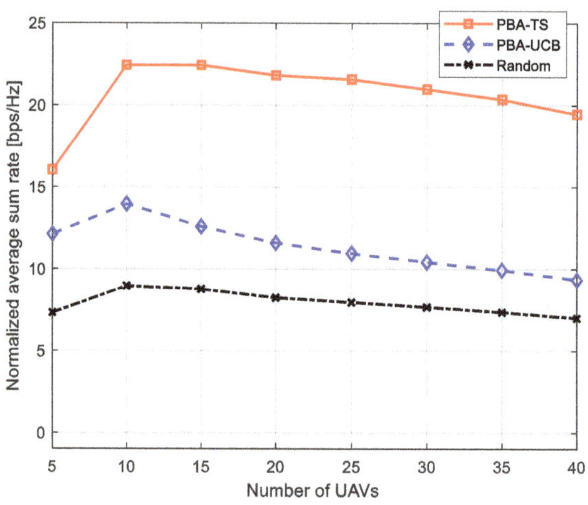

Figure 4. Normalized average sum rate against number of UAVs using 10 ETC gates.

5.2. Convergence Rate

The convergence rate is considered one of the most important parameters to judge the efficiency of online learning algorithms such as MAB algorithms; the faster the algorithm can converge, the better the reward that can be gained in just a few attempts. Hence, this section studies the convergence rate of the achievable total average system rate for our proposed PBA-MAB algorithms with different settings. Figures 5 and 6 show the convergence rate of the achievable total average system rate using 10 ETC gates while changing the number of UAVs to be 10 and 30. This can show the convergence rate for each algorithm under different network setup and different interference values. As shown in these figures, the horizontal axis indicates the count for rounds. Each algorithm runs its iterative process over counts till the algorithm converges toward a higher data rate. The proposed PBA-TS algorithm can converge faster than the PBA-UCB algorithm due to the fact that it uses Bayesian strategy over the posterior distributions of the reward. On the other hand, the PBA-UCB fluctuates during the few beginning rounds, and it takes more time to converge than the PBA-TS algorithm. Furthermore, it has a less convergence rate that the PBA-TS algorithm when both of the algorithms saturate by the end of the simulation rounds. These results can be concluded that both proposed PBA-MAB algorithms can deal with the adversarial network setup and selfish behavior of the UAVs. Hence, it means that every UAV learns how to select the most suitable transmission power value to enhance the overall system performance at every round. Furthermore, without loss of generality, it keeps an eye on the interference level while choosing this most suitable action.

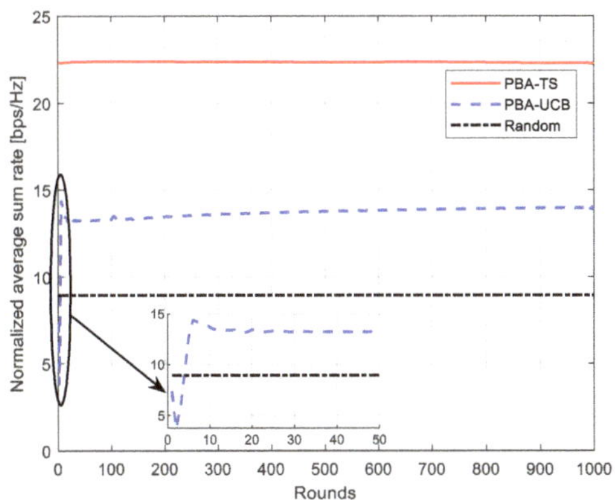

Figure 5. Convergence of normalized average sum rate using 10 ETC gates and 10 UAVs.

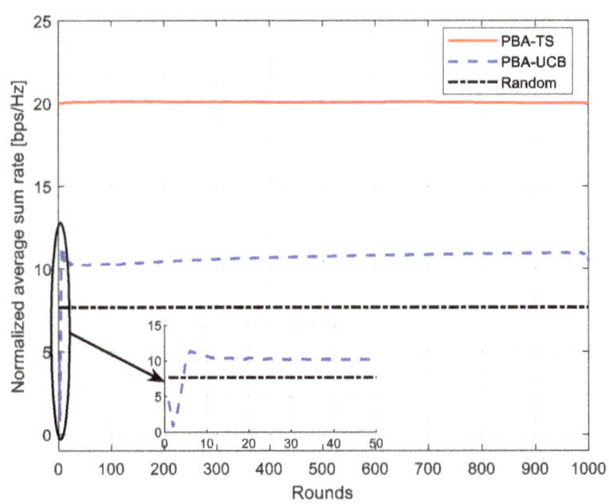

Figure 6. Convergence of normalized average sum rate using 10 ETC gates and 30 UAVs.

6. Conclusions

In this paper, we have investigated the radio resource allocation for a CRN through DSA system to support a disaster surveillance system using UAVs wireless networks. To tackle this problem, we proposed two MAB algorithms, i.e., the PBA-UCB algorithm and the PBA-TS algorithm. The idea behind deploying MAB algorithms, as a class of RL algorithms, is the ability of MAB algorithms to solve online optimization problems with conflicting parameters that need to be jointly optimized. Since there is no information exchange between all UAVs, multi-player PBA-MAB algorithms were introduced to deal with this selfish configuration. Proposed PBA-MAB algorithms show outstanding performance over transmission using a random power value selection. Furthermore, the proposed algorithms showed a moderate convergence rate. The obtained results showed the capability of different MAB algorithms to deal with such problems with a high degree of randomness. Therefore, it can open the way for applying ML algorithms and more precise MAB algorithms to handle various wireless communication problems.

Author Contributions: Conceptualization, A.A. and E.M.M.; methodology, A.A. and G.K.T.; software, A.A.; validation, A.A. and G.K.T.; formal analysis, A.A.; investigation, A.A.; resources, A.A.; data curation, A.A.; writing—original draft preparation, A.A.; writing—review and editing, A.A. and G.K.T.; visualization, A.A.; supervision, E.M.M., G.K.T. and K.S.; project administration, G.K.T. and K.S.; funding acquisition, G.K.T. and K.S. All authors have read and agreed to the published version of the manuscript.

Funding: This research received no external funding.

Institutional Review Board Statement: Not applicable.

Informed Consent Statement: Not applicable.

Data Availability Statement: Not applicable.

Acknowledgments: We would like to acknowledge the KDDI Foundation International Students Scholarship and the Telecommunications Advancement Foundation for the financial support to complete this research.

Conflicts of Interest: The authors declare no conflict of interest.

Abbreviations

The following abbreviations are used in this manuscript:

DSA	Dynamic Spectrum Access
UAV	Unmanned Aerial Vehicle
CRN	Cognitive Radio Network
ETC	Electronic Toll Gate
MAB	Multi-armed Bandit
PBA-MAB	Power-Budget-Aware Multi-armed Bandit
UCB	Upper Confidence Bound
TS	Thompson Sampling
PBA-UCB	Power-Budget-Aware Upper Confidence Bound
PBA-TS	Power-Budget-Aware Thompson Sampling
PU	Primary User
SU	Secondary User
QoS	Quality of Service
ML	Machine Learning
RL	Reinforcement Learning
SINR	Signal-to-Interference-Plus-Noise Ratio
WSN	Wireless Sensor Network
CP	Control Plane
DP	Data Plane

VANET	Vehicle Ad hoc NETwork
iid	independent and identical distribution
WLAN	Wireless Local Area Network
LoS	Line of Sight
AWGN	Additive White Gaussian Noise
HetNets	Heterogeneous Networks
PHY	Physical layer
MAC	Medium Access Control layer
IoT	Internet of Things
VHF	Very High Frequency
UHF	Ultra High Frequency
mmWave	millimeter-wave
D2D	Device to Device

References

1. Mkiramweni, M.E.; Yang, C.; Li, J.; Zhang, W. A Survey of Game Theory in Unmanned Aerial Vehicles Communications. *IEEE Commun. Surv. Tutor.* **2019**, *21*, 3386–3416. [CrossRef]
2. Mozaffari, M.; Saad, W.; Bennis, M.; Nam, Y.H.; Debbah, M. A Tutorial on UAVs for Wireless Networks: Applications, Challenges, and Open Problems. *IEEE Commun. Surv. Tutor.* **2019**, *21*, 2334–2360. [CrossRef]
3. Panda, K.G.; Das, S.; Sen, D.; Arif, W. Design and Deployment of UAV-Aided Post-Disaster Emergency Network. *IEEE Access* **2019**, *7*, 102985–102999. [CrossRef]
4. Hu, F.; Chen, B.; Zhu, K. Full Spectrum Sharing in Cognitive Radio Networks Toward 5G: A Survey. *IEEE Access* **2018**, *6*, 15754–15776. [CrossRef]
5. Zhao, Q.; Swami, A. A Survey of Dynamic Spectrum Access: Signal Processing and Networking Perspectives. In Proceedings of the 2007 IEEE International Conference on Acoustics, Speech and Signal Processing—ICASSP '07, Honolulu, HI, USA, 15–20 April 2007; Volume 4, pp. IV-1349–IV-1352. [CrossRef]
6. Akyildiz, I.F.; Lee, W.y.; Vuran, M.C.; Mohanty, S. A survey on spectrum management in cognitive radio networks. *IEEE Commun. Mag.* **2008**, *46*, 40–48. [CrossRef]
7. Haykin, S. Cognitive radio: Brain-empowered wireless communications. *IEEE J. Sel. Areas Commun.* **2005**, *23*, 201–220. [CrossRef]
8. Drozd, A.L.; Mohan, C.K.; Varshney, P.K.; Werner, D.D. Multiobjective joint optimization and frequency diversity for efficient utilization of the RF transmission hyperspace. In Proceedings of the 2004 International Waveform Diversity Design Conference, Edinburgh, UK, 8–10 November 2004; pp. 1–7. [CrossRef]
9. Sutton, R.S.; Barto, A.G. *Reinforcement Learning: An Introduction*; MIT Press, Cambridge, MA, USA, 2018.
10. Kwasinski, A.; Wang, W.; Mohammadi, F.S. Reinforcement learning for resource allocation in cognitive radio networks. *Mach. Learn. Future Wirel. Commun.* **2020**, 2020, 27–44. doi:10.1002/9781119562306.ch2. [CrossRef]
11. Katehakis, M.N.; Veinott, A.F., Jr. The multi-armed bandit problem: Decomposition and computation. *Math. Oper. Res.* **1987**, *12*, 262–268. [CrossRef]
12. Bubeck, S.; Cesa-Bianchi, N. Regret analysis of stochastic and nonstochastic multi-armed bandit problems. *arXiv* **2012**, arxiv:1204.5721v2.
13. Auer, P.; Cesa-Bianchi, N.; Fischer, P. Finite-time analysis of the multiarmed bandit problem. *Mach. Learn.* **2002**, *47*, 235–256.:1013689704352. [CrossRef]
14. Audibert, J.Y.; Munos, R.; Szepesvári, C. Exploration–exploitation tradeoff using variance estimates in multi-armed bandits. *Theor. Comput. Sci.* **2009**, *410*, 1876–1902. [CrossRef]
15. Francisco-Valencia, I.; Marcial-Romero, J.R.; Valdovinos-Rosas, R.M. A comparison between UCB and UCB-Tuned as selection policies in GGP. *J. Intell. Fuzzy Syst.* **2019**, *36*, 5073–5079. [CrossRef]
16. Olivieri, M.; Barnett, G.; Lackpour, A.; Davis, A.; Ngo, P. A scalable dynamic spectrum allocation system with interference mitigation for teams of spectrally agile software defined radios. In Proceedings of the First IEEE International Symposium on New Frontiers in Dynamic Spectrum Access Networks (DySPAN 2005), Baltimore, MD, USA, 8–11 November 2005; pp. 170–179. [CrossRef]
17. Agrawal, S.; Goyal, N. Further optimal regret bounds for thompson sampling. *Artif. Intell. Stat.* **2013**, *31*, 99–107.
18. Sun, Y.; Peng, M.; Zhou, Y.; Huang, Y.; Mao, S. Application of Machine Learning in Wireless Networks: Key Techniques and Open Issues. *IEEE Commun. Surv. Tutor.* **2019**, *21*, 3072–3108. [CrossRef]
19. Huang, Y.; Xu, C.; Zhang, C.; Hua, M.; Zhang, Z. An Overview of Intelligent Wireless Communications using Deep Reinforcement Learning. *J. Commun. Inf. Netw.* **2019**, *4*, 15–29. [CrossRef]
20. Wilhelmi, F.; Cano, C.; Neu, G.; Bellalta, B.; Jonsson, A.; Barrachina-Muñoz, S. Collaborative spatial reuse in wireless networks via selfish multi-armed bandits. *Ad Hoc Netw.* **2019**, *88*, 129–141. [CrossRef]
21. Zhao, Q.; Sadler, B.M. A Survey of Dynamic Spectrum Access. *IEEE Signal Process. Mag.* **2007**, *24*, 79–89. 361604. [CrossRef]

22. Yucek, T.; Arslan, H. A survey of spectrum sensing algorithms for cognitive radio applications. *IEEE Commun. Surv. Tutor.* **2009**, *11*, 116–130. [CrossRef]
23. Wang, J.; Ghosh, M.; Challapali, K. Emerging cognitive radio applications: A survey. *IEEE Commun. Mag.* **2011**, *49*, 74–81. [CrossRef]
24. Liang, Y.C.; Chen, K.C.; Li, G.Y.; Mahonen, P. Cognitive radio networking and communications: An overview. *IEEE Trans. Veh. Technol.* **2011**, *60*, 3386–3407. [CrossRef]
25. Srinivasa, S.; Jafar, S.A. Cognitive Radios for Dynamic Spectrum Access—The Throughput Potential of Cognitive Radio: A Theoretical Perspective. *IEEE Commun. Mag.* **2007**, *45*, 73–79. [CrossRef]
26. Akan, O.B.; Karli, O.B.; Ergul, O. Cognitive radio sensor networks. *IEEE Netw.* **2009**, *23*, 34–40. [CrossRef]
27. Xing, Y.; Chandramouli, R.; Mangold, S.; N, S. Dynamic spectrum access in open spectrum wireless networks. *IEEE J. Sel. Areas Commun.* **2006**, *24*, 626–637. [CrossRef]
28. Theis, N.C.; Thomas, R.W.; DaSilva, L.A. Rendezvous for Cognitive Radios. *IEEE Trans. Mob. Comput.* **2011**, *10*, 216–227. [CrossRef]
29. Homssi, B.A.; Al-Hourani, A.; Krusevac, Z.; Rowe, W.S.T. Machine Learning Framework for Sensing and Modeling Interference in IoT Frequency Bands. *IEEE Internet Things J.* **2021**, *8*, 4461–4471. [CrossRef]
30. Xiao, Z.; Li, F.; Jiang, H.; Bai, J.; Xu, J.; Zeng, F.; Liu, M. A Joint Information and Energy Cooperation Framework for CR-Enabled Macro–Femto Heterogeneous Networks. *IEEE Internet Things J.* **2020**, *7*, 2828–2839. [CrossRef]
31. Farag, H.M.; Mohamed, E.M. Soft decision cooperative spectrum sensing with noise uncertainty reduction. *Pervasive Mob. Comput.* **2017**, *35*, 146–164. [CrossRef]
32. Erdelj, M.; Natalizio, E. UAV-assisted disaster management: Applications and open issues. In Proceedings of the 2016 International Conference on Computing, Networking and Communications (ICNC), Kauai, HI, USA, 15–18 February 2016; pp. 1–5. [CrossRef]
33. Merwaday, A.; Tuncer, A.; Kumbhar, A.; Guvenc, I. Improved Throughput Coverage in Natural Disasters: Unmanned Aerial Base Stations for Public-Safety Communications. *IEEE Veh. Technol. Mag.* **2016**, *11*, 53–60. [CrossRef]
34. Lee, S.; Har, D.; Kum, D. Drone-Assisted Disaster Management: Finding Victims via Infrared Camera and Lidar Sensor Fusion. In Proceedings of the 2016 3rd Asia-Pacific World Congress on Computer Science and Engineering (APWC on CSE), Nadi, Fiji, 4–6 December 2016; pp. 84–89. [CrossRef]
35. Rivera, A.; Villalobos, A.; Monje, J.; Mariñas, J.; Oppus, C. Post-disaster rescue facility: Human detection and geolocation using aerial drones. In Proceedings of the 2016 IEEE Region 10 Conference (TENCON), Singapore, 22–26 November 2016; pp. 384–386. [CrossRef]
36. Kobayashi, T.; Matsuoka, H.; Betsumiya, S. Flying Communication Server in case of a Largescale Disaster. In Proceedings of the 2016 IEEE 40th Annual Computer Software and Applications Conference (COMPSAC), Atlanta, GA, USA, 10–14 June 2016; Volume 2, pp. 571–576. [CrossRef]
37. Sánchez-García, J.; García-Campos, J.; Toral, S.L.; Reina, D.G.; Barrero, F. A Self Organising Aerial Ad Hoc Network Mobility Model for Disaster Scenarios. In Proceedings of the 2015 International Conference on Developments of E-Systems Engineering (DeSE), Dubai, United Arab Emirates, 13–14 December 2015; pp. 35–40. [CrossRef]
38. Ejaz, W.; Ahmed, A.; Mushtaq, A.; Ibnkahla, M. Energy-efficient task scheduling and physiological assessment in disaster management using UAV-assisted networks. *Comput. Commun.* **2020**, *155*, 150–157. [CrossRef]
39. Zhan, P.; Yu, K.; Swindlehurst, A.L. Wireless Relay Communications with Unmanned Aerial Vehicles: Performance and Optimization. *IEEE Trans. Aerosp. Electron. Syst.* **2011**, *47*, 2068–2085. [CrossRef]
40. Zhang, S.; Shi, S.; Gu, S.; Gu, X. Power Control and Trajectory Planning Based Interference Management for UAV-Assisted Wireless Sensor Networks. *IEEE Access* **2020**, *8*, 3453–3464. [CrossRef]
41. Baek, J.; Han, S.I.; Han, Y. Energy-Efficient UAV Routing for Wireless Sensor Networks. *IEEE Trans. Veh. Technol.* **2020**, *69*, 1741–1750. [CrossRef]
42. Seliem, H.; Shahidi, R.; Ahmed, M.H.; Shehata, M.S. Drone-Based Highway-VANET and DAS Service. *IEEE Access* **2018**, *6*, 20125–20137. [CrossRef]
43. Lai, T.L.; Robbins, H. Asymptotically efficient adaptive allocation rules. *Adv. Appl. Math.* **1985**, *6*, 4–22. [CrossRef]
44. Thompson, W.R. On the likelihood that one unknown probability exceeds another in view of the evidence of two samples. *Biometrika* **1933**, *25*, 285–294. [CrossRef]
45. Toldov, V.; Clavier, L.; Loscrí, V.; Mitton, N. A Thompson sampling approach to channel exploration-exploitation problem in multihop cognitive radio networks. In Proceedings of the 2016 IEEE 27th Annual International Symposium on Personal, Indoor, and Mobile Radio Communications (PIMRC), Valencia, Spain, 4–7 September 2016; pp. 1–6. [CrossRef]
46. Bonnefoi, R.; Moy, C.; Palicot, J. Advanced metering infrastructure backhaul reliability improvement with cognitive radio. In Proceedings of the 2016 IEEE International Conference on Smart Grid Communications (SmartGridComm), Sydney, Australia, 6–9 November 2016; pp. 230–236. [CrossRef]
47. Kalathil, D.; Nayyar, N.; Jain, R. Decentralized Learning for Multiplayer Multiarmed Bandits. *IEEE Trans. Inf. Theory* **2014**, *60*, 2331–2345. [CrossRef]

48. Liu, K.; Zhao, Q.; Krishnamachari, B. Decentralized multi-armed bandit with imperfect observations. In Proceedings of the 2010 48th Annual Allerton Conference on Communication, Control, and Computing (Allerton), Monticello, IL, USA, 29 September–1 October 2010; pp. 1669–1674. [CrossRef]
49. Anandkumar, A.; Michael, N.; Tang, A.K.; Swami, A. Distributed Algorithms for Learning and Cognitive Medium Access with Logarithmic Regret. *IEEE J. Sel. Areas Commun.* **2011**, *29*, 731–745. [CrossRef]
50. Mohamed, E.M.; Hashima, S.; Aldosary, A.; Hatano, K.; Abdelghany, M.A. Gateway Selection in Millimeter Wave UAV Wireless Networks Using Multi-Player Multi-Armed Bandit. *Sensors* **2020**, *20*, 3947. [CrossRef]
51. Takeuchi, S.; Hasegawa, M.; Kanno, K.; Uchida, A.; Chauvet, N.; Naruse, M. Dynamic channel selection in wireless communications via a multi-armed bandit algorithm using laser chaos time series. *Sci. Rep.* **2020**, *10*, 1574. [CrossRef]
52. Oshima, K.; Onishi, T.; Kim, S.J.; Ma, J.; Hasegawa, M. Efficient wireless network selection by using multi-armed bandit algorithm for mobile terminals. *Nonlinear Theory Its Appl. IEICE* **2020**, *11*, 68–77. [CrossRef]
53. Mohamed, E.M.; Hashima, S.; Hatano, K.; Fouda, M.M.; Fadlullah, Z.M. Sleeping Contextual/Non-Contextual Thompson Sampling MAB for mmWave D2D Two-Hop Relay Probing. *IEEE Trans. Veh. Technol.* **2021**, *70*, 12101–12112. [CrossRef]
54. Hashima, S.; Hatano, K.; Takimoto, E.; Mahmoud Mohamed, E. Neighbor Discovery and Selection in Millimeter Wave D2D Networks Using Stochastic MAB. *IEEE Commun. Lett.* **2020**, *24*, 1840–1844. [CrossRef]
55. Bhardwaj, P.; Panwar, A.; Ozdemir, O.; Masazade, E.; Kasperovich, I.; Drozd, A.L.; Mohan, C.K.; Varshney, P.K. Enhanced Dynamic Spectrum Access in Multiband Cognitive Radio Networks via Optimized Resource Allocation. *IEEE Trans. Wirel. Commun.* **2016**, *15*, 8093–8106. [CrossRef]
56. Miao, G.; Himayat, N.; Li, G.Y.; Talwar, S. Low-Complexity Energy-Efficient Scheduling for Uplink OFDMA. *IEEE Trans. Commun.* **2012**, *60*, 112–120. [CrossRef]
57. Gupta, P.; Kumar, P. The capacity of wireless networks. *IEEE Trans. Inf. Theory* **2000**, *46*, 388–404. [CrossRef]
58. Arnold, P.; Bayer, N.; Belschner, J.; Zimmermann, G. 5G radio access network architecture based on flexible functional control/user plane splits. In Proceedings of the 2017 European Conference on Networks and Communications (EuCNC), Oulu, Finland, 12–15 June 2017; pp. 1–5. [CrossRef]
59. Chapelle, O.; Li, L. An empirical evaluation of thompson sampling. *Adv. Neural Inf. Process. Syst.* **2011**, *24*, 2249–2257.

sensors

MDPI

Article

Optimal UAV Deployment and Resource Management in UAV Relay Networks

Sang Ik Han [1,*] and Jaeuk Baek [2]

1 School of Smart IT, Semyung University, 65 Semyeong-ro, Jecheon-si 27136, Korea
2 Electronics and Telecommunications Research Institute (ETRI), 218 Gajeong-ro, Yuseong-gu, Daejeon 34129, Korea; jubaek@etri.re.kr
* Correspondence: sihan@semyung.ac.kr

Abstract: UAV equipped three-dimensional (3D) wireless networks can provide a solution for the requirements of 5G communications, such as enhanced Mobile Broadband (eMBB) and massive Machine Type Communications (mMTC). Especially, the introduction of an unmanned aerial vehicle (UAV) as a relay node can improve the connectivity, extend the terrestrial base station (BS) coverage and enhance the throughput by taking advantage of a strong air-to-ground line of sight (LOS) channel. In this paper, we consider the deployment and resource allocation of UAV relay network (URN) to maximize the throughput of user equipment (UE) within a cell, while guaranteeing a reliable transmission to UE outside the coverage of BS. To this end, we formulate joint UAV deployment and resource allocation problems, whose analytical solutions can be hardly obtained, in general. We propose a fast and practical algorithm to provide the optimal solution for the number of transmit time slots and the UAV relay location in a sequential manner. The transmit power at BS and UAV is determined in advance based on the availability of channel state information (CSI). Simulation results demonstrate that the proposed algorithms can significantly reduce the computational effort and complexity to determine the optimal UAV location and transmit time slots over an exhaustive search.

Keywords: UAV relay networks; UAV positioning; resource management; transmit time allocation

Citation: Han, S.I.; Baek, J. Optimal UAV Deployment and Resource Management in UAV Relay Networks. *Sensors* **2021**, *21*, 6878. https://doi.org/10.3390/s21206878

Academic Editor: Margot Deruyck

Received: 15 September 2021
Accepted: 8 October 2021
Published: 16 October 2021

1. Introduction

As one of the diverse emerging applications of unmanned aerial vehicles (UAV), it can be utilized as an aerial base station (BS) or an aerial relay node in three-dimensional (3D) wireless networks to satisfy the service requirement of the fifth generation (5G) communication [1–4], such as enhanced Mobile Broadband (eMBB) and massive Machine Type Communications (mMTC). Due to their mobility, versatile UAVs can adjust their locations to improve the connectivity among user equipment (UEs). Easy deployment of UAV enables to construct 3D networks efficiently with terrestrial networks, which can extend the service coverage or accommodate a large number of devices. By introducing a strong air-to-ground line of sight (LOS) channel, UAV can improve the capability of networks through diverse applications such as (1) emergency supports where communication services are unavailable [5], (2) Internet of Things (IoT) platforms where UAV can collect data from distributed IoT devices by saving their transmit power [6,7], (3) terrestrial network supports where UAVs can assist terrestrial BS transmission or device-to-device (D2D) transmissions [8].

In UAV networks, UAV positioning and radio resource allocation are key factors to extend cell coverage and to improve network performance. The locations of UAV BSs determine the coverage area and the number of UEs within its service area, whereas the resource allocation affects the overall performance of networks. Likewise for UAVs as relay node, its location and resource management (e.g., power control and transmit time allocation) are critical to guarantee seamless connectivity to UEs outside BS service areas without performance degradation.

Many studies on UAV BS scenario (UBS) [9–13] have focused on finding the optimal 3D UAV location to take advantage of strong air-to-ground LOS channels. The study [9] uses a circle packing theory to determine optimal locations of UAVs, and maximizes energy efficiency of UAVs. Another study [10], adopts an optimal transfer theory to minimize total transmit power at UAVs, and investigates the effect of UAV height on power efficiency. The authors of [11] propose a spiral algorithm that sequentially determines the locations of multiple UAVs, and their UAV deployment algorithm is shown to outperform other heuristic schemes in terms of performance and computational time. Study [12] analyzes the effect of interference between UAVs and derives the optimal height of UAV that can maximize the coverage of UAVs. The authors of [13] assume a disaster scenario and propose UAV deployment considering the coexistence of aerial and terrestrial BS. Also, recent works [6,14,15] address more complicated problems of optimizing both UAV deployment and resource allocation to improve network performance. In [6], optimal locations of UAVs, cell association and power controls are provided to maximize energy efficiency in IoT communications. The study [14] optimizes both UAV locations and cell association to minimize network delay. Study [15] achieves capacity enhancement in heterogeneous networks by optimizing UAV deployment, load balancing and traffic offload.

Compared to studies on UBS, research on UAV relay network (URN) is in its infancy, and more efforts are required to optimize both UAV deployment and resource allocation for reliable transmission. Especially, the transmit period of each relay transmission link is one of the most crucial factors on URN because it affects both optimal UAV location and the network performance. Studies [16,17] analyze the performance of URN during two transmit time slots and find the optimal height of UAV [16] and UAV operation range numerically [17] to guarantee a reliable relay transmission. However, a relay transmission during two transmit time slots is sometimes insufficient when UE stays far from BS or requires a high level of quality of service (QoS). To deal with this, Ref. [18] adopts multiple transmit time slots (>2) and derives the maximum distance between UAVs to achieve a reliable relay transmission. However, it does not consider the optimal height of UAV and the performance analysis may not be applicable in all circumstances due to a fixed height of UAVs.

Research on URN to optimize both UAV deployment and resource allocation during multiple transmit time slots (>2) can be rarely found due to the following two main reasons; a relay transmission under time-varying channels and a difficulty on joint optimization of UAV deployment and resource allocation. It is impractical to optimize UAV deployment and resource allocation reflecting channel variations within a single time slot. So, joint optimization of UAV deployment and resource allocation for multiple transmit time slots is required even though they depend on each other.

In this paper, we consider with no constraint on the number of overall transmit time slots in URN to joint optimization for UAV deployment and resource allocation. Especially, the throughput of UE within a cell is maximized while guaranteeing a reliable relay transmission to UE in its extended service area. Multiple transmit time slots are utilized in URN, but the minimum number of overall transmit time slots is considered in a relay transmission for efficient resource management and without performance degradation of UEs within its original coverage due to reduced service opportunity by the BS. The formulated joint UAV deployment and transmit time allocation problem is a mixed-integer nonlinear problem, which is difficult to solve and requires huge computational effort to achieve global optimality. To tackle this, a time-varying channel condition is approximated to the channel expectation in URN. The joint optimization problem is decomposed in a sequential manner. As a solution, we propose the fast and practical UAV deployment and transmit time allocation (UDTA) algorithm, which consists of a novel time slot determination (TSD) algorithm and UAV deployment (UD) algorithm that determines the optimal number of transmit time slots and optimal UAV location, respectively. Transmit power at

BS and UAV is determined based on channel state information (CSI). To the best of our knowledge, no such work on URN to optimize UAV deployment and resource allocation for generalized multiple transmit time slots is conducted.

The paper is organized as follows. Section 2 describes a URN system model. In Section 3, the joint UAV deployment and transmit time allocation problem for throughput maximization of UEs is formulated. Section 4 optimizes the UAV location for given transmit time slots, and the optimal number of transmit time slots is determined in Section 5. Computational complexity of the proposed algorithm is analyzed in Section 6. Simulation results in Section 7 demonstrate the optimality and low complexity of the proposed algorithm, followed by the conclusion in Section 8.

2. System Model

We consider a downlink URN, where UAV is used as an aerial relay node to assist BS transmission in the networks, as shown in Figure 1. Two UEs are considered in URN; a UE at the cell edge, denoted as CU, and an isolated UE, denoted as IU. CU can receive a signal from BS through BS-to-CU link, whereas IU can only receive a signal by the relay transmission through BS-to-UAV-to-IU link due to severe pathloss attenuation or blockage between BS and IU. We assume that UAV operates in a half-duplex mode, and hence two transmission phases are considered. UAV receives data from BS in the first transmission phase, and forwards it to IU in the second transmission phase. Multiple time slots are allocated to each transmission phase to guarantee a reliable signal reception at both UAV and IU. Full channel state information (CSI) is assumed at BS, but not at UAV.

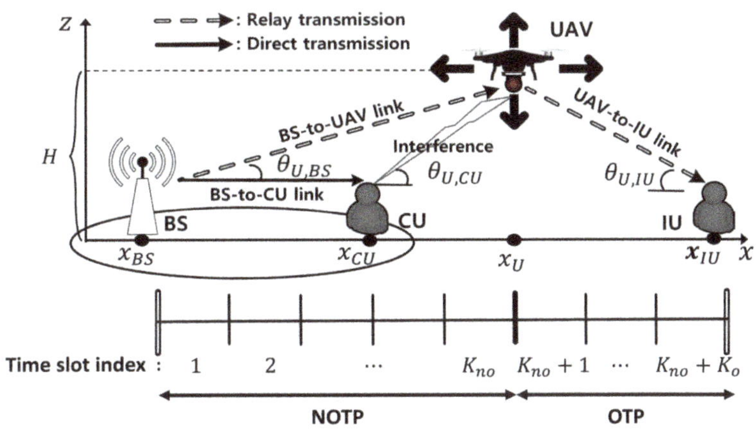

Figure 1. UAV relay network.

We assume that UAV is located at the height of H over the line between BS and IU to avoid unnecessary signal attenuation in relay transmissions. In addition, to investigate the effect of interference from UAV to the cell (especially, the worst case of maximum interference to the cell), CU is assumed to be located on the same line between BS and IU for analytical simplicity. Thereby, UAV and ground nodes can be projected onto a plane (i.e., x-z plane), which reduces to the line between BS and IU (i.e., x-axis). The location of ground node v ($v \in \{BS, IU, CU\}$) can be represented by its x-coordinate x_v, and the UAV location, denoted as $\mathbf{U}$, can be expressed as $\mathbf{U} = \{x_U, H\}$, where x_U is the x-coordinate of UAV. $x_{CU} \leq x_U \leq x_{IU}$ is assumed to set a strong UAV-to-IU link. Note that the projected two-dimensional (2D) space includes the information on UAV height, so it can clearly reflect the air-to-ground LOS channel characteristics in URN.

2.1. Channel Modeling and Assumption

Conventional relay systems where all nodes are located on the ground consider only a ground-to-ground link to characterize channels between nodes, whereas URN consists of not only ground nodes (i.e., BS, CU and IU), but also an aerial node (i.e., UAV). Therefore, an air-to-ground link should be considered along with a ground-to-ground link to characterize channels in URN.

For a ground-to-ground link, a small-scale fading with a pathloss dependent large-scale fading can be used to reflect a rich-scattering environment and a signal attenuation [19]. In URN, the channel between BS and CU is modeled as $h_{BS,CU}d_{BS,CU}^{-\beta_G}$, where $h_{BS,CU}{\sim}exp(1)$ denotes the small-scale fading modeled by Rayleigh distribution, and $d_{BS,CU}^{-\beta_G}$ denotes the pathloss dependent large-scale fading. $d_{BS,CU}$ is the distance between BS and CU, and β_G denotes a pathloss exponent in a ground-to-ground link.

For an air-to-ground link, strong signals in LOS and Non-LOS (NLOS) links dominate the channel characteristics and reduce the randomness of channel fluctuations. Hence, a small-scale fading can be neglected, and only a pathloss dependent large-scale fading in LOS and NLOS links is considered to model an air-to-ground channel in URN [8]. Ref. [20] derives LOS probability of an air-to-ground link between UAV U and ground node v as

$$\mathbb{P}_{los}^{v} = F(\theta_{U,v}) = \frac{1}{1 + C\exp(-B[\theta_{U,v} - C])}, \tag{1}$$

where $\theta_{U,v}$ is an elevation angle between UAV U and ground node v, as shown in Figure 1. B and C are coefficients that reflect the characteristics of the environment, such as rural, suburban and urban areas. Compared to a LOS link, an NLOS link experiences an additional signal attenuation of ς [dB]. Therefore, an air-to-ground channel between UAV U and ground node v can be modeled as $d_{v,U}^{-\beta_A}(\mathbb{P}_{los}^{v} + \varsigma\mathbb{P}_{nlos}^{v})$, where $d_{v,U}$ is the distance between UAV U and ground node v, β_A is a pathloss exponent of air-to-ground link, and $\mathbb{P}_{los}^{v}$ and $\mathbb{P}_{nlos}^{v}$ are the LOS and NLOS probabilities of the link between UAV U and ground node v with $\mathbb{P}_{nlos}^{v} = 1 - \mathbb{P}_{los}^{v}$.

We assume that the channel condition between BS and CU is better than that between BS and UAV (i.e., $h_{BS,CU}d_{BS,CU}^{-\beta_G} > d_{BS,U}^{-\beta_A}(\mathbb{P}_{los}^{BS} + \varsigma\mathbb{P}_{nlos}^{BS})$). The distance between BS and UAV is much longer than that between BS and CU (i.e., $d_{BS,U} \geq d_{BS,CU}$), because UAV should be located close to IU for a reliable UAV-to-IU link. Due to a long distance between BS and UAV, a pathloss attenuation becomes dominant in the channel condition of LOS link between BS and UAV. Hence, the channel condition between BS and UAV gets worse than that between BS and CU [17].

2.2. Transmission Schemes in URN

Based on the result of [17] that a non-orthogonal transmission at BS outperforms an orthogonal transmission in URN in terms of overall throughput of UEs in the cell, we adopt the non-orthogonal transmission at BS in the first transmission phase, where BS transmits a superposition-coded signal to CU and UAV simultaneously [21]. On the other hands, in the second transmission phase, the orthogonal transmission is used at BS and UAV, where BS transmits a signal to CU, and UAV forwards the received data from BS in the first transmission phase to IU. In the rest of this paper, the non-orthogonal transmission phase (NOTP) and the orthogonal transmission phase (OTP) are used to represent the first and second transmission phase, respectively.

2.3. Power Control Strategy and Overall Transmit Time Slots

A pairwise power control [22] is adopted at BS during entire transmit time slots of URN to guarantee a required QoS in the cell while supporting a relay transmission to IU. In NOTP, BS allocates $P_{BS,CU} = \rho h_{BS,CU}^{-1} d_{BS,CU}^{\beta_G}$ transmit power to BS-to-CU link to guarantee a received signal power of ρ at CU, and the remaining transmit power at BS, $P_{BS,U}$ (i.e., $P_{BS,U} = P_{BS}^{\max} - P_{BS,CU}$), is allocated to BS-to-UAV link, where $P_{BS}^{\max}$ is a maximum transmit power at BS. As a full CSI is available at BS, BS can determine $P_{BS,CU}$ depending on the channel condition in a BS-to-CU link, and then $P_{BS,U}$ can be determined. Similarly, in OTP, BS holds a pairwise power control to guarantee the received signal power of ρ at CU. On the other hand, since UAV has no CSI, UAV uses its maximum transmit power, $P_U^{\max}$, to provide seamless communication service to IU.

As illustrated in Figure 1, URN consists of overall $n = K_{no} + K_o$ time slots, where NOTP is composed of K_{no} time slots with a time index $k_{no} \in \mathbf{K_{no}} = \{1, \ldots, K_{no}\}$ and OTP has K_o time slots with a time index $k_o \in \mathbf{K_o} = \{K_{no}+1, \ldots, K_{no}+K_o\}$. (K_{no}, K_o) denotes a pair of time slots for each transmission phase.

3. URN: UAV Relay Network

3.1. Throughput of CU and IU

At a time slot k_{no} ($\forall k_{no} \in \mathbf{K_{no}}$) in NOTP, BS transmits a superposition-coded signal to CU and UAV simultaneously with transmit power $P_{BS,CU}(k_{no}) = \rho h_{BS,CU}^{-1}(k_{no}) d_{BS,CU}^{\beta_G}$ and $P_{BS,U}(k_{no}) = P_{BS}^{\max} - P_{BS,CU}(k_{no})$, as explained in Section 2.2. CU can perform the successive interference cancellation (SIC) [21] to eliminate an interference from BS-to-UAV link due to the channel assumption in Section 2.1 (i.e., $h_{BS,CU}(k_{no}) d_{BS,CU}^{-\beta_G} \geq d_{U,BS}^{-\beta_A}(\mathbb{p}_{los}^{BS} + \varsigma \mathbb{p}_{nlos}^{BS})$). On the other hand, UAV cannot eliminate an interference from BS-to-CU link. Hence, the corresponding signal to interference plus noise ratios (SINRs) at CU, $\psi_{CU}^{no}(k_{no})$, and UAV, $\psi_U(k_{no})$, at a time slot k_{no} in NOTP can be expressed as

$$\psi_{CU}^{no}(k_{no}) = \frac{\rho}{\sigma_{CU}^2}, \tag{2}$$

$$\psi_U(k_{no}) = \frac{P_{BS,U}(k_{no}) d_{BS,U}^{-\beta_A}(\mathbb{p}_{los}^{BS}(1-\varsigma)+\varsigma)}{P_{BS,CU}(k_{no}) d_{BS,U}^{-\beta_A}(\mathbb{p}_{los}^{BS}(1-\varsigma)+\varsigma)+\sigma_U^2}, \tag{3}$$

where $d_{BS,U}^{-\beta_A}(\mathbb{p}_{los}^{BS}(1-\varsigma)+\varsigma)$ in (3) represents the channel in BS-to-UAV link with LOS probability $\mathbb{p}_{los}^{BS}$. σ_i^2 indicates the variance of additive white Gaussian noise (AWGN) at a node i.

At a time slot k_o ($\forall k_o \in \mathbf{K_o}$) in OTP, BS transmits a signal to CU, and UAV relays the received data from BS in NOTP to IU. Therefore, the SINRs at CU, $\psi_{CU}^o(k_o)$, and IU, $\psi_{IU}(k_o)$, at a time slot k_o in OTP can be given by

$$\psi_{CU}^o(k_o) = \frac{\rho}{I_{U,CU} + \sigma_{CU}^2}, \tag{4}$$

$$\psi_{IU}(k_o) = \frac{P_U^{\max} d_{U,IU}^{-\beta_A}(\mathbb{p}_{los}^{IU}(1-\varsigma)+\varsigma)}{\sigma_{IU}^2}, \tag{5}$$

where $I_{U,CU} \triangleq P_U^{\max} d_{U,CU}^{-\beta_A}(\mathbb{p}_{los}^{CU}(1-\varsigma)+\varsigma)$ in (4) represents the interference from UAV to CU. IU does not receive any interference from BS in OTP due to severe pathloss attenuation in BS-to-IU link. $\mathbb{p}_{los}^{CU}$ and $\mathbb{p}_{los}^{IU}$ are the LOS probabilities of UAV-to-CU link and UAV-to-IU link, respectively. Note that we assume that the adjacent cell utilizes different frequency bands from that of the cell of interest to avoid the inter-cell interference, and that other interference received at CU is negligible except that from the link between UAV and IU in OTP, which is dominant.

From (2), (4) and (5), we can find that the SINR at CU in both transmission phases and that at IU in OTP are time-invariant (i.e., $\psi_{CU}^{no}(k_{no}) = \psi_{CU}^{no}$, $\psi_{CU}^o(k_o) = \psi_{CU}^o$ and $\psi_{IU}(k_o) = \psi_{IU}, \forall k_{no}, k_o$) due to the pairwise power control and channel characteristics of air-to-ground LOS link. However, the SINR at UAV in NOTP (i.e., (3)) is time-varying for each time slot k_{no} because $P_{BS,U}(k_{no})$ and $P_{BS,CU}(k_{no})$ vary with the channel condition of BS-to-CU link.

Based on the Shannon capacity theorem [19], the amount of received data at CU, $r_{CU}^{\Sigma}(|\mathbf{K_{no}}|)$, and at UAV, $r_{U}^{\Sigma}(|\mathbf{K_{no}}|)$, in NOTP ($\forall k_{no} \in \mathbf{K_{no}}$) can be obtained using (2) and (3) as

$$r_{CU}^{\Sigma}(|\mathbf{K_{no}}|) = \sum_{k_{no}=1}^{K_{no}} f(\psi_{CU}^{no}) = f(\psi_{CU}^{no})K_{no}, \tag{6}$$

$$r_{U}^{\Sigma}(|\mathbf{K_{no}}|) = \sum_{k_{no}=1}^{K_{no}} f(\psi_{U}(k_{no})), \tag{7}$$

respectively, where $f(x) \triangleq \log(1+x)$. (6) follows that each time slot has a unit length and ψ_{CU}^{no} is a time-invariant.

Similarly, the amount of received data at CU, $r_{CU}^{\Sigma}(|\mathbf{K_{o}}|)$, and at IU, $r_{IU}^{\Sigma}(|\mathbf{K_{o}}|)$, in OTP ($\forall k_{o} \in \mathbf{K_{o}}$) can be obtained using (4) and (5) as

$$r_{CU}^{\Sigma}(|\mathbf{K_{o}}|) = \sum_{k_{o}=1}^{K_{o}} f(\psi_{CU}^{o}) = f(\psi_{CU}^{o})K_{o}, \tag{8}$$

$$r_{IU}^{\Sigma}(|\mathbf{K_{o}}|) = \sum_{k_{o}=1}^{K_{o}} f(\psi_{IU}) = f(\psi_{IU})K_{o}. \tag{9}$$

For the overall time slots n, the average data rate of CU, R_{CU} [bps/Hz], can be defined by (6) and (8) as

$$R_{CU} = \frac{1}{n}(r_{CU}^{\Sigma}(|\mathbf{K_{no}}|) + r_{CU}^{\Sigma}(|\mathbf{K_{o}}|)), \tag{10}$$

and the total amount of received data at IU via relay transmission, D_{IU} [bit/Hz], can be obtained by (7) and (9) as

$$D_{IU} = \min(r_{U}^{\Sigma}(|\mathbf{K_{no}}|), r_{IU}^{\Sigma}(|\mathbf{K_{o}}|)), \tag{11}$$

where (11) follows that the amount of transmitted data through a forwarding link (i.e., UAV-to-IU link) cannot exceed that of received data at UAV via backhaul link (i.e., BS-to-UAV link) in a relay transmission.

3.2. Problem Formulation: *JUDTAP*

The throughput maximization of UEs in URN is equivalent to maximizing R_{CU} while delivering the required amount of data to IU, D_{req}, during the minimum number of overall time slots n with respect to UAV location $\mathbf{U} = \{x_U, H\}$ and transmit time slots $\mathbf{K} = \{K_{no}, K_o\}$. Hence, the multi-objective optimization problem, denoted as joint UAV deployment and transmit time allocation problem (**JUDTAP**), can be formulated as (12)

$$\textbf{JUDTAP:} \max_{\mathbf{U},\mathbf{K}} \quad [R_{CU}, \frac{1}{n}] \tag{12}$$

$$\text{s.t.} \quad D_{IU} \geq D_{req} \tag{12a}$$

$$n = K_{no} + K_o \tag{12b}$$

$$K_{no} \geq 1, K_o \geq 1 \tag{12c}$$

$$k_{no} \in \mathbf{K_{no}}, k_o \in \mathbf{K_o} \tag{12d}$$

$$x_{CU} < x_U < x_{IU}, H \geq 0 \tag{12e}$$

where multi-objective function implies that the overall number of time slots n should be minimized before the average data rate of R_{CU} is maximized, as explained in Section 1. (12a) shows the requirement on the amount of received data at IU. (12b)–(12d) represent the constraints on the number of time slots in URN, and (12e) indicates the possible operation range that UAV can be deployed.

The **JUDTAP** is a mixed-integer nonlinear programming [23] and its combinatorial nature makes the bulk of computational load to find a global optimal solution (i.e., $\mathbf{U^{opt}}, \mathbf{K^{opt}}$). In addition, mutual-influence between UAV location and transmit periods in both transmission phases makes it more difficult to be solved. For example, to maximize R_{CU}, UAV should be located close to IU to reduce interference from UAV to CU (i.e., $I_{U,CU}$ in (4)). However, it may increase K_{no}, eventually n, to guarantee a data transmission in BS-to-UAV link (i.e., to satisfy (12a)). Therefore, the **JUDTAP** cannot be solved by optimizing **U** and **K** independently due to their close relationships.

One approach to solve the **JUDTAP** is updating **U** and **K** iteratively. However, these procedures are not practical and cannot guarantee a convergence to global optimal solution. Therefore, in this paper, we propose a fast and practical algorithm that finds $\mathbf{U^{opt}}$ and $\mathbf{K^{opt}}$ in a sequential manner;

- Step 1: Find the optimal pair of (K_{no}, K_o), $\mathbf{K^{opt}}$, that leads to a minimum n.
- Step 2: Determine the optimal location of UAV, $\mathbf{U^{opt}}$.

The details on each step will be presented in Sections 4 and 5.

3.3. Analysis on Relay Transmission during Multiple Time Slots

As explained in the previous section, the relay transmission during multiple transmit time slots makes it difficult to analyze the constraint on BS-to-UAV link in (12a) (i.e., $r_U^{\Sigma}(|\mathbf{K_{no}}|) \geq D_{req}$). More specifically, a time-varying small-scale fading in BS-to-CU link changes $\psi_U(k_{no})$ in (3) and $r_U^{\Sigma}(|\mathbf{K_{no}}|)$ in (7) for each time slot k_{no}, so it is challenging to find optimal **U** and **K** that satisfy $r_U^{\Sigma}(|\mathbf{K_{no}}|) \geq D_{req}$. To cope with this issue, we introduce the expected channel model in a ground-to-ground link (i.e., BS-to-CU link in URN) because the effect of random fluctuation by small-scale fading during multiple time slots is negligible and it is impractical to adjust the location of UAV for the short period of each time slot. Therefore, $\mathbb{E}[h_{BS,CU}]d_{BS,CU}^{-\beta_G}$ is used to model the channel condition of BS-to-CU link, where $\mathbb{E}[h_{BS,CU}]$ represents the expectation of small-scale fading $h_{BS,CU}(k), \forall k \in \mathbf{K_{no}} \cup \mathbf{K_o}$. The expected channel model affects the pairwise power control at BS, and the SINR at UAV in (3) as follows.

The pairwise power control at BS is simplified to a fixed power control. In NOTP, BS allocates $\bar{P}_{BS,CU} = \rho\mathbb{E}[h_{BS,CU}]^{-1}d_{BS,CU}^{\beta_G}$ and $\bar{P}_{BS,U} = P_{BS}^{\max} - \bar{P}_{BS,CU}$ to BS-to-CU link and BS-to-UAV link, respectively. Similarly, in OTP, BS allocates $\bar{P}_{BS,CU} = \rho\mathbb{E}[h_{BS,CU}]^{-1}d_{BS,CU}^{\beta_G}$ to BS-to-CU link.

The time-varying SINR at UAV in NOTP (i.e., $\psi_U(k_{no}), \forall k_{no}$ in (3)) can be replaced into a time-invariant $\bar{\psi}_U$ due to $\bar{P}_{BS,CU}$ and $\bar{P}_{BS,U}$. The amount of received data at UAV in (7) can be simplified as $\bar{r}_U^{\Sigma}(|\mathbf{K_{no}}|)$, where $\bar{r}_U^{\Sigma}(|\mathbf{K_{no}}|) = f(\bar{\psi}_U)K_{no}$.

Therefore, the constraint (12a) can be expressed as

$$\begin{aligned} & D_{IU} \geq D_{req} \\ \Leftrightarrow\ & \frac{\bar{P}_{BS,U}d_{BS,U}^{-\beta_A}(\mathbb{P}_{los}^{BS}(1-\varsigma)+\varsigma)}{\bar{P}_{BS,CU}d_{BS,U}^{-\beta_A}(\mathbb{P}_{los}^{BS}(1-\varsigma)+\varsigma)+\sigma_U^2} \geq 2^{\frac{D_{req}}{K_{no}}} - 1, \end{aligned} \tag{13}$$

$$\text{and } P_U^{\max}\frac{d_{U,IU}^{-\beta_A}}{\sigma_{IU}^2}(\mathbb{P}_{los}^{IU}(1-\varsigma)+\varsigma) \geq 2^{\frac{D_{req}}{K_o}} - 1, \tag{14}$$

where (13) and (14) are obtained from time-invariant $\bar{r}_U^{\Sigma}(|\mathbf{K_{no}}|)$ and (9), respectively.

After rearranging above two inequalities, the left terms of (13) and (14) can be expressed with respect to LOS probability, which are given as

$$F(\theta_{U,BS}) \geq X_{BS}(d_{U,BS}, K_{no}), \tag{15}$$

$$F(\theta_{U,IU}) \geq X_{IU}(d_{U,IU}, K_o), \tag{16}$$

where

$$X_{BS}(d_{U,BS}, K_{no}) \triangleq \frac{(2^{\frac{D_{req}}{K_{no}}} - 1)\frac{\sigma_U^2}{d_{U,BS}^{-\beta_A}(1-\varsigma)}}{(P_{BS}^{\max} - 2^{\frac{D_{req}}{K_{no}}}\bar{P}_{BS,CU})} - \frac{\varsigma}{1-\varsigma}, \tag{17}$$

$$X_{IU}(d_{U,IU}, K_o) \triangleq \frac{(2^{\frac{D_{req}}{K_o}} - 1)\frac{\sigma_{IU}^2}{d_{U,IU}^{-\beta_A}(1-\varsigma)}}{P_U^{\max}} - \frac{\varsigma}{1-\varsigma}. \tag{18}$$

Based on (15)–(18), a sequential algorithm for Steps 1 and 2 is derived in the following sections.

4. UAV Deployment

In this section, we investigate the effect of UAV location on the network performance, and derive the optimal UAV location, $\mathbf{U}^{\mathbf{opt}}$. For this purpose, we assume that the transmit time allocation is given (i.e., (K_{no}, K_o)) and guarantees the existence of UAV locations that can provide a reliable relay transmission to IU. We utilize distance and elevation angle in x-z plane to reflect the channel characteristics between UAV and ground node in Section 2.1. $\mathbf{\Theta} = \{\theta_{U,v} | v \in \{BS, CU, IU\}\}$ and $\mathbf{D} = \{d_{i,j} | i,j \in \{BS, U, IU, CU\}\}$ represent sets of elevation angles and distances respectively, and the UAV location $\mathbf{U} = \{x_U, h\}$ in (12) can be expressed as $\mathbf{U} = \{d_{U,v}, \theta_{U,v}\}$.

4.1. UAV Deployment Problem

Based on (2) and (4), the maximization of multi-objective function for a given time allocation in **JUDTAP** is equivalent to the minimization of interference from UAV to CU (i.e., minimization of $I_{U,CU}$ in (4)). The constraint (12a) can be replaced by (15) and (16), and the constraints (12b)–(12d) can be omitted because the transmit time allocation is given.

Therefore, for a given time allocation, **JUDTAP** reduces to UAV deployment problem (**UDP**), which can be formulated as (19) with respect to $\mathbf{\Theta}$ and $\mathbf{D}$.

$$\textbf{UDP:} \min_{\mathbf{\Theta},\mathbf{D}} \quad d_{U,CU}^{-\beta_A}(F(\theta_{U,CU})(1-\varsigma)+\varsigma) \tag{19}$$

$$\text{s.t.} \quad F(\theta_{U,BS}) \geq X_{BS}(d_{U,BS}|K_{no}), \tag{19a}$$

$$F(\theta_{U,IU}) \geq X_{IU}(d_{U,IU}|K_o), \tag{19b}$$

$$d_{BS,CU} \leq d_{U,BS}\cos(\theta_{U,BS}) \leq d_{BS,IU}. \tag{19c}$$

The objective function is given by $I_{U,CU}/P_U^{\max}$. (19a) and (19b) are the constraints of BS-to-UAV link and UAV-to-IU link, respectively, and derived from (15)–(18) by replacing $X_{BS}(d_{U,BS}, K_{no})$ and $X_{IU}(d_{U,IU}, K_o)$ with $X_{BS}(d_{U,BS}|K_{no})$ and $X_{IU}(d_{U,IU}|K_o)$ due to the assumption on time allocation. (19c) represents the UAV operation range that UAV can be deployed (i.e., (12e) of **JUDTAP**).

The optimal solution of **UDP** (i.e., $\mathbf{\Theta}^{\mathbf{opt}}, \mathbf{D}^{\mathbf{opt}}$) determines the optimal UAV location, $\mathbf{U}^{\mathbf{opt}}$, for a given time allocation. However, all the elements of $\mathbf{\Theta}$ and $\mathbf{D}$ should be considered simultaneously to find $\mathbf{U}^{\mathbf{opt}}$, so no closed-form solution to **UDP** exists. Therefore, we propose UAV deployment (UD) algorithm, which updates the UAV location iteratively to reach $\mathbf{U}^{\mathbf{opt}}$ based on search areas and directions. In the following section, we define search areas and directions for a given UAV location, and investigate them to update the UAV location toward $\mathbf{U}^{\mathbf{opt}}$.

4.2. Search Areas and Directions

For a given location of UAV U, we define *search areas* and *directions* using lines and circles as shown in Figure 2, where UAV and ground nodes are placed on the x-z plane as explained in Section 2. UAVs on a line have the same elevation angle of $\theta_{U,v}$ from ground node v (i.e., $v \in \{CU, BS, IU\}$), while those on a circle have equal distance of $d_{U,v}$ from ground node v to UAV. The line and circle inside dashed-rectangle (i.e., UAV operation range) in Figure 2b define the search areas considering the interference from UAV to CU, while those in Figure 2c,d represent the search directions based on each relay transmission link. All the search areas and directions for a given UAV location are integrated in Figure 2a. The search areas and directions change when a given UAV location is updated. Therefore, we investigate search areas and directions for the given location of UAV U, $\mathbf{U}^{U}$, to find the updated location of UAV U', $\mathbf{U}^{U'}$.

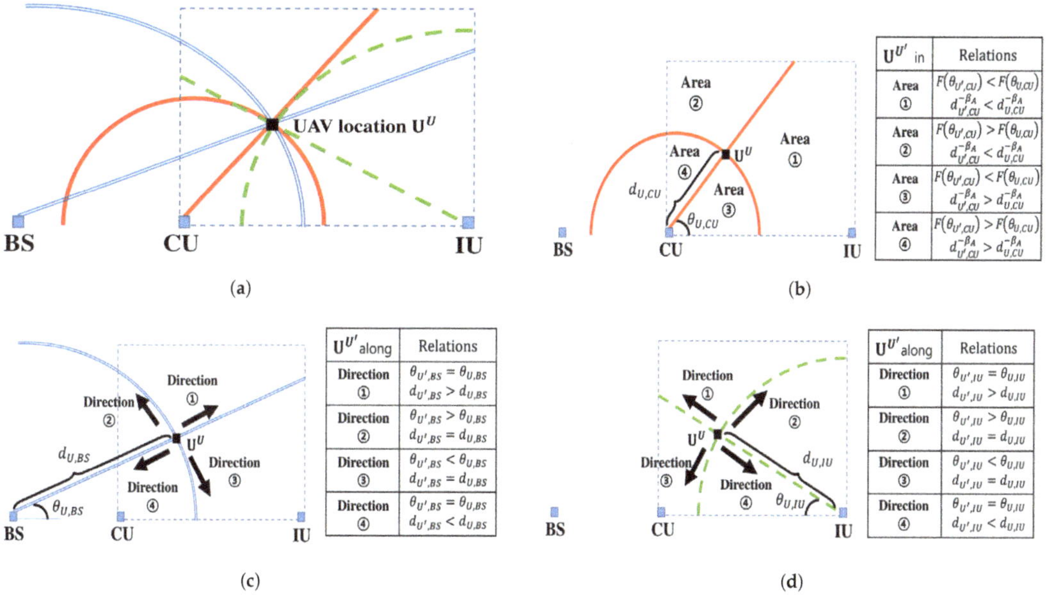

$\mathbf{U}^{U'}$ in	Relations
Area ①	$F(\theta_{U',CU}) < F(\theta_{U,CU})$, $d_{U',CU}^{-\beta_A} < d_{U,CU}^{-\beta_A}$
Area ②	$F(\theta_{U',CU}) > F(\theta_{U,CU})$, $d_{U',CU}^{-\beta_A} < d_{U,CU}^{-\beta_A}$
Area ③	$F(\theta_{U',CU}) < F(\theta_{U,CU})$, $d_{U',CU}^{-\beta_A} > d_{U,CU}^{-\beta_A}$
Area ④	$F(\theta_{U',CU}) > F(\theta_{U,CU})$, $d_{U',CU}^{-\beta_A} > d_{U,CU}^{-\beta_A}$

$\mathbf{U}^{U'}$ along	Relations
Direction ①	$\theta_{U',BS} = \theta_{U,BS}$, $d_{U',BS} > d_{U,BS}$
Direction ②	$\theta_{U',BS} > \theta_{U,BS}$, $d_{U',BS} = d_{U,BS}$
Direction ③	$\theta_{U',BS} < \theta_{U,BS}$, $d_{U',BS} = d_{U,BS}$
Direction ④	$\theta_{U',BS} = \theta_{U,BS}$, $d_{U',BS} < d_{U,BS}$

$\mathbf{U}^{U'}$ along	Relations
Direction ①	$\theta_{U',IU} = \theta_{U,IU}$, $d_{U',IU} > d_{U,IU}$
Direction ②	$\theta_{U',IU} > \theta_{U,IU}$, $d_{U',IU} = d_{U,IU}$
Direction ③	$\theta_{U',IU} < \theta_{U,IU}$, $d_{U',IU} = d_{U,IU}$
Direction ④	$\theta_{U',IU} = \theta_{U,IU}$, $d_{U',IU} < d_{U,IU}$

Figure 2. Search areas and directions for updating $\mathbf{U}^{U}$ toward $\mathbf{U}^{U'}$. (**a**) The integrated search areas and directions for $\mathbf{U}^{U}$. (**b**) Four search areas for $\mathbf{U}^{U'}$ based on the objective function of (19) (i.e., link between UAV and CU). (**c**) Four search directions for $\mathbf{U}^{U'}$ based on the constraint (19a) (i.e., link between BS and UAV). (**d**) Four search directions for $\mathbf{U}^{U'}$ based on the constraint (19b) (i.e., link between UAV and IU).

4.2.1. Search Area

The search areas ①–④ in Figure 2b are divided by the line and circle based on CU location and $\mathbf{U}^{U}$. UAV U should move towards the area where the interference from UAV U' to CU (i.e., objective function in (19)) decreases. When UAV U moves into Area ①, both $F(\theta_{U,CU})$ and $d_{U,CU}^{-\beta_A}$ in (19) decreases due to smaller $\theta_{U,CU}$ and longer $d_{U,CU}$. Any UAV locations within Area ① always reduce the UAV-to-CU interference, hence, Area ① is a potential search area for $\mathbf{U}^{U'}$. On the other hand, all UAV locations in Area ④ increases both $F(\theta_{U,CU})$ and $d_{U,CU}^{-\beta_A}$. Therefore, they cannot decrease the objective function in (19), thereby excluding Area ④ from potential search areas.

The search areas ② and ③ possess the uncertainty on the interference from UAV U' to CU. UAV locations within Area ② decrease $d_{U,CU}^{-\beta_A}$, but increase the $F(\theta_{U,CU})$, while those within Area ③ results in the opposite. However, when UAV U' is within Area ②, a pathloss attenuation dominates the LOS connection in BS-to-CU link. More specifically, $F(\theta_{U',CU})$ is close to one due to large $\theta_{U',CU}$ [20], but $d_{U',CU}$ can be sufficiently large so that $d_{U',CU}^{-\beta_A}$ becomes a dominant factor in the objective function of (19). Therefore, UAV locations within Area ② can reduce the UAV-to-CU interference compared to given UAV location $\mathbf{U}^{U}$. On the other hand, UAV locations within Area ③ makes more severe UAV-to-CU interference due to proximity of their locations, hence, Area ③ cannot be the potential search area.

- Observation 1: The objective function in (19) can be decreased by moving UAV into Area ① or ② in Figure 2b.

The Observation 1 is directly applicable for a feasible UAV location (which satisfies the constraints (19a) and (19b)) to reduce the UAV-to-CU interference. On the other hands, when UAV location cannot satisfy the constraints (i.e., infeasible UAV location), Observation 1 and the channel condition of relay transmission links should be considered simultaneously to find a feasible UAV location and to reduce the objective function in (19). In the following subsection, we examine the search directions to satisfy the constraints and decrease the objective function in (19) simultaneously.

4.2.2. Search Directions

Although the potential search areas that can be used to find $\mathbf{U}^{u'}$ from $\mathbf{U}^{u}$ are described on Observation 1, there is no clue on $\mathbf{U}^{u'}$ within the potential area. Hence, the points on a line or a circle within the potential search area are utilized to determine $\mathbf{U}^{u'}$. In particular, search directions in Figure 2c,d are examined to move UAV U into the feasible UAV location $\mathbf{U}^{u'}$ when $\mathbf{U}^{u}$ cannot satisfy the constraints (19a) or (19b).

From the constraints, there are four cases (i.e., C_1, $\bar{C_1}$, C_2 and $\bar{C_2}$) to be considered at $\mathbf{U}^{u}$. C_1 and C_2 indicate that $\mathbf{U}^{u}$ satisfies (19a) and (19b) respectively, while $\bar{C_1}$ and $\bar{C_2}$ represent that it does not. Each case follows a different search direction in Figure 2c for C_1 and $\bar{C_1}$ and in Figure 2d for C_2 and $\bar{C_2}$.

C_1 and C_2 indicate that $\mathbf{U}^{u'}$ can be found based on Observation 1 to decrease the objective function in (19). In the case of C_1, Direction ① or ② in Figure 2c should be selected because they are within the potential search areas ① and ② in Figure 2b (see Figure 2a). For the same reason, Direction ①, ② or ④ of Figure 2d should be selected in case of C_2.

$\bar{C_1}$ and $\bar{C_2}$ represent that $\mathbf{U}^{u}$ cannot satisfy the constraints (19a) and (19b) due to poor channel conditions in BS-to-UAV and UAV-to-IU link, hence resulting in $F(\theta_{U,BS}) < X_{BS}(d_{U,BS}|K_{no})$ and $F(\theta_{U,IU}) < X_{IU}(d_{U,IU}|K_o)$ respectively. Therefore, to find the feasible UAV location $\mathbf{U}^{u'}$, $\theta_{U,v}$ should be increased along a circle or $d_{U,v}$ should be decreased along a line in Figure 2c (when $v = BS$) and Figure 2d (when $v = IU$). This is because $F(\theta_{U,v})$ $v \in \{BS, IU\}$, $X_{BS}(d_{U,BS}|K_{no})$ and $X_{IU}(d_{U,IU}|K_o)$ are increasing functions of $\theta_{U,v}$, $d_{U,BS}$ and $d_{U,IU}$, respectively.

In the case of $\bar{C_1}$, the movement of UAV U along Direction ④ in Figure 2c decreases $d_{U,BS}$, while that along Direction ② increases $\theta_{U,BS}$. Direction ④ always provides the feasible UAV location $\mathbf{U}^{u'}$ that satisfies (19a), while Direction ② could find it only when $X_{BS}(d_{U,BS}|K_{no}) \leq 1$ because $\max F(\theta_{U',BS}) = 1$. Similarly, in the case of $\bar{C_2}$, the movement of UAV U along Direction ② in Figure 2d increases $\theta_{U,IU}$, while that along Direction ④ in Figure 2d decreases $d_{U,IU}$. Direction ② could find the feasible UAV location $\mathbf{U}^{u'}$ that satisfies (19b) only when $X_{IU}(d_{U,IU}|K_o) \leq 1$.

- Observation 2 (C_1): When $\mathbf{U}^{u}$ satisfies the constraint (19a), Direction ① along a line or Direction ② along a circle in Figure 2c should be selected to determine $\mathbf{U}^{u'}$.
- Observation 3 ($\bar{C_1}$): When $\mathbf{U}^{u}$ cannot satisfy the constraint (19a), Direction ④ along a line in Figure 2c always provides the feasible UAV location $\mathbf{U}^{u'}$, while Direction ② along a circle in Figure 2c could find $\mathbf{U}^{u'}$ only when $X_{BS}(d_{U,BS}|K_{no}) \leq 1$.
- Observation 4 (C_2): When $\mathbf{U}^{u}$ satisfies the constraint (19b), Direction ① or ④ along a line, or Direction ② along a circle in Figure 2d should be selected to determine $\mathbf{U}^{u'}$.
- Observation 5 ($\bar{C_2}$): When $\mathbf{U}^{u}$ cannot satisfy the constraint (19b), Direction ④ along a line in Figure 2d always provides the feasible UAV location $\mathbf{U}^{u'}$, while Direction ② along a circle in Figure 2d could find $\mathbf{U}^{u'}$ only when $X_{IU}(d_{U,IU}|K_o) \leq 1$.

4.2.3. Combined Search Directions

The Observations 2-5 should be integrated to consider the constraints (19a) and (19b) together. First, Observations 2 and 4 can be used to decrease the objective function in (19) for the case of $C_1 \cap C_2$ where $C_1 \cap C_2$ indicates that $\mathbf{U}^U$ is a feasible UAV location and satisfies both (19a) and (19b). As in Figure 2, the movement along Direction ① in Figure 2c and Direction ② in Figure 2d decreases both $F(\theta_{U,CU})$ and $d_{U,CU}^{-\beta_A}$ in (19). They have the same properties, but differ on moving along line and circle, respectively. Similarly, the movement along Direction ② in Figure 2c and Direction ① in Figure 2d decreases $d_{U,CU}^{-\beta_A}$ and achieves large $\theta_{U,CU}$, resulting in $F(\theta_{U,CU}) \approx 1$. Either direction that has same properties can be selected for the movement towards $\mathbf{U^{opt}}$. However, it is preferable to select search direction moving along a line (i.e., Direction ① in Figure 2c and Direction ① in Figure 2d) to reduce computation time, which will be discussed in Section 4.3. Note that, UAV locations along Direction ④ in Figure 2d could break the constraint (19a) due to insufficient height of UAV and small $\theta_{U,BS}$, therefore, it is not an option for $C_1 \cap C_2$.

- Observation 6 ($C_1 \cap C_2$): When $\mathbf{U}^U$ satisfies both the constraints (19a) and (19b), $\mathbf{U}^{U'}$ will be found along Direction ① in Figure 2c or Direction ① in Figure 2d.

When $\mathbf{U}^U$ is infeasible location, there are three cases (i.e., $\bar{C}_1 \cap C_2$, $C_1 \cap \bar{C}_2$ and $\bar{C}_1 \cap \bar{C}_2$) to be considered. However, it is clear that $\bar{C}_1 \cap C_2 \subset \bar{C}_1$ and $C_1 \cap \bar{C}_2 \subset \bar{C}_2$, therefore, Observations 3 and 5 will be solutions for each case.

- Observation 7 ($\bar{C}_1 \cap C_2$): When $\mathbf{U}^U$ satisfies the constrains (19b), but (19a), Direction ② with the condition on $X_{BS}(d_{U,BS}|K_{no})$ or Direction ④ in Figure 2c should be selected to find $\mathbf{U}^{U'}$ that satisfies (19a).
- Observation 8 ($C_1 \cap \bar{C}_2$): When $\mathbf{U}^U$ satisfies the constraints (19a), but (19b), Direction ② with the condition on $X_{IU}(d_{U,IU}|K_o)$ or Direction ④ in Figure 2d should be selected to find $\mathbf{U}^{U'}$ that satisfies (19b).

Lastly, $\bar{C}_1 \cap \bar{C}_2$ indicates that $\mathbf{U}^U$ cannot satisfy both constraints on relay transmission links. Unfortunately, there is no solution based on Observations 3 and 5. For example, if UAV U moves along Direction ④ in Figure 2c to make (19a) satisfied (which is opposite to Direction ② in Figure 2d suggested in Observation 5 for the satisfaction of (19b)), it causes $d_{U,IU} < d_{U',IU}$ and $\theta_{U,IU} > \theta_{U',IU}$, thereby resulting in $F(\theta_{U',IU}) < F(\theta_{U,IU}) < X_{IU}(d_{U,IU}|K_o) < X_{IU}(d_{U',IU}|K_o)$ (i.e., (19b) is still not satisfied). Similarly, other search directions on Observations 3 and 5 also cannot simultaneously improve both relay transmission links, so we declare that no feasible UAV location exists for the case of $\bar{C}_1 \cap \bar{C}_2$. To deal with this issue, more transmit time slots should be allocated to the relay transmission, which will be discussed in Section 5.

- Observation 9 ($\bar{C}_1 \cap \bar{C}_2$): When $\mathbf{U}^U$ cannot satisfy both constraints (19a) and (19b), no feasible UAV location $\mathbf{U}^{U'}$ exists without allocating more transmit time slots to relay transmission.

4.3. *UAV Deployment (UD) Algorithm*

In this section, we propose a novel UAV deployment (UD) algorithm for a given time allocation based on search directions. The constraints (19a) and (19b) are described graphically in Figure 3a as parabolic curves C_a and C_b, which are drawn with an equality in (19a) and (19b) respectively. The UAV locations inside C_a and C_b satisfy the constraints (19a) and (19b) respectively, therefore, areas for $C_1 \cap C_2$, $C_1 \cap \bar{C}_2$, $\bar{C}_1 \cap C_2$, and $\bar{C}_1 \cap \bar{C}_2$ (i.e., $A_{C_1 \cap C_2}$, $A_{C_1 \cap \bar{C}_2}$, $A_{\bar{C}_1 \cap C_2}$, and $A_{\bar{C}_1 \cap \bar{C}_2}$) can be defined as in Figure 3a. In particular, $A_{C_1 \cap C_2}$ (see dashed area in Figure 3a) is of special interest to find the optimal UAV location $\mathbf{U^{opt}}$ because it indicates the feasible UAV locations and always includes $\mathbf{U^{opt}}$. The $\mathbf{U^{opt}}$ will be determined to be on either C_a or C_b within $A_{C_1 \cap C_2}$ (refer to Section 4.6 in [24]), especially near the upper point of intersection of C_a and C_b to minimize the interference between UAV and CU. Note that $A_{C_1 \cap C_2}$ always exists due to the assumption at the beginning of Section 4 that the given transmit time allocation (K_{no}, K_o) guarantees the existence of feasible UAV locations.

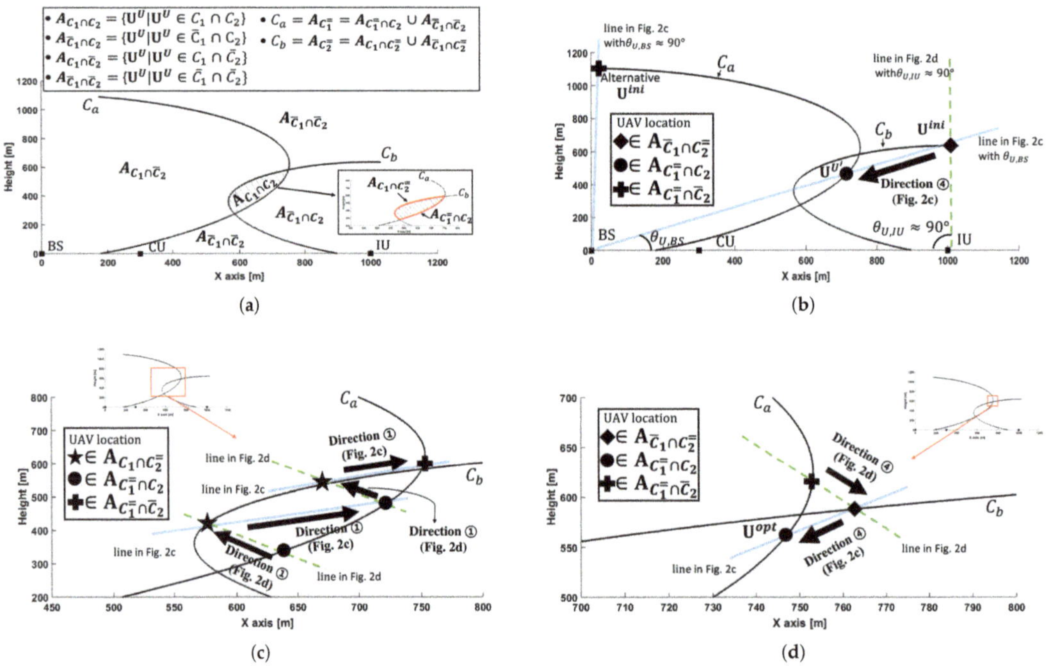

Figure 3. The UAV deployment (UD) algorithm. (**a**) Three areas for $\mathbf{U}^{U}$ with respect to constraints (19a) and (19b). $\mathbf{U}^{opt}$ is within $A_{C_1 \cap C_2}$ (specifically, either $A_{C_1^{=} \cap C_2}$ or $A_{C_1 \cap C_2^{=}}$). (**b**) [Step 1] on UD algorithm to determine $\mathbf{U}^{U'} = \mathbf{U}^{f}$ from $\mathbf{U}^{U} = \mathbf{U}^{ini}$. (**c**) [Step 2] on UD algorithm to update $\mathbf{U}^{f}$ until $\mathbf{U}^{U'}$ is determined outside $A_{C_1 \cap C_2}$. (**d**) [Step 3] on UD algorithm to determine $\mathbf{U}^{opt}$ and terminate the algorithm.

We introduce two more cases $C_1^{=}$ and $C_2^{=}$, which represent that current UAV location satisfies the constraints (19a) and (19b) with an equality, respectively. Hence, it is clear that curve C_a consists of $A_{C_1^{=} \cap C_2}$ and $A_{C_1^{=} \cap \bar{C}_2}$, whereas C_b is composed of $A_{C_1 \cap C_2^{=}}$ and $A_{\bar{C}_1 \cap C_2^{=}}$. In addition, $A_{C_1 \cap C_2}$ includes $A_{C_1^{=} \cap C_2}$ and $A_{C_1 \cap C_2^{=}}$ (see Figure 3a).

The UD algorithm consists of three steps: [Step 1] for finding a feasible UAV location $\mathbf{U}^{f}$ from an initial UAV location $\mathbf{U}^{ini}$, [Step 2] for updating $\mathbf{U}^{f}$ towards $\mathbf{U}^{\mathbf{opt}}$, and [Step 3] for determining $\mathbf{U}^{\mathbf{opt}}$ and terminating the algorithm. Figure 3b,d represent three steps respectively, and search directions along a line or a circle have the same properties as those in Figure 2.

4.3.1. [Step 1] Finding $\mathbf{U}^{f}$ from $\mathbf{U}^{ini}$

To utilize Observations for search directions, it is essential to place an initial UAV at an arbitrary location. We suggest that the initial UAV location $\mathbf{U}^{ini} = \mathbf{U}^{U}$ be at the point of intersection of C_b and line in Figure 2d with $\theta_{U,IU} \approx 90°$ (e.g., '◆' in Figure 3b). This point belongs to $A_{\bar{C}_1 \cap C_2^{=}}$, hence Direction ④ on Observation 7 can be applied to find $\mathbf{U}^{f} = \mathbf{U}^{U'}$ by decreasing $d_{U,BS}$ along the line between BS and $\mathbf{U}^{ini}$. Since this line *always* passes through $A_{C_1 \cap C_2}$, the feasible UAV location $\mathbf{U}^{f}$ can be found within $A_{C_1 \cap C_2}$, specifically at the intersection of $A_{C_1^{=} \cap C_2}$ and the line (e.g., '●' in Figure 3b) to minimize interference from

UAV to CU. Therefore, $\mathbf{U}^f = \mathbf{U}^{\mathcal{U}'} = \{d_{U',BS}, \theta_{U,BS}\}$ can be obtained by directly calculating $d_{U',BS}$ from (19a) as

$$d_{U',BS} = X_{BS}^{-1}(F(\theta_{U,BS})\,|K_{no}), \tag{20}$$

where $X_{BS}^{-1}(\cdot\,|K_{no})$ is the inverse function of $X_{BS}(\cdot|K_{no})$ and $\theta_{U,BS} = \theta_{U',BS}$ is the elevation angle between BS and $\mathbf{U}^{ini}$.

Alternatively, $\mathbf{U}^{ini}$ at the intersection of C_a and line in Figure 2c with $\theta_{U,BS} \approx 90°$ (e.g., '✚' in Figure 3b) can be considered, and it is within $A_{C_1^{=} \cap \bar{C}_2} \subset A_{C_1 \cap \bar{C}_2}$. Based on Direction ④ on Observation 8, $\mathbf{U}^f = \mathbf{U}^{\mathcal{U}'} = \{d_{U',IU}, \theta_{U,IU}\}$ can be determined to be on C_b, where

$$d_{U',IU} = X_{IU}^{-1}(F(\theta_{U,IU})\,|K_o). \tag{21}$$

It is obtained by taking the inverse function of $X_{IU}(\cdot|K_o)$, $X_{IU}^{-1}(\cdot\,|K_o)$, to (19b) with $\theta_{U,IU} = \theta_{U',IU}$.

4.3.2. [Step 2] Updating $\mathbf{U}^f$

[Step 1] finds the feasible UAV location $\mathbf{U}^f$ ($= \mathbf{U}^{\mathcal{U}}$ in this step) within $A_{C_1^{=} \cap C_2}$ on C_a, or $A_{C_1 \cap C_2^{=}}$ on C_b. [Step 2] updates $\mathbf{U}^{\mathcal{U}}$ iteratively towards $\mathbf{U}^{\mathbf{opt}}$ and near upper point of intersection of C_a and C_b as shown in Figure 3c. $A_{C_1^{=} \cap C_2}$ and $A_{C_1 \cap C_2^{=}}$ belong to $A_{C_1 \cap C_2}$, hence, Observation 6 can be applied. When $\mathbf{U}^{\mathcal{U}}$ is within $A_{C_1^{=} \cap C_2}$ on C_a (e.g., '●' in Figure 3c), $\mathbf{U}^{\mathcal{U}'} = \{d_{U',IU}, \theta_{U,IU}\}$ will be on C_b by increasing $d_{U,IU}$ along Direction ① in Figure 2d, according to (21). On the other hand, when $\mathbf{U}^{\mathcal{U}}$ is within $A_{C_1 \cap C_2^{=}}$ on C_b (e.g., '★' in Figure 3c), $\mathbf{U}^{\mathcal{U}'} = \{d_{U',BS}, \theta_{U,BS}\}$ will be obtained on C_a by increasing $d_{U,BS}$ along Direction ① in Figure 2c based on (20). The newly obtained $\mathbf{U}^{\mathcal{U}'}$ becomes $\mathbf{U}^{\mathcal{U}}$ for the next procedure to draw a line for search directions. These procedures iterate until $\mathbf{U}^{\mathcal{U}'}$ locates on either C_a or C_b outside $A_{C_1 \cap C_2}$, but near the upper intersection point of C_a and C_b (e.g., '✚' in Figure 3c).

Note that, search directions along the line (i.e., Direction ① in Figure 2c,d) are selected on Observation 6 rather than those along the circle (i.e., Direction ② in Figure 2c,d). Since (19a) is not a function of $\theta_{U,IU}$, Direction ② in Figure 2d cannot find $\mathbf{U}^{\mathcal{U}'}$ within $A_{C_1^{=} \cap C_2}$ (e.g., '●' in Figure 3c) from $\mathbf{U}^{\mathcal{U}}$ within $A_{C_1 \cap C_2^{=}}$ directly, but numerically by searching $\theta_{r,iu}$ as

$$\theta_{U',IU} = \theta_{U,IU} + \min \Delta_\theta \tag{22}$$
$$s.t.\ \mathbf{U}^{\mathcal{U}'} \text{satisfies (19a) with an equality,}$$

where Δ_θ is an increment of $\theta_{U,IU}$ and $d_{U',IU} = d_{U,IU}$. Similarly, Direction ② in Figure 2c requires numerical updating $\theta_{U,BS}$ to find $\mathbf{U}^{\mathcal{U}'}$ within $A_{C_1 \cap C_2^{=}}$ (e.g., '★' in Figure 3c) from $\mathbf{U}^{\mathcal{U}}$ within $A_{C_1^{=} \cap C_2}$. These numerical updates increase computation time on Step 2, therefore, it is preferable to select Direction ① in Figure 2c and Direction ① in Figure 2d on Observation 6, thereby determining $\mathbf{U}^{\mathcal{U}'}$ from (20) or (21) directly.

4.3.3. [Step 3] Determining $\mathbf{U}^{\mathbf{opt}}$

[Step 2] places $\mathbf{U}^{\mathcal{U}'}$ at '✚' in Figure 3c, which is outside $A_{C_1 \cap C_2}$. [Step 3] puts it back at a point either on C_a or C_b within $A_{C_1 \cap C_2}$, and then declares the optimal UAV location $\mathbf{U}^{\mathbf{opt}}$.

Observations 7 and 8 can be utilized for this step because $\mathbf{U}^{\mathcal{U}'}$ from [Step 2] (=$\mathbf{U}^{\mathcal{U}}$ in this step) is on either C_a or C_b outside $A_{C_1 \cap C_2}$, specifically $A_{C_1^{=} \cap \bar{C}_2}$ or $A_{\bar{C}_1 \cap C_2^{=}}$. When $\mathbf{U}^{\mathcal{U}}$ is within $A_{C_1^{=} \cap \bar{C}_2}$ (e.g., '✚' in Figure 3d), Direction ④ on Observation 8 can be applied using (21) to put $\mathbf{U}^{\mathcal{U}'}$ on C_b, while Direction ④ on Observation 7 can be utilized to place $\mathbf{U}^{\mathcal{U}'}$ on C_a using (20) when $\mathbf{U}^{\mathcal{U}}$ is within $A_{\bar{C}_1 \cap C_2^{=}}$ (e.g., '◆' in Figure 3d). If newly obtained $\mathbf{U}^{\mathcal{U}'}$ is within $A_{C_1 \cap C_2}$, more accurately $A_{C_1^{=} \cap C_2}$ or $A_{C_1 \cap C_2^{=}}$, the UD algorithm declares that it is

$\mathbf{U^{opt}}$, and terminates. If not, it repeats [Step 3] until $\mathbf{U}^{\mathcal{U}'}$ is within $A_{C_1^{=} \cap C_2}$ or $A_{C_1 \cap C_2^{=}}$. The details of UD algorithm are summarized in the Algorithm 1.

Algorithm 1 UD algorithm.

Input: (K_{no}, K_o).
Output: Optimal UAV location $\mathbf{U^{opt}}$.
Step 1. Find $\mathbf{U}^f$ from $\mathbf{U}^{ini}$:
1: initial UAV location is determined as $\mathbf{U}^{ini} = \{d_{U,IU}, \theta_{U,IU}\}$
$= \{X_{IU}^{-1}(F(\theta_{U,IU})\,|K_o), \theta_{U,IU}\}$ with $\theta_{U,IU} \approx 90°$.
2: update $\mathbf{U}^{ini}$ following Direction ④ in Figure 2c based on (20).
Then, $\mathbf{U}^f$ within $A_{C_1^{=} \cap C_2}$ is obtained, and go to Step 2.
Step 2. Update $\mathbf{U}^f$ towards $\mathbf{U^{opt}}$:
3: **if** $\mathbf{U}^f$ is within $A_{C_1^{=} \cap C_2}$ **then**
4: find $\mathbf{U}^{\mathcal{U}'}$ following Direction ① in Figure 2d based on (21).
5: **else if** $\mathbf{U}^f$ is within $A_{C_1 \cap C_2^{=}}$ **then**
6: find $\mathbf{U}^{\mathcal{U}'}$ following Direction ① in Figure 2c based on (20).
7: **end if**
8: **if** $\mathbf{U}^{\mathcal{U}'}$ is within $A_{C_1^{=} \cap \bar{C}_2}$ **then**
9: $\mathbf{U}^{\mathcal{U}} \leftarrow \mathbf{U}^{\mathcal{U}'}$, and go to Step 3.
10: **else**
11: $\mathbf{U}^f \leftarrow \mathbf{U}^{\mathcal{U}'}$, and go to line 3.
12: **end if**
Step 3. Determine $\mathbf{U^{opt}}$:
13: **if** $\mathbf{U}^{\mathcal{U}}$ is within $A_{C_1^{=} \cap \tilde{C}_2}$ **then**
14: find $\mathbf{U}^{\mathcal{U}'}$ following Direction ④ in Figure 2d based on (21).
15: **else if** $\mathbf{U}^{\mathcal{U}}$ is within $A_{\tilde{C}_1 \cap C_2^{=}}$ **then**
16: find $\mathbf{U}^{\mathcal{U}'}$ following Direction ④ in Figure 2c based on (20).
17: **end if**
18: **if** $\mathbf{U}^{\mathcal{U}'}$ is within $A_{C_1^{=} \cap C_2}$ or $A_{C_1 \cap C_2^{=}}$ **then**
19: $\mathbf{U}^{opt} \leftarrow \mathbf{U}^{\mathcal{U}'}$, and terminates the algorithm.
20: **else**
21: go to line 13.
22: **end if**

5. Optimal Number of Transmit Time Slots

To maximize the multi-objective function of **JUDTAP** in (12), the overall number of time slots n that guarantees a reliable relay transmission to IU should be minimized before UD algorithm is performed. Hence, in this section, we propose the time slot determination (TSD) algorithm to determine an optimal pair of time slots $(K_{no}^{opt}, K_o^{opt})$, equivalently the minimum (optimal) number of overall time slots of $n^{opt} = K_{no}^{opt} + K_o^{opt}$.

5.1. Existence of Feasible UAV Locations

In Section 4, UD algorithm is proposed to find $\mathbf{U^{opt}}$ with the assumption that the given time allocation (K_{no}, K_o) guarantees the existence of feasible UAV locations, equivalently $A_{C_1 \cap C_2}$, that satisfies both constraint (19a) and (19b). However, when K_{no} and K_o are not enough for reliable relay transmissions, $A_{C_1 \cap C_2}$ in Figure 3a does not exist for $\mathbf{U^{opt}}$. Therefore, it is critical to select a proper pair of (K_{no}, K_o), but we consider the minimum K_{no} and K_o as an optimum in the resource-efficiency aspect. In addition, scanning all UAV operation range (i.e., inside dashed rectangular in Figure 2) to check the existence of $A_{C_1 \cap C_2}$ for each possible (K_{no}, K_o) is impractical. Therefore, a time-efficient determination for the existence of $A_{C_1 \cap C_2}$ should be considered. This can be realized by utilizing [Step 1] of UD algorithm. If $\mathbf{U}^f$ within $A_{C_1^{=} \cap C_2}$ can be found from $\mathbf{U}^{ini}$, $A_{C_1 \cap C_2}$ exists for (K_{no}, K_o) to guarantee a reliable relay transmission. On the other hand, if [Step 1] cannot find $\mathbf{U}^f$ within $A_{C_1^{=} \cap C_2}$, (K_{no}, K_o) is not enough to satisfy both constraints on relay transmission

(12a), thereby resulting in absence of $A_{C_1 \cap C_2}$. Hence, more time slots should be allocated for a reliable relay transmission.

5.2. Time Slot Determination (TSD) Algorithm

In this section, we propose a novel time slot determination (TSD) algorithm to derive the minimum number of overall time slots n^{opt} along with (K_{no}^{opt}, K_{o}^{opt}) for a reliable relay transmission. First, it defines l_{min} and l_{max} with the assumption of $K_{no} = K_o$, and utilizes them to reduce the search range for K_{no}^{opt} and K_{o}^{opt}. l_{min} is the minimum number of time slots where $(K_{no}, K_o) = (l_{min}, l_{min})$ could provide $A_{C_1 \cap C_2}$, but not necessarily guarantee it, whereas l_{max} is the number of time slots that does guarantee $A_{C_1 \cap C_2}$ for $(K_{no}, K_o) = (l_{max}, l_{max})$. l_{min} is determined by the upper bounds of $d_{U,BS}$ and $d_{U,IU}$, denoted as $d_{U,BS}^{max}(K_{no})$ and $d_{U,IU}^{max}(K_o)$, while l_{max} is obtained based on l_{min} and utilized to find (K_{no}^{opt}, K_{o}^{opt}) and n^{opt}.

5.2.1. Determination of l_{min}

The l_{min} aims at restricting search range, thereby reducing a computation time for (K_{no}^{opt}, K_{o}^{opt}). It does not require guaranteeing $A_{C_1 \cap C_2}$ necessarily, but provides a lower bound for l_{max}. It can be derived from maximum distances of $d_{U,BS}$ and $d_{U,IU}$, $d_{U,BS}^{max}(K_{no})$ and $d_{U,IU}^{max}(K_o)$, for each K_{no} and K_o. From (15) and (16), $d_{U,BS} \leq X_{BS}^{-1}(F(\theta_{U,BS}), K_{no})$ and $d_{U,IU} \leq X_{IU}^{-1}(F(\theta_{U,IU}), K_o)$ can be obtained, and the right terms of both inequalities are increasing functions of $F(\theta_{U,BS})$ and $F(\theta_{U,IU})$ respectively. Hence, $d_{U,BS}^{max}(K_{no})$ and $d_{U,IU}^{max}(K_o)$ can be defined as

$$d_{U,BS}^{max}(K_{no}) \triangleq X_{BS}^{-1}(F(\theta_{U,BS}), K_{no})_{|F(\theta_{U,BS})=1} = X_{BS}^{-1}(1, K_{no}), \tag{23}$$

$$d_{U,IU}^{max}(K_o) \triangleq X_{IU}^{-1}(F(\theta_{U,IU}), K_o)_{|F(\theta_{U,IU})=1} = X_{IU}^{-1}(1, K_o), \tag{24}$$

where $F(\theta_{U,i}) = 1, i \in \{BS, IU\}$ because $\max F(\theta_{U,I}) = 1$. Note that $d_{U,BS}^{max}(K_{no})$ and $d_{U,IU}^{max}(K_o)$ are increasing functions of K_{no} and K_o, respectively.

Using (23) and (24), l_{min} is defined as

$$l_{\min} = \underset{z \in \mathbb{Z}_{>0}}{\arg\min}\ d_{U,BS}^{max}(z) + d_{U,IU}^{max}(z) \geq d_{BS,IU}, \tag{25}$$

where z is a positive integer indicating $z = K_{no} = K_o$, and $d_{BS,IU}$ is a distance between BS and IU. The $l_{\min}$ given by (25) makes two circles drawn at ground node i with radius $d_{U,i}^{max}(l_{\min})$, $i \in \{BS, IU\}$ (i.e., blue and dashed green circles in Figure 2a) overlap each other. As aforementioned, the feasible UAV locations could exist within an overlapped region by two circles, but not guaranteed because the derivation of l_{min} starts with the assumption of $F(\theta_{U,i}) = 1, i \in \{BS, IU\}$ in (23) and (24).

5.2.2. Determination of l_{max}

If a feasible UAV location exists for $(K_{no}, K_o) = (l_{\min}, l_{\min})$, l_{max} can be determined as $l_{max} \triangleq l_{min}$ by the definition. Otherwise, l_{max} should be greater than l_{min} such that a feasible UAV location exists for $(K_{no}, K_o) = (l_{\max}, l_{\max})$. In addition, K_{no} and K_o, both less than l_{min}, do not need to be considered because it can not provide $A_{C_1 \cap C_2}$. Hence, l_{max} can be obtained as

$$\begin{aligned} l_{\max} = \quad & \underset{l_{\min}+z}{\arg\min}\ (l_{\min} + z, l_{\min} + z), \\ s.t \quad & z \in \mathbb{Z}_{\geq 0} \text{ and} \\ & A_{C_1 \cap C_2} \text{ exists for } (K_{no}, K_o) = (l_{\min}+z, l_{\min}+z), \end{aligned} \tag{26}$$

where z is a non-negative integer value. If a feasible UAV location exists for $(K_{no}, K_o) = (l_{\min}, l_{\min})$, $z = 0$, and otherwise, $z > 0$. Note that it is obvious that $n^{opt} \le 2l_{\max}$ because a feasible UAV location always exists for $(K_{no}, K_o) = (l_{\max}, l_{\max})$.

5.2.3. Determination of $(K_{no}^{opt}, K_o^{opt})$

Based on l_{max}, the initial number of overall time slots is defined as $n_{ini} = 2l_{max}$. The TSD algorithm targets to the minimum number of overall time slots, so K_{no} and K_o can be different even though n_{ini} is derived from the assumption of $K_{no} = K_o$. In addition, for each number of overall time slot n, it is preferable to maximize K_{no} to achieve maximum R_{CU}, because the SINR at CU in NOTP (i.e., (2)) is larger than that in OTP (i.e., (4)). Therefore, n^{opt} will be determined by any pair of (K_{no}, K_o), which leads to the minimum $K_{no} + K_o$. However, when n^{opt} is given, $(K_{no}^{opt}, K_o^{opt})$ is preferred to be (K_{no}, K_o) with $K_{no} \ge K_o$ if exists.

Figure 4 represents a way to update K_{no} and K_o towards K_{no}^{opt} and K_o^{opt}. The row and column of the table indicate values of K_o and K_{no}. K_o is upper bounded by l_{max} because $(K_{no}, K_o) = (l_{\max} - z, l_{\max} + z)$, $z \in \mathbb{Z}_{>0}$ for $n_{ini} = 2l_{\max}$ cannot achieve $K_{no} \ge K_o$. Each element of table denotes the sum of row and column values (i.e., the number of overall time slots), and pairs of K_{no} and K_o resulting in $K_{no} + K_o > 2l_{max}$ are out of interest (i.e., upper triangle in Figure 4). To find $(K_{no}^{opt}, K_o^{opt})$, an initial time allocation is set to $(K_{no}, K_o) = (l_{max}, l_{max})$, $n^{opt} = n_{ini}$, and two rules for updating (K_{no}, K_o) are defined as follows:

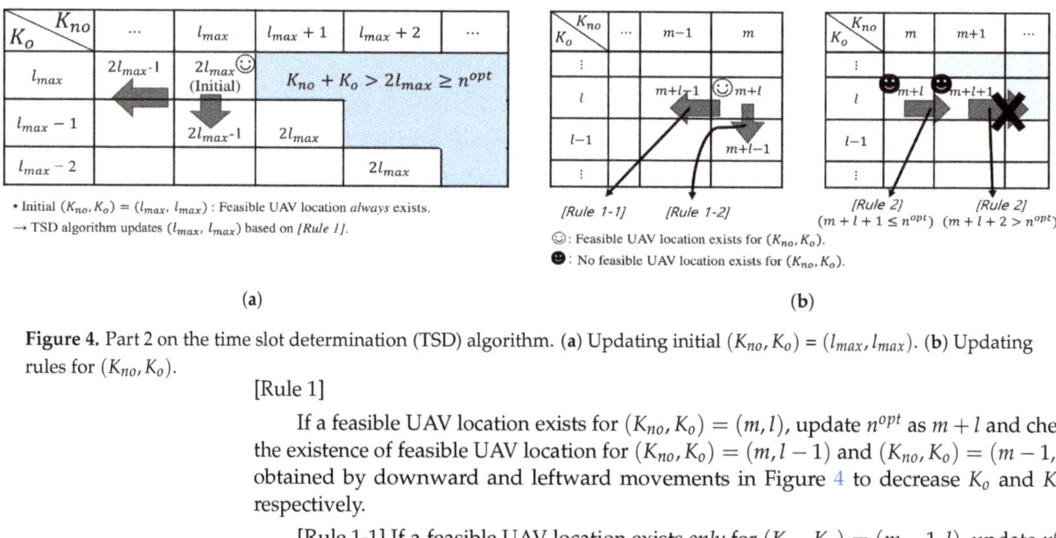

Figure 4. Part 2 on the time slot determination (TSD) algorithm. (**a**) Updating initial $(K_{no}, K_o) = (l_{max}, l_{max})$. (**b**) Updating rules for (K_{no}, K_o).

[Rule 1]

If a feasible UAV location exists for $(K_{no}, K_o) = (m, l)$, update n^{opt} as $m + l$ and check the existence of feasible UAV location for $(K_{no}, K_o) = (m, l-1)$ and $(K_{no}, K_o) = (m-1, l)$, obtained by downward and leftward movements in Figure 4 to decrease K_o and K_{no}, respectively.

[Rule 1-1] If a feasible UAV location exists *only* for $(K_{no}, K_o) = (m-1, l)$, update n^{opt} as $m + l - 1$ and repeat [Rule 1] at $(K_{no}, K_o) = (m-1, l)$.
[Rule 1-2] If a feasible UAV location exists for $(K_{no}, K_o) = (m, l-1)$, update n^{opt} as $m + l - 1$ and repeat [Rule 1] at $(K_{no}, K_o) = (m, l-1)$.

[Rule 2]

If no feasible UAV location exists for $(K_{no}, K_o) = (m, l)$ and $m + l + 1 \le n^{opt}$, move right in Figure 4 to increase K_{no} and check the existence of feasible UAV location for $(K_{no}, K_o) = (m+1, l)$.

[Rule 1] aims at checking the availability of smaller n^{opt}, while [Rule 2] is to investigate the existence of (K_{no}, K_o) with $K_{no} > K_o$ for the candidate n^{opt}. When feasible UAV locations exist for both $(K_{no}, K_o) = (m-1, l)$ and $(K_{no}, K_o) = (m, l-1)$ in [Rule 1], $(K_{no}, K_o) = (m, l-1)$ in [Rule 1-2] is selected to maximize K_{no} for $n^{opt} = m + l - 1$. TSD algorithm terminates when $(K_{no}, K_o) = (m+1, l)$ in [Rule 2] results in $m + l + 1 > n^{opt}$. The last updated n^{opt} is minimum (optimal) number of overall time slots, and (K_{no}, K_o), which leads to n^{opt}, becomes $(K_{no}^{opt}, K_o^{opt})$.

5.2.4. TSD Algorithm

Details of the TSD algorithm are summarized in Algorithm 2. Part 1 determines l_{min} and l_{max} from (25) and (26), respectively. Part 2 derives n^{opt} and $(K_{no}^{opt}, K_o^{opt})$ from (K_{no}, K_o) = (l_{max}, l_{max}) by the updating rules for (K_{no}, K_o). Note that only a few iterations are required on Part 2 of the TSD algorithm. From (26), it is clear that $(K_{no}, K_o) = (l_{\max} - 1, l_{\max} - 1)$ with $n = 2l_{\max} - 2$ does not provide a feasible UAV location. This results from the assumption of $K_{no} = K_o$, hence, $n = 2l_{\max} - 2$ could give a feasible UAV location when K_{no} is different from K_o. However, there is little chance for such a case to obtain a valid (K_{no}, K_o) with $n = 2l_{\max} - 2$ because K_{no} or K_o may be too small to set a reliable relay link between BS and UAV or between UAV and IU. In other words, a leftward or a downward movement in Figure 4 may be enough once or twice to reach n^{opt}, and so may a rightward movement to maximize K_{no} for the same reason. Therefore, the TSD algorithm reduces a search time dramatically compared to exhaustive algorithm or others, hence delivers $\mathbf{K^{opt}} = (K_{no}^{opt}, K_o^{opt})$ quickly.

Algorithm 2 TSD algorithm.

Part 1. Calculate of $l_{\min}$ and $l_{\max}$.
1: Determine $l_{\min}$ using (25).
2: Determine $l_{\max}$ using (26).
Part 2: Find $(K_{no}^{opt}, K_o^{opt})$.
Input: $(l_{\max}, l_{\max})$.
Output: $(K_{no}^{opt}, K_o^{opt}), n^{opt}$.
Initialization: $(K_{no}, K_o) \leftarrow (l_{\max}, l_{\max})$, $n^{opt} \leftarrow 2l_{\max}$.
3: **if** feasible UAV location exists for (K_{no}, K_o) **then**
4:   **if** feasible UAV location exists for $(K_{no}, K_o - 1)$ **then**
5:     $(K_{no}^{opt}, K_o^{opt}) \leftarrow (K_{no}, K_o - 1)$, $n^{opt} \leftarrow K_{no} + K_o - 1$.
6:     $(K_{no}, K_o) \leftarrow (K_{no}, K_o - 1)$, and go to line 3.
7:   **else if** feasible UAV location exists for $(K_{no} - 1, K_o)$ **then**
8:     $(K_{no}^{opt}, K_o^{opt}) \leftarrow (K_{no} - 1, K_o)$, $n^{opt} \leftarrow K_{no} + K_o - 1$.
9:     $(K_{no}, K_o) \leftarrow (K_{no} - 1, K_o)$, and go to line 3.
10:   **end if**
11: **else**
12:   **if** $K_{no} + K_o + 1 \leq n^{opt}$ **then**
13:     $(K_{no}, K_o) \leftarrow (K_{no} + 1, K_o)$, and go to line 3.
14:   **else**
15:     Terminate the algorithm.
16:   **end if**
17: **end if**

6. UAV Deployment and Time Allocation Algorithm

The TSD and UD algorithms are presented to determine $\mathbf{K^{opt}}=\{K_{no}^{opt}, K_o^{opt}\}$ and to decide $\mathbf{U^{opt}}$ for $\mathbf{K^{opt}}$, respectively. The **JUDTAP** can be solved by UAV deployment and transmit time allocation (UDTA) algorithm, which consists of TSD and UD algorithms and runs them in a sequential manner. Details of UDTA algorithm are summarized in Algorithm 3. As mentioned in Sections 4 and 5, TSD and UD algorithms reduce search range for $\mathbf{K^{opt}}$ and search area for $\mathbf{U^{opt}}$ respectively, thereby requiring much less computation time over exhaustive search algorithm. In the following subsection, a computational complexity is analyzed with respect to the total number of computations, considered for searching and determining $\mathbf{K^{opt}}$ and $\mathbf{U^{opt}}$.

Algorithm 3 UDTA algorithm.

Input: D_{req}, x_{BS}, x_{CU}, and x_{IU}.
Output: $\mathbf{K}^{\mathbf{opt}}$, $\mathbf{U}^{\mathbf{opt}}$.
Part 1: Optimal number of time slots.
1: Find $l_{\min}$ and $l_{\max}$ from Part 1 on TSD algorithm.
2: Find $\mathbf{K}^{\mathbf{opt}}$ $=(K_{no}^{opt}, K_{o}^{opt})$ from Part 2 on TSD algorithm.
Part 2: Optimal UAV location.
3: Based on $(K_{no}^{opt}, K_{o}^{opt})$, find $\mathbf{U}^{\mathbf{opt}}$ using UD algorithm.

Complexity Analysis

To determine the number of UAV locations for an exhaustive search, we consider that a grid is superimposed over the operation range (i.e., dashed rectangular in Figure 2a) with lines separated by Δ_d [m]. As a result, the number of UAV locations to be considered increases as Δ_d decreases. Total computations for an exhaustive search is derived as $\mathcal{O}(\Delta_d) \triangleq \left\lceil \frac{S}{\Delta_d^2} \right\rceil^{\frac{n^{opt}(n^{opt}-1)}{2}}$. $\left\lceil \frac{S}{\Delta_d^2} \right\rceil$ is the number of UAV locations within the operation range of S [m^2], where $\lceil \cdot \rceil$ is rounding up to the nearest integer. $\frac{n^{opt}(n^{opt}-1)}{2}$ is the number of combinations for (K_{no}, K_o) to find $(K_{no}^{opt}, K_{o}^{opt})$ from initial (K_{no}, K_o) = (1,1). Hence, $\mathcal{O}(\Delta_d)$ indicates that $\left\lceil \frac{S}{\Delta_d^2} \right\rceil$ locations for their feasibility need to be considered for each time allocation (K_{no}, K_o). On the other hand, the UDTA algorithm requires at most $l_{min} + 2^{|l_{max}-l_{min}|} + 2^{2\left|l_{max}-K_o^{opt}\right|+\left|l_{max}-K_{no}^{opt}\right|} + 10$ number of computations to find $\mathbf{K}^{\mathbf{opt}}$ and $\mathbf{U}^{\mathbf{opt}}$. On TSD algorithm, l_{min} computations are required in (25) to determine l_{min} from $(K_{no}, K_o) = (1,1)$. To determine l_{max}, $2^{|l_{max}-l_{min}|}$ computations are required in (26) from l_{min}, where only two UAV locations (i.e., initial and first feasible locations in [Step 1] on UD algorithm) are considered for each time allocation (K_{no}, K_o). Similarly, there are at most $2\left|l_{max} - K_o^{opt}\right| + \left|l_{max} - K_{no}^{opt}\right|$ combinations for (K_{no}, K_o) to find $(K_{no}^{opt}, K_{o}^{opt})$ from (l_{max}, l_{max}), hence resulting in $2^{2\left|l_{max}-K_o^{opt}\right|+\left|l_{max}-K_{no}^{opt}\right|}$ computations. In the UD algorithm, the total number of computations in (20) or (21) is less than 10, which is derived from simulations and reasonable due to dramatically reduced search area within $A_{C_1 \cap C_2}$ by the proposed algorithm. As a result, the UDTA algorithm requires much fewer UAV locations and (K_{no}, K_o) combinations to be considered for $\mathbf{K}^{\mathbf{opt}}$ and $\mathbf{U}^{\mathbf{opt}}$ over exhaustive search algorithm, thereby reducing computational time and effort significantly.

7. Numerical Results

In this section, we compare the optimal UAV location and transmit time allocations by proposed algorithms with those from an exhaustive search, and demonstrate that the UDTA algorithm achieves optimality while significantly reducing computational complexity. For simulations, we assume that P_{BS}^{max} = 30 [dBm], P_{U}^{max} = 25 [dBm], $\beta_A = 3$, $\beta_G = 2$, and ς = 20 [dB] [25–28]. An urban environment is assumed with $B = 0.136$ and $C = 11.95$ [8]. $\lambda \triangleq \frac{d_{BS,IU}}{d_{BS,CU}}$ is a relative location of IU with respect to CU. In order to evaluate the optimality of proposed algorithm, the throughput gap (%) is defined by R_{CU} difference from exhaustive search of $\Delta_d = 1$ because $\Delta_d = 1$ is sufficiently small to find the global optimal UAV location for exhaustive search.

Figure 5 represents feasible combinations of (K_{no}, K_o) and formation of $A_{C_1 \cap C_2}$, and compares $\mathbf{U}^{\mathbf{opt}}$ by UD algorithm with that by exhaustive search of $\Delta_d = 1$, where $x_{BS} = 0$, $x_{CU} = 300$ [m], $x_{IU} = 1000$ [m] and $D_{req} = 3$ [bit/Hz]. When $(K_{no}, K_o) = (4,4)$, $A_{C_1 \cap C_2}$ starts to appear, but TSD algorithm concludes $(K_{no}^{opt}, K_{o}^{opt}) = (4,3)$ resulting in $n^{opt} = 7$, even though $(K_{no}, K_o) = (3,4)$ also provides $A_{C_1 \cap C_2}$. This is because $(K_{no}^{opt}, K_{o}^{opt}) = (4,3)$ maximizes K_{no}^{opt} for given n^{opt}. TSD algorithm reduces the feasible UAV locations $A_{C_1 \cap C_2}$ significantly, and UD algorithm successfully determines $\mathbf{U}^{\mathbf{opt}}$ close to that from the exhaustive search of $\Delta_d = 1$.

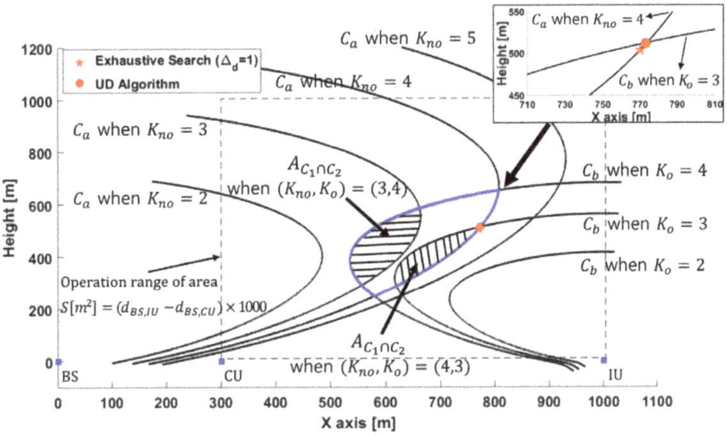

Figure 5. Feasible combinations of (K_{no}, K_o) and comparison of $\mathbf{U}^{\mathbf{opt}}$ from different algorithms.

Figure 6 represents $(K_{no}^{opt}, K_o^{opt})$, n^{opt}, UAV location, and throughput gap with respect to D_{req} for same $x_v, v \in \{BS, CU, IU\}$ in Figure 5. The UDTA algorithm finds $(K_{no}^{opt}, K_o^{opt})$ and UAV location close to optimum with negligible throughput gap. n^{opt} increases as D_{req} increases to set a reliable relay connection. For a given n^{opt}, UAV should be placed lower and close to CU as D_{req} increases. Even though this UAV movement increases the interference to CU, it is necessary for reliable relay transmission in BS-to-UAV link. For example, Figure 6a represents that $n^{opt} = 2$ is required for $0.5 \leq D_{req} \leq 0.9$. When $D_{req} = 0.5$, UAV can be placed very high and remote from CU. As D_{req} increases, however, UAV moves towards BS to set a reliable BS-to-UAV link. Lastly, the throughput gap between the proposed algorithm and exhaustive search is less than 0.1 (%) for entire range of D_{req}, hence, it demonstrates that UDTA algorithm successfully determines $\mathbf{K}^{\mathbf{opt}}$ and $\mathbf{U}^{\mathbf{opt}}$ with negligible throughput gap from the exhaustive search of $\Delta_d = 1$.

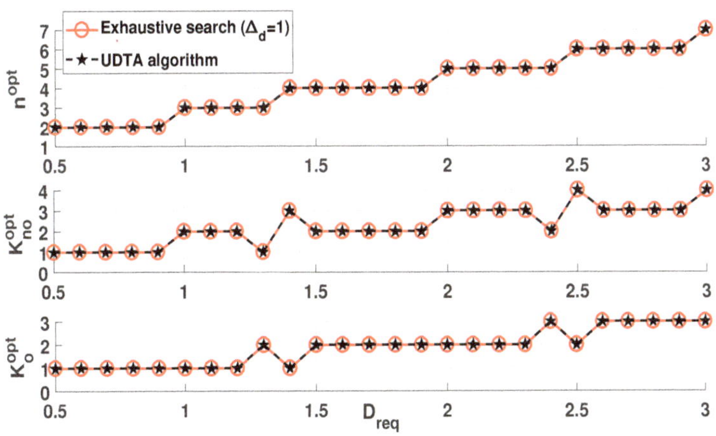

(a)

Figure 6. *Cont.*

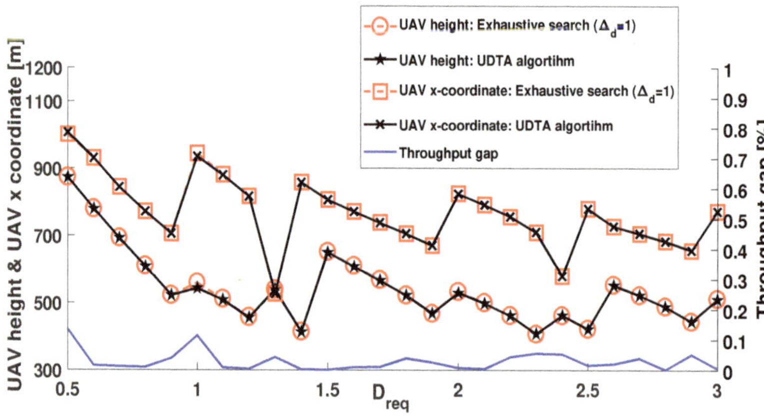

(**b**)

Figure 6. Performance comparison of different algorithms and their throughput gap with respect to D_{req}. (**a**) The number of transmit time slots. (**b**) UAV location and throughput gap.

URN with multiple transmit time slots (n^{opt} >2) on each relay transmission link can reduce redundant usages of transmit time slots. For example, URN consisting of two time slots for a single relay transmission (i.e., single transmit time slot allocation to each relay transmission link) requires at least three repetitions of relay transmissions to provide $D_{req} = 1.9$ at IU since $D_{req} = 0.9$ is a maximum delivered data to IU by URN with $n^{opt} = 2$, as shown in Figure 6a. Therefore, six time slots are required for URN that utilizes two transmit time slots for a single relay transmission, while URN that allocates multiple transmit time slots on each relay transmission link only requires four time slots for $D_{req} = 1.9$. Hence, multiple transmit time slots should be adopted in URN to efficiently utilize the transmit time slots.

Figure 7 represents $(K_{no}^{opt}, K_{o}^{opt})$, n^{opt}, UAV location, and throughput gap with respect to λ for $x_{CU} = 300$ [m]. As λ increases, IU moves away from BS, hence, larger n^{opt} is required in URN for a reliable relay transmission. In addition, UAV height should be increased for large $\theta_{U,v}$ to set a strong LOS connection between UAV and ground nodes v. In order to guarantee the minimum number of transmit time slots, K_{no}^{opt} can be smaller than K_{o}^{opt} as explained in Section 5.2.3. For example, when $2.43 \leq \lambda \leq 2.83$, n^{opt} is equal to 5 with $(K_{no}^{opt}, K_{o}^{opt}) = (3,2)$ or $(2,3)$. Especially, $(K_{no}^{opt}, K_{o}^{opt}) = (2,3)$ is selected when $2.76 \leq \lambda \leq 2.83$ to achieve the minimum number of overall time slots, however, it requires UAV to move towards BS for reliable BS-to-UAV link due to smaller K_{no}^{opt}. Similar to Figure 6, UDTA algorithm achieves negligible throughput gap, less than 0.2 (%), over the exhaustive search of $\Delta_d = 1$.

Figure 8 shows the throughput gap of exhaustive searches with respect to the number of computations. It is obvious that the throughput gap increases as Δ_d increases, due to the reduction on the number of UAV locations considered for searching $\mathbf{K^{opt}}$ and $\mathbf{U^{opt}}$, compared with $\Delta_d = 1$. As aforementioned, the UDTA algorithm significantly reduces computational time to find $\mathbf{K^{opt}}$ and $\mathbf{U^{opt}}$ due to the time-efficient determination of $(K_{no}^{opt}, K_{o}^{opt})$ based on l_{min} and l_{max}, and small $A_{C_1 \cap C_2}$ derived from $(K_{no}^{opt}, K_{o}^{opt})$. Therefore, it provides optimal solution for **JUDTAP**, and achieves much less computations of $l_{min} + 2^{|l_{max}-l_{min}|} + 2^{2\left|l_{max}-K_{o}^{opt}\right|+\left|l_{max}-K_{no}^{opt}\right|} + 10 \ll O(1)$ with negligible throughput gap over the exhaustive search of even $\Delta_d \leq 1$.

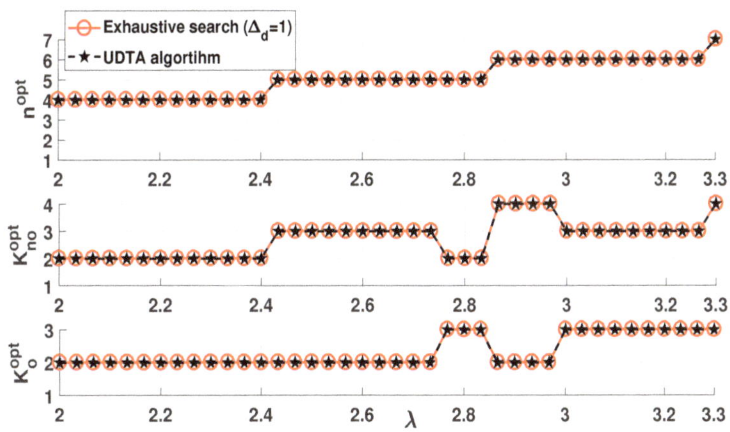

(a)

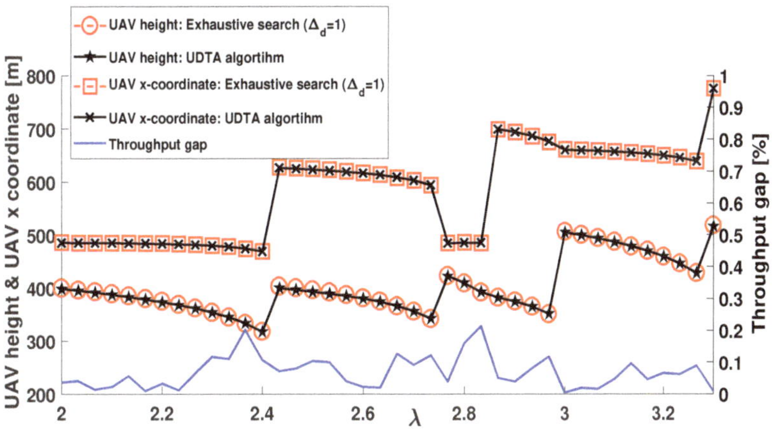

(b)

Figure 7. Performance comparison of different algorithms and their throughput gap with respect to λ. (**a**) The number of transmit time slots. (**b**) UAV location and throughput gap.

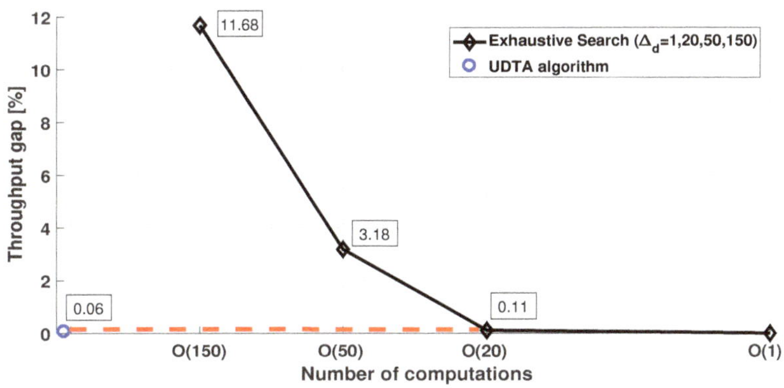

Figure 8. Throughput gap of exhaustive searches of different Δ_d.

8. Conclusions

In this paper, we have investigated URN with multiple transmit time slots, and proposed algorithms to maximize the throughput of UE in a cell while guaranteeing a reliable transmission to UE in its extended service area. The formulated multi-objective joint UAV deployment and transmit time allocation optimization problem (**JUDTAP**) is solved by TSD and UD algorithms to determine the optimal number of overall transmit time slots $\mathbf{K}^{\mathrm{opt}}$ and optimal UAV location $\mathbf{U}^{\mathrm{opt}}$ in a sequential manner. Simulation results demonstrate that $\mathbf{K}^{\mathrm{opt}}$ and $\mathbf{U}^{\mathrm{opt}}$ are critical to URN for a reliable relay transmission. $\mathbf{K}^{\mathrm{opt}}$ and $\mathbf{U}^{\mathrm{opt}}$ by the proposed algorithm match well those from exhaustive search, but with significantly reduced computation complexity to determine them over the exhaustive search. In addition, URN allocating multiple transmit time slots on relay transmission links is better than that utilizing two transmit time slots for a single relay transmission in terms of resource efficiency.

Author Contributions: Conceptualization, S.I.H. and J.B.; methodology, S.I.H. and J.B.; validation, S.I.H. and J.B.; formal analysis, S.I.H. and J.B.; investigation, J.B.; writing—original draft preparation, S.I.H.; writing—review and editing, S.I.H.; supervision, S.I.H.; project administration, S.I.H. All authors have read and agreed to the published version of the manuscript.

Funding: This research received no external funding.

Institutional Review Board Statement: Not applicable.

Informed Consent Statement: Not applicable.

Data Availability Statement: Data is contained within the article.

Conflicts of Interest: The authors declare no conflict of interest.

References

1. Wong, V.W.; Schober, R.; Ng, D.W.K.; Wang, L.-C. *Key Technologies for 5G Wireless Systems;* Cambridge University Press: Cambridge, UK, 2017.
2. Zeng, Y.; Zhang, R.; Lim, T.J. Wireless communications with unmanned aerial vehicles: Opportunities and challenges. *IEEE Commun. Mag.* **2016**, *54*, 36–42. [CrossRef]
3. Gupta, L.; Jain, R.; Vaszkun, G. Survey of important issues in UAV communication networks. *IEEE Commun. Surv. Tutor.* **2016**, *18*, 1123–1152. [CrossRef]
4. Marchese, M.; Moheddine, A.; Patrone, F. IoT and UAV Integration in 5G Hybrid Terrestrial-Satellite Networks. *Sensors* **2019**, *19*, 3704. [CrossRef] [PubMed]
5. Castellanos, G.; Deruyck, M.; Martens, L.; Joseph, W. Performance Evaluation of Direct-Link Backhaul for UAV-Aided Emergency Networks. *Sensors* **2019**, *19*, 3342. [CrossRef] [PubMed]
6. Mozaffari, M.; Saad, W.; Bennis, M.; Debbah, M. Mobile unmanned aerial vehicles (UAVs) for energy-efficient internet of things communications. *IEEE Trans. Wirel. Commun.* **2017**, *16*, 7574-7589. [CrossRef]
7. Fu, S.; Zhao, L.; Su, Z.; Jian, X. UAV Based Relay for Wireless Sensor Networks in 5G Systems. *Sensors* **2018**, *18*, 2413. [CrossRef] [PubMed]
8. Mozaffari, M.; Saad, W.; Bennis, M.; Debbah, M. Unmanned aerial vehicle with underlaid device-to-device communications: Performance and tradeoffs. *IEEE Trans. Wirel. Commun.* **2016**, *15*, 3949–3963. [CrossRef]
9. Mozaffari, M.; Saad, W.; Bennis, M.; Debbah, M. Efficient deployment of multiple unmanned aerial vehicles for optimal wireless coverage. *IEEE Commun. Lett.* **2016**, *20*, 1647–1650. [CrossRef]
10. Mozaffari, M.; Saad, W.; Bennis, M.; Debbah, M. Optimal transport theory for power-efficient deployment of unmanned aerial vehicles. In Proceedings of the 2016 IEEE International Conference on Communications (ICC), Kuala Lumpur, Malaysia, 23–27 May 2016.
11. Lyu, J.; Zeng, Y.; Zhang, R.; Lim, T. J. Placement optimization of UAV-mounted mobile base stations. *IEEE Commun. Lett.* **2017**, *21*, 604–607. [CrossRef]
12. Mozaffari, M.; Saad, W.; Bennis, M.; Debbah, M. Drone small cells in the clouds: Design, deployment and performance analysis. In Proceedings of the 2015 IEEE Global Communications Conference (GLOBECOM), San Diego, CA, USA, 6–10 December 2015.
13. Komerl, J.; Vilhar, A. Base stations placement optimization in wireless networks for emergency communications. In Proceedings of the 2014 IEEE International Conference on Communications Workshops (ICC), Sydney, Australia, 10–14 June 2014.
14. Mozaffari, M.; Saad, W.; Bennis, M.; Debbah, M. Optimal transport theory for cell association in UAV-enabled cellular networks. *IEEE Commun. Lett.* **2017**, *21*, 2053–2056. [CrossRef]

15. Sharma, V.; Bennis, M.; Kumar, R. UAV-assisted heterogeneous networks for capacity enhancement. *IEEE Commun. Lett.* **2016**, *20*, 1207–1210. [CrossRef]
16. Chen, Y.; Feng, W.; Zheng, G. Optimum placement of UAV as relays. *IEEE Commun. Lett.* **2017**, *22*, 248–251. [CrossRef]
17. Baek, J.; Han, S. I.; Han, Y. Optimal resource allocation for non-orthogonal transmission in UAV relay systems. *IEEE Wirel. Commun. Lett.* **2017**, *7*, 356–359. [CrossRef]
18. Han, S. I.; Baek, J.; Han, Y. Deployment of multi-layer UAV relay system. In Proceedings of the 2018 IEEE Wireless Communications and Networking Conference (WCNC), Barcelona, Spain, 15–18 April 2018.
19. Goldsmith, A. *Wireless Communications*; Cambridge University Press: Cambridge, UK, 2005.
20. Al-Hourani, A.; Kandeepan, S.; Lardner, S. Optimal LAP altitude for maximum coverage. *IEEE Wirel. Commun. Lett.* **2014**, *3*, 569–572. [CrossRef]
21. Saito, Y.; Benjebbour, A.; Kishiyama, Y.; Nakamura, T. System-level performance evaluation of downlink non-orthogonal multiple access (NOMA). In Proceedings of the IEEE 24th Annual International Symposium on Personal, Indoor, and Mobile Radio Communications (PIMRC), London, UK, 8–11 September 2013.
22. Weber, S. P.; Yang, X.; Andrews, J.G.; de Veciana, G. Transmission capacity of wireless ad hoc networks with outage constraints. *IEEE Trans. Inf. Theory* **2005**, *51*, 4091–4102. [CrossRef]
23. Boyd, S.; Vandenberghe, L. *Convex Optimization*; Cambridge University Press: Cambridge, UK, 2004.
24. Arora, J. *Introduction to Optimum Design*; Elsevier: Amsterdam, The Netherlands, 2016.
25. Islam, M.M.; Funabiki, N.; Sudibyo, R.W.; Munene, K.I.; Kao, W.-C. A dynamic access-point transmission power minimization method using PI feedback control in elastic WLAN system for IoT applications. *Internet Things* **2019**, *8*, 100089. [CrossRef]
26. Alzenad, M.; El-Keyi, A.; Yanikomeroglu, H. 3-D Placement of an Unmanned Aerial Vehicle Base Station for Maximum Coverage of Users With Different QoS Requirements. *IEEE Wirel. Commun. Lett.* **2018**, *7*, 38–41. [CrossRef]
27. Sodhro, A. H.; Li, Y.; Shah, M. A. Energy-efficient adaptive transmission power control for wireless body area networks. *IET Commun.* **2016**, *10*, 81–90. [CrossRef]
28. Feng, Q.; McGeehan, J.; Tameh, E.K.; Nix, A.R. Path Loss Models for Air-to-Ground Radio Channels in Urban Environments. In Proceedings of the 2006 IEEE 63rd Vehicular Technology Conference, Melbourne, Australia, 7–10 May 2006.

sensors MDPI

Article

Analytical Blind Beamforming for a Multi-Antenna UAV Base-Station Receiver in Millimeter-Wave Bands

Pingchuan Liu [1,2], Kuangang Fan [2,3,*] and Yuhang Chen [2,3]

1 School of Mechanical and Electrical Engineering, Jiangxi University of Science and Technology, Hongqi Street No. 86, Ganzhou 34100, China; wsnpingchuan@163.com
2 Magnetic Suspension Technology Key Laboratory of Jiangxi Province, Jiangxi University of Science and Technology, Hongqi Street No. 86, Ganzhou 34100, China; chenyuhang@jxust.edu.cn
3 School of Electrical Engineering and Automation, Jiangxi University of Science and Technology, Hongqi Street No. 86, Ganzhou 34100, China
* Correspondence: fankuangang@jxust.edu.cn

Abstract: Over the last decade, unmanned aerial vehicles (UAVs) with antenna arrays have usually been employed for the enhancement of wireless communication in millimeter-wave bands. They are commonly used as aerial base stations and relay platforms in order to serve multiple users. Many beamforming methods for improving communication quality based on channel estimation have been proposed. However, these methods can be resource-intensive due to the complexity of channel estimation in practice. Thus, in this paper, we formulate an MIMO blind beamforming problem at the receivers for UAV-assisted communications in which channel estimation is omitted in order to save communication resources. We introduce one analytical method, which is called the analytical constant modulus algorithm (ACMA), in order to perform blind beamforming at the UAV base station; this relies only on data received by the antenna. The feature of the constant modulus (CM) is employed to restrict the target user signals. Algebraic operations, such as singular value decomposition (SVD), are applied to separate the user signal space from other interferences. The number of users in the region served by the UAV can be detected by exploring information in the measured data. We seek solutions that are expressible as one Kronecker product structure in the signal space; then, the beamformers that correspond to each user can be successfully estimated. The simulation results show that, by using this analytically derived blind method, the system can achieve good signal recovery accuracy, a reasonable system sum rate, and acceptable complexity.

Keywords: UAV base station; MIMO; millimeter-wave band; blind beamforming; signal recovery

Citation: Liu, P.; Fan, K.; Chen, Y. Analytical Blind Beamforming for a Multi-Antenna UAV Base-Station Receiver in Millimeter-Wave Bands. *Sensors* **2021**, *21*, 6561. https://doi.org/10.3390/s21196561

Academic Editor: Margot Deruyck

Received: 7 September 2021
Accepted: 28 September 2021
Published: 30 September 2021

1. Introduction

Over the last few decades, wider bandwidths and more robust data transformations have always been the trends of the development of wireless communication. As one new and promising technology, millimeter-wave (mmWave) techniques have provided the possibility of applying upcoming 5G technologies, such as massive multiple-input–multiple-output (MIMO) technologies, to wireless systems [1–4].

One of the highlighted application scenarios of mmWave communication is in unmanned aerial vehicle (UAVs); such scenarios are presented with a base station, communication relay platform, or other communication enhancement equipment [1–12]. For instance, Z. Xiao et al. declared the advantages and challenges of mmWave UAV base stations (UAV-BSs) in cellular networks and explored different ways of counteracting signal blockage [2]. J. Du et al. studied UAV-BSs using multi-user massive MIMO hybrid beamforming and took an energy-saving strategy into consideration due to practical constraints, such as the size and weight of UAVs [3]. In addition to energy and power constraints, the mobility and positions of ground users are also crucial factors for UAV-assisted communication systems. In Reference [4], by exploiting the mobility and position information of a UAV system and ground users, W. Miao et al. proposed a lightweight beamforming algorithm in order to enhance the transmission performance of 5G UAV broadcasting system. In Reference [5], by using a UAV as a relay platform, F. Jiang and A. L. Swindlehurst studied the collection of messages from co-channel users on the ground and derived an algorithm for adjusting the UAV heading to maximize the approximate ergodic sum rate of the uplink channel. In some application scenarios, single-antenna UAVs are regarded as connection points or moving antennas for the enhancement of data transmission. For example, J. Ouyang et al. used a single-antenna UAV as a data transmission relay to connect access points and base stations [6]. M. Mozaffari et al. designed a novel framework for deploying a single-antenna UAV as an antenna array in order to minimize the wireless transmission time [8]. Some other recent research using single-antenna UAV-BSs has considered other scenarios [13–15].

In the research mentioned above and in other related research, usually, linear or planar antenna arrays are employed on the UAV-BSs. It is clear that adopting MIMO systems and mmWave communication in UAV-BSs has some convincing advantages. One aspect is that mmWave systems have rich spectral resources and flexible beamforming, which can meet the great demand of the data transfer rate. On the other hand, at particular positions, multi-antenna UAV-BSs can achieve high beamforming gain towards different line-of-sight (LoS) users by using mmWave communication, thus resulting in a more satisfying system throughput and communication efficiency [16–22]. However, due to some practical system constraints, such as multipath effects, blockage, and co-channel interference from different users, a UAV-BS system will sometimes have the issues of a low signal-to-noise ratio (SNR) in data acquisition and a low communication efficiency when serving multiple users.

To overcome these shortcomings and better utilize the advantages of mmWave, various beamforming techniques have been intensively investigated in UAV communication systems [16–21,23–26]. For example, Q. Song et al. studied the joint design of beamforming and power allocation to maximize the instantaneous data rate based on an efficient sub-optimal solution based on the block-coordinate descent method [23]. Some location-based beamforming algorithms were designed to improve the secrecy outage probability performance, maximize the network throughput, and provide flexible coverage [24–26]. In Reference [16], L. Liu et al. proposed a cooperative interference cancellation strategy motivated by the Xhaul structure to mitigate the uplink multi-beam UAV communication interference. In general communication systems, LoS transmission can efficiently improve the throughput of the UAV-BS. However, in urban areas, LoS links are prone to severe deterioration due to the complex propagation environments. In Reference [17,18], reconfigurable intelligent reflecting surfaces were used to reflect the received signals in order to enhance the data transmission quality, and passive beamforming was performed to better maximize the average achievable rate and received power. Combined with machine learning and a mean field game (MFG) control scheme, L. Li et al. proposed a joint beamforming and beam-steering method in order to build a reliable and steady connection between a UAV-BS and ground users in Reference [19]. Z. Xiao et al. utilized the artificial bee colony (ABC) algorithm to find the near-optimal beamforming vector for ground users when deploying a multi-antenna UAV-BS [20]. In Reference [21], W. Zhang et al. designed a beam-training code book for moving ground users, which can be regarded as a prior work of UAV-BS beamforming.

In this research and with the most recent beamforming techniques in UAV-BS communication systems, obtaining channel state information (CSI) is usually regarded as a crucial part of prior work. The effectiveness of channel estimation/identification may greatly affect the performance of most existing methods. In practical scenarios, especially in urban areas, specific CSI is difficult to acquire due to the complex propagation environment and multi-user interference. Another factor that greatly affects the channel estimation is the mobility of transmitters and receivers. The transmitting distance can also affect the performance of channel estimation. Accurate time-varying MIMO channel estimation can be very difficult to realize [4,7,20,27]. Moreover, the positions and mobility of ground users and UAV-BSs further increase the difficulty of channel estimation [28]. This puts extra pressure on the timeliness of a system, which causes bad beamforming performance. For this reason, we would like to formulate a blind beamforming problem for a UAV-BS that does not depend on channel estimation or CSI.

Recently, some works explored beamforming methods that needed less specific CSI. For example, X. Li et al. derived a beamforming method for different users under the assumption of only statistical CSI, i.e., the LoS component and Rician K-factor [29]. L. Du et al. proposed a robust pre-coder using imperfect CSI [30]. A beamforming scheme considering limited CSI was proposed with the aid of full-diversity rotation matrices [31]. However, these beamforming methods still partially rely on channel estimation, which can be time-consuming in complicated propagation environments. To tackle the thorny problem mentioned above that is caused by complicated channel estimation, in this paper, we will employ a method called the analytical constant modulus algorithm (ACMA) in order to perform blind beamforming in UAV-BS communication systems. Differently from methods that require detailed CSI, this blind method needs only the antenna output signals (the received signals) and some of their statistical information. In other words, it saves the computational costs caused by complex channel estimation. Another advantage of the blind beamforming method is that no training sequence is needed in the data transmission process, which saves the bandwidth and UAV's battery life. The task of blind beamforming is to calculate proper weight vectors for each ground user without detailed signal and channel information [32–34].

Compared with a traditional blind beamforming method called the constant modulus algorithm (CMA), the ACMA scheme can overcome the CMA's disadvantages of a lengthy iteration process and irregular convergence to local minima [32]. The analytical scheme can successfully derive a proper beamformer for the UAV-BS system without CSI. It is, in essence, a space separation method, which needs no lengthy iterations and achieves near-optimal accuracy. Using algebraic operations, such as singular value decomposition and diagonalization, we can successfully separate the user signal space (signals originally transmitted by ground users) from interference space. By properly investigating the properties of antenna outputs, this method can achieve near-optimal calculations of the weight vectors. These well-calculated weight vectors will finally formulate blind beamformers in order to assign the power of each antenna towards the corresponding ground users. At the same time, co-channel interference among different users will be efficiently compensated, and the desired signals can be successfully separated from other signals. We will evaluate this method's performance by applying it to a UAV-BS communication system in later sections. Different propagation settings are taken into consideration. The simulation results show that, through this analytical blind method, UAV-BS beamforming in an MIMO situation can provide a reasonable system sum rate while, at the same time, guaranteeing the signal reconstruction accuracy of each signal. Its complexity is comparable to that of the CMA and other classic methods under proper settings.

The main contributions of this paper are listed below:

- In this paper, we solve the UAV-BS beamforming problem in a blind way, which means that channel estimation is not necessary in our work. We only explore information from antenna outputs at the UAV-BS. This will save computation resources for the whole communication system.
- The number of UAV-serving users (ground users) can be detected in the scheme for sufficient snapshots and a reasonable signal–noise ratio. Algebraic operations can successfully separate the user signal space from other interference spaces, thus causing the beamformers to concentrate on user signals instead of other interference signals.
- With sufficient snapshots, this analytical blind method is robust for different channel settings. It can suppress distortions caused by both interference and additive noise. A near-optimal beamformer can be rapidly derived even when the number of snapshots is not very large. The simulation results show that the derived beamformer can achieve good MSE performance, a reasonable system sum rate, and acceptable complexity.

One potential UAV-BS application scenario can be a situation in which some specific users/target signals need higher-quality communications.

The rest of this paper is organized as follows: Section 2 introduces the basic models for the UAV-BS communication system. In Section 3, we formulate a blind beamforming problem. Section 4 demonstrates the way in which we solve the formulated problem. In Section 5, we evaluate the performance of our introduced method under different settings.

Notations: We denote scalars, vectors, and matrices by using lowercase a, bold lowercase $\boldsymbol{a}$, and bold uppercase letters $\boldsymbol{A}$, respectively. We further denote the transpose, element-wise complex conjugation, and conjugate transpose of a matrix $\boldsymbol{A}$ by $\boldsymbol{A}^{\mathrm{T}}$, $\boldsymbol{A}^*$, and $\boldsymbol{A}^{\mathrm{H}}$, respectively. $\mathbf{0}$ denotes an all-zero-element vector or matrix of appropriate size. Moreover, $\otimes$ denotes a Kronecker product, whereas $\mathrm{pinv}(\boldsymbol{A})$, $\mathrm{SVD}(\boldsymbol{A})$, and $\mathrm{rowspan}(\boldsymbol{A})$ represent a pseudo-inverse, singular value decomposition, and row span of matrix $\boldsymbol{A}$. Finally, $(\boldsymbol{A})_i$ denotes the i-th row, $\boldsymbol{A}^j$ the j-th column, and $\boldsymbol{A}_{ij}$ the element in row i and column j of matrix $\boldsymbol{A}$, respectively.

2. Basic Model

2.1. System Model

In this paper, we consider an MIMO uplink scenario in which a multi-antenna UAV-BS serves ground users in mmWave bands, as shown in Figure 1. In this scenario, the UAV-BS is usually equipped with a half-wavelength-spaced uniform linear array (ULA) or uniform rectangular array (URA), while each ground user is equipped with a single antenna. The ground users are distributed among distinct locations, transmitting signals of the same frequency. Due to the existence of scattering and a reflective environment, a multi-path effect is also considered in this scenario. We first introduce the well-known model for describing this signal transmission process:

$$\boldsymbol{X} = \boldsymbol{H}\boldsymbol{S} + \boldsymbol{N}, \tag{1}$$

where $\boldsymbol{X} \in \mathbb{C}^{m\times n}$ is the output of the m-element receiver on the UAV-BS over n symbol snapshots, $\boldsymbol{H} \in \mathbb{C}^{m\times d}$ is the channel matrix between users and the UAV-BS, and $\boldsymbol{S} \in \mathbb{C}^{d\times n}$ is the source signal matrix transmitted by ground users. Each $\boldsymbol{s}_i^{\mathrm{T}} = (\boldsymbol{S})_i \in \mathbb{C}^n, i \in \{1,\ldots,d\}$ corresponds to d ground users. $\boldsymbol{N} \in \mathbb{C}^{m\times n}$ denotes the channel noise matrix.

To fight against the impacts of the channels on the signals, beamforming is usually applied at the transmitters or receivers (sometimes both).

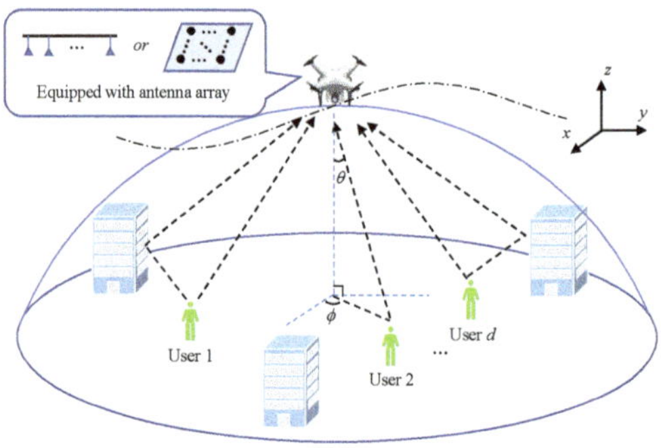

Figure 1. Application scenario of a UAV-BS serving multiple users.

In this paper, we focus on beamforming (we can also call it beam-combining) at the UAV-BS receivers, which can be modeled as

$$\hat{\boldsymbol{S}} = \boldsymbol{W}\boldsymbol{X}. \tag{2}$$

The beamforming matrix $W \in \mathbb{C}^{d\times m}$ is applied to compensate for the channels' impacts, especially for interference cancellation, power allocation, further signal recovery processes, and so on. The rows of W, which are denoted as $w_i^{\mathsf{T}} = (W)_i$, $i \in \{1,\ldots,d\}$, correspond to the beamforming vectors for each ground user.

2.2. Channel Model

We then establish a 3D rectangular coordinate system to better illustrate the positions of the UAV-BS and ground users. The UAV-BS is located at (x, y, h_{uav}), where h_{uav} represents the UAV's flying altitude. The position of ground user i is $(x_i, y_i, 0)$.

The channel matrix H consists of channel response vectors between users and the UAV-BS. In the classic 2D situation, due to the existence of multi-path components (MPCs), each channel vector h_i can be expressed as

$$h_i = \sum_{l=1}^{L_i} \kappa_{i,l}\, a(m, \theta_{i,l}), \tag{3}$$

where $\kappa_{i,l}$ represents the channel gain coefficient corresponding to the l-th MPC of user i, $a(m, \theta_{i,l})$ is the steering vector, m is the number of antenna elements, L_i is the number of existing MPCs, and $\theta_{i,l}$ is the elevation angle of arrival (AoA) of the $l-$th MPC. Due to the spatial sparsity of the angles of arrival in the mmWave channel, different MPCs will have distinct AoAs. The steering vector can then be derived as

$$a(m, \theta_{i,l}) = \left[1, e^{j\pi \sin\theta_{i,l}}, e^{j\pi 2 \sin\theta_{i,l}} \ldots, e^{j\pi(m-1)\sin\theta_{i,l}}\right]^{\mathsf{T}}. \tag{4}$$

Equation (4) illustrates the 2D situation, though when applying 3D beamforming by using a uniform rectangular array (URA), the steering vector will be slightly different. For a half-spaced URA, the steering vector can be derived as

$$a(m_x, m_y, \theta_{i,l}, \phi_{i,l}) = a_x(m_x, \theta_{i,l}, \phi_{i,l}) \otimes a_y(m_y, \theta_{i,l}, \phi_{i,l}), \tag{5}$$

where m_x and m_y are the URA dimensions along the $x-$axis and $y-$axis directions, and the total array elements are $m = m_x * m_y$. $\phi_{i,l}$ is the azimuth angle of the $l-$th MPC. The geometry denoted with a_x and a_y can further be described as

$$a_x(m_x, \theta_{i,l}, \phi_{i,l}) = \left[1, e^{j\pi \sin\theta_{i,l}\cos\phi_{i,l}}, \ldots, e^{j\pi(m_x-1)\sin\theta_{i,l}\cos\phi_{i,l}}\right]^{\mathsf{T}}, \tag{6a}$$

$$a_y(m_y, \theta_{i,l}, \phi_{i,l}) = \left[1, e^{j\pi \sin\theta_{i,l}\sin\phi_{i,l}}, \ldots, e^{j\pi(m_y-1)\sin\theta_{i,l}\sin\phi_{i,l}}\right]^{\mathsf{T}}. \tag{6b}$$

Generally, the MPCs between the UAV-BS and ground users are composed of the LoS part and non-LoS (NLoS) part under the condition of no blockage. For the LoS component, the elevation angle $\theta_{i,1}$ and azimuth angle $\phi_{i,1}$ of user i can be expressed as

$$\begin{cases} \theta_{i,1} = \arctan \dfrac{\sqrt{(x-x_i)^2 + (y-y_i)^2}}{h_{uav}} \\ \phi_{i,1} = \arctan \dfrac{y-y_i}{x-x_i} \end{cases}, \tag{7}$$

which are based on the UAV-BS position and user position. For the NLoS components, these angles will be some random values for which the elevation angles $\theta_{i,l}$ generally vary from –90° to 90°, while the azimuth angles $\phi_{i,l}$ vary from –180° to 180°. In the scenario of blockage, all MPCs will be NLoS components, and the channel H is usually described as a classic Rayleigh fading channel.

The next crucial part in determining the channel response vector h_i is its channel gain coefficient $\kappa_{i,l}$. For the LoS component, $\kappa_{i,1}$ mainly depends on the path loss, which is affected by the transmitting distance D_i and carrier frequency f. The channel gain coefficient for LoS can be described as

$$\kappa_{i,1} = \frac{1}{(\frac{4\pi f}{c}) \cdot D_i^{\alpha}} = \frac{1}{(4\pi\lambda) \cdot D_i^{\alpha}}, \tag{8}$$

in which λ is the carrier wavelength, $D_i = \sqrt{(x - x_i)^2 + (y - y_i)^2 + h_{uav}^2}$ is the linear distance between the UAV-BS and ground user i, and α is the LoS path-loss factor. For the NLoS components, we have the following channel gain:

$$\kappa_{i,l} = \frac{\sigma_f}{(4\pi\lambda) \cdot D_i^{\beta}}, \tag{9}$$

where σ_f is a small-scale Rayleigh fading factor, and β is the NLoS path-loss factor. Generally, with the existence of LoS, the channel gain will be mainly dominated by the LoS part, and other NLoS parts will sometimes be considered as interferences.

In practice, due to the mobility of the UAV and ground users, some new features, such as non-stationarity and the Doppler shift effect, should also be considered in UAV communications [27,28]. Thus, the time-varying channel response vectors for a URA-equipped UAV-BS should further be expressed as

$$\boldsymbol{h}_i(t) = \sum_{l=1}^{L_i} \kappa_{i,l}(t)\boldsymbol{a}(m_x, m_y, \theta_{i,l}(t), \phi_{i,l}(t)) \cdot e^{j\frac{2\pi}{\lambda}\int_{t_0}^{t} f_{i,l}(t)dt}, \tag{10}$$

where $f_{i,l}$ is the Doppler frequency. These time-varying factors in Equation (10) are, in essence, mainly determined by the status of the UAV-BS and ground users. As mentioned in Reference [27], only a few mmWave channel sounders have the ability to measure MIMO channels in time-variant environments. MIMO channels will change due to environmental variations, even when transmitters and receivers are static, not to mention in situations, such as a moving UAV-BS serving moving ground users. Due to all these difficulties, which can greatly impact the quality of the channel estimation and the latency of the system, we focus on blind methods to compensate for signal distortion caused by the channel.

3. Problem Formulation

In general, in MIMO uplink wireless communications, the basic task of beamforming is to introduce proper weight vectors $\boldsymbol{w}_i$ in order to achieve signal recovery against interference. Generally, non-blind beamforming methods will rely on the channel information of $\boldsymbol{H}$ and set the channel response matrix $\boldsymbol{H}$ as prior knowledge that which can be obtained through channel estimation, but this consumes resources. In this paper, the problem that we aim to solve is to derive a blind beamforming scheme at the UAV-BS receivers in order to improve the communication quality of the communication systems. In other words, we aim to calculate the proper beamforming (beam-combining) matrix $\boldsymbol{W}$ for optimal signal recovery at the UAV receivers without detailed information about $\boldsymbol{H}$. It was shown in Reference [35] that a proper $\boldsymbol{W}$ usually means that the mean squared error (MSE) of the recovered signals can be minimized. Thus, we formulate the blind UAV-BS beamforming problem as a signal recovery MSE minimization problem. We denote by $\boldsymbol{WX} = \hat{\boldsymbol{S}}$ the recovered signals; then, the problem can be expressed as finding a proper $\boldsymbol{W}$ that gives

$$\min \frac{1}{d \cdot n} \sum_{i=1}^{d} \sum_{j=1}^{n} \left(|\boldsymbol{S}_{ij}|^2 - |\hat{\boldsymbol{S}}_{ij}|^2\right)^2, \tag{11}$$

where d is the total user number, and n is the number of snapshots. This problem illustrates that we want to minimize the signal recovery MSE for the whole system.

In the noiseless case with perfect CSI, we can directly use $\boldsymbol{W} = \mathrm{pinv}(\boldsymbol{H})$ as a proper beamformer, resulting in

$$\boldsymbol{W}\boldsymbol{X} = \hat{\boldsymbol{S}} = \boldsymbol{S}. \quad (12)$$

The recovered signals of the UAV-BS are exactly the original signals sent by ground users. However, in blind situations, we have no clues about what exactly the channel $\boldsymbol{H}$ is. Thus, we try to solve this problem in another way, which is to explore the properties of the UAV's received signal matrix $\boldsymbol{X}$.

Generally, some signals in communication systems have a constant modulus (CM) feature, such as signals using BPSK or QPSK modulations. This CM feature can provide useful information for $\boldsymbol{X}$. We set the user signals with the CM feature as the target signals. For the convenience of illustration, we assume that all the target user signals have unit modulus elements via an appropriate scaler. The classic constant modulus algorithm (CMA) usually regards this signal recovery problem as a constant modulus factorization problem, that is, it makes $\boldsymbol{X} = \boldsymbol{H}\boldsymbol{S}$ or $\boldsymbol{S} = \boldsymbol{W}\boldsymbol{X}$ factorizable under ideal conditions. In essence, the recovered user signals are derived from combinations of vectors in $\boldsymbol{X}$, and each recovered user signal can be expressed as $\hat{\boldsymbol{s}_i}^{\mathsf{T}} = \boldsymbol{w}_i^{\mathsf{T}}\boldsymbol{X}$. Consequently, each $\hat{\boldsymbol{s}}_i^{\mathsf{T}}$ can be seen as a projection onto the row span of $\boldsymbol{X}$. As we now have the signal CM property, for a more distinct illustration, the problem mentioned in Equation (11) can be transformed into

$$\begin{aligned} &\min \ \frac{1}{d \cdot n}\sum_{i=1}^{d}\sum_{j=1}^{n}\left(1 - |\hat{\boldsymbol{s}}_{ij}|^2\right)^2 \\ &\text{s.t.} \quad \hat{\boldsymbol{s}}_i^{\mathsf{T}} = (\hat{\boldsymbol{S}})_i \in \mathrm{rowspan}(\boldsymbol{X}),\ \forall i \in \{1,\ldots,d\}. \end{aligned} \quad (13)$$

This equation guarantees that the derived beamformer explores the CM feature in order to achieve optimal beamforming for signal recovery accuracy, and simultaneously makes sure that the recovered signals belong to specific UAV-BS receivers. We will introduce an analytical blind method called the analytical constant modulus algorithm (ACMA) in order to solve this problem in the following sections.

4. Solution of the Problem

We first assume that the ground users are transmitting independent signals with enough phase richness, which means that $\boldsymbol{S}$ is a full rank and the CM factorization can be unique. In practice, some trivial transformations, such as phase invariance, will sometimes cause admissible non-uniqueness, but the uniqueness of recovered signals $\hat{\boldsymbol{S}}$ can be guaranteed for $n \to \infty$. Once the uniqueness is guaranteed, we can make sure that the recovered user signals at the UAV-BS site are the target signals originally sent by ground users [32]. To achieve a minimal signal recovery MSE for the whole system, we can pay attention to each user signal. Ideally, each recovered user signal should have unit modulus elements, as we assumed before, that satisfy

$$\hat{\boldsymbol{s}}_i^{\mathsf{T}} = [1,\ldots,1],\ \forall i \in \{1,\ldots,d\} \quad (14a)$$

$$\text{s.t.} \quad \hat{\boldsymbol{s}}_i^{\mathsf{T}} = (\hat{\boldsymbol{S}})_i \in \mathrm{rowspan}(\boldsymbol{X}),\ \forall i \in \{1,\ldots,d\}. \quad (14b)$$

As mentioned in the previous sections, the UAV-BS's received signal matrix $\boldsymbol{X}$ is the only data that we can acquire in the system. For an antenna-given $\boldsymbol{X} \in \mathbb{C}^{m \times n}$, the first step in the ACMA scheme is to separate the signal space from the interference space. To divide spaces in $\boldsymbol{X}$, we need to apply singular value decomposition (SVD) to $\boldsymbol{X}$.

Let $\mathrm{SVD}(\boldsymbol{X}) = \boldsymbol{U}\boldsymbol{\Sigma}\boldsymbol{V}^*$, where $\boldsymbol{U} \in \mathbb{C}^{m \times m}$, $\boldsymbol{\Sigma} \in \mathbb{R}^{m \times n}$, and $\boldsymbol{V} \in \mathbb{C}^{n \times n}$. By estimating the rank of $\boldsymbol{X}$, we can first obtain the number of received signals. One direct way is to check how many singular values are non-zero in $\boldsymbol{\Sigma}$; the number of non-zero values is equal to the number of transmitted signals. Remember that ground users are transmitting d

independent signals, and the received signal matrix $\boldsymbol{X}$ is, in essence, the combination of transmitted signals $\boldsymbol{s}_i$ through channels, so there are actually d independent row vectors in the row space of $\boldsymbol{X}$. The SVD operation gives us the singular non-zero values and their corresponding vectors, which can actually form the basis of rowspan $(\boldsymbol{X})$. This SVD of $\boldsymbol{X}$ first gives us the number of transmitted signals; then, we extract the signal space from $\boldsymbol{X}$, which is helpful for the following operations. Notice that the antenna array will sometimes receive all of the signals that appear in the system, including non-CM (non-user) ones, so the rank of $\boldsymbol{X}$ is not always precisely d in the presence of non-CM signals. We use δ as the total signal number; then, we have δ non-zero values in $\boldsymbol{\Sigma}$. In fact, the SVD of $\boldsymbol{X}$ can only extract the signal space, not the target CM signal space. To obtain the CM signal space, more operations are needed.

Let $\hat{\boldsymbol{V}} \in \mathbb{C}^{\delta \times n}$ be the submatrix of $\boldsymbol{V}$ that contains the singular vectors corresponding to δ singular non-zero values; then, we have that the rows of $\hat{\boldsymbol{V}}$ are actually the orthogonal basis of rowspan$(\boldsymbol{X})$, which can be seen as the signal space. Thus, we have

$$\boldsymbol{s}_i^{\mathsf{T}} \in \text{rowspan}(\boldsymbol{X}) \Leftrightarrow \boldsymbol{s}_i^{\mathsf{T}} = \boldsymbol{\omega}_i^{\mathsf{T}} \hat{\boldsymbol{V}}, \quad \forall i \in \{1, \ldots, \delta\}, \tag{15}$$

where the weight vectors $\boldsymbol{w}_i$ are acting on the signal space. Notice that, here, $\boldsymbol{s}_i^{\mathsf{T}}$, $\forall i \in \{1, \ldots, \delta\}$ are not yet the recovered target user signals, and some non-CM signals are still in the subspace. In other words, the SVD operation of $\boldsymbol{X}$ extracts the d user signals together with $\delta - d$ interference signals. Clearly, this cannot yet meet the requirements for user signal recovery. In the following, we will separate user signals from interference signals by using the CM property.

We know that $\hat{\boldsymbol{V}} = [\boldsymbol{v}_1, \ldots, \boldsymbol{v}_n]$, where $\boldsymbol{v}_j, \forall j \in \{1, \ldots, n\}$ is the j−th column of $\hat{\boldsymbol{V}}$; then, we can rewrite Equation (14a) as

$$\begin{aligned} \hat{\boldsymbol{s}}_i^{\mathsf{T}} &= [1, \ldots, 1], \forall i \in \{1, \ldots, d\} \\ &\Leftrightarrow \left[\left|(\hat{\boldsymbol{s}}_i^{\mathsf{T}})_1\right|^2 \ldots \left|(\hat{\boldsymbol{s}}_i^{\mathsf{T}})_n\right|^2\right] = [1, \ldots, 1], \ \forall i \in \{1, \ldots, d\} \\ &\Leftrightarrow \boldsymbol{w}_i^{\mathsf{T}} \boldsymbol{v}_j \boldsymbol{v}_j^{\mathsf{H}} \boldsymbol{w}_i^* = 1, \quad \forall i \in \{1, \ldots, d\}, \forall j \in \{1, \ldots, n\}. \end{aligned} \tag{16}$$

We first define $\boldsymbol{P}_j = \boldsymbol{v}_j \boldsymbol{v}_j^{\mathsf{H}} \in \mathbb{C}^{\delta \times \delta}, \forall j \in \{1, \ldots, n\}$; then, the CM property in Equation (16) can be expressed as

$$\hat{\boldsymbol{s}}_i = \begin{bmatrix} \boldsymbol{w}_i^{\mathsf{T}} \boldsymbol{P}_1 \boldsymbol{w}_i^* \\ \vdots \\ \boldsymbol{w}_i^{\mathsf{T}} \boldsymbol{P}_j \boldsymbol{w}_i^* \\ \vdots \\ \boldsymbol{w}_i^{\mathsf{T}} \boldsymbol{P}_n \boldsymbol{w}_i^* \end{bmatrix} = \begin{bmatrix} 1 \\ \vdots \\ 1 \end{bmatrix}. \tag{17}$$

Now, in the extracted signal space, our goal is to find linearly independent beamformers $\boldsymbol{w}_i$ for each user that can enforce this signal CM property. Once we have found all of the $\boldsymbol{w}_i$, we will have separated the user signal space (the target CM signal space) from the non-user signal space, and the blind beamforming problem for the UAV-BS can be solved.

For a better illustration, we can transform Equation (17) into a matrix computation form. In the following, we will temporarily drop the index i of $\boldsymbol{w}$ for clarity, i.e., $\boldsymbol{w}$ stands for all solutions $\boldsymbol{w}_i, \forall i \in \{1, \ldots, d\}$. We define two operators, $vec(\cdot)$ and $mat(\cdot)$, where $vec(\cdot)$ will stack all elements of a matrix into a vector by column, and $mat(\cdot)$ is the inverse operator of $vec(\cdot)$.

Let $\boldsymbol{p}_j = vec(\boldsymbol{P}_j) \in \mathbb{C}^{\delta^2}$; with the structure of $\boldsymbol{y} = vec(\boldsymbol{w}^* \boldsymbol{w}^{\mathsf{T}}) = \boldsymbol{w} \otimes \boldsymbol{w}^* \in \mathbb{C}^{\delta^2}$, we have the original CM elements in which $\boldsymbol{w}^{\mathsf{T}} \boldsymbol{P}_j \boldsymbol{w}^* = 1$ is equal to $\boldsymbol{p}_j^{\mathsf{T}} \boldsymbol{y} = 1$; then, Equation (17) can be expressed as

$$
Py = \begin{bmatrix} 1 \\ \vdots \\ 1 \end{bmatrix}, where\ P = \begin{bmatrix} p_1^{\top} \\ \vdots \\ p_j^{\top} \\ \vdots \\ p_n^{\top} \end{bmatrix}, y = w \otimes w^*. \tag{18}
$$

For greater convenience of computation, we can try to use some linear algebraic methods to transform Equation (18) into a more manageable form. One reasonable choice can be Householder transformation (HT) [32,36], though other options that can achieve the same goal are also acceptable. Let

$$
Q = I - 2\frac{qq^*}{q^*q}, \ q = \begin{bmatrix} 1 \\ \vdots \\ 1 \end{bmatrix} - \begin{bmatrix} \sqrt{n} \\ 0 \\ \vdots \\ 0 \end{bmatrix}. \tag{19}
$$

Then, by applying this Q to P, we have separated the system into two parts:

$$
QPy = \begin{bmatrix} \hat{p}_1 \\ \hat{P} \end{bmatrix} y \Leftrightarrow \begin{cases} (1)\ \hat{p}_1 y = \sqrt{n} \\ (2)\ \hat{P}y = \mathbf{0} \end{cases}, \begin{array}{l} \hat{p}_1 : 1 \times \delta^2 \\ \hat{P} : (n-1) \times \delta^2 \end{array}. \tag{20}
$$

After this transformation, we can directly focus on the second part of this system, which is easier to solve than the system $Py = 1$. The first part corresponds to the scaling of the vectors, which will be addressed later. Now, we try to find a solution y that satisfies

$$
\hat{P}y = \begin{bmatrix} 0 \\ \vdots \\ 0 \end{bmatrix}, y = w \otimes w^*. \tag{21}
$$

Due to the existence of multiple ground users, there will be more than one solution to the system. We require the number of snapshots n to be larger than δ^2; otherwise, there will be no solutions to the system. One simple and direct way to calculate these solutions is to employ SVD for $\hat{P}$ because the solutions that make $\hat{P}y = \mathbf{0}$ can be seen as the orthogonal basis of the null-space ($\hat{P}$). Thus, the singular vectors corresponding to singular zero values can be seen as the solutions. In the case of noisy channels, we retrieve the singular vectors corresponding to the smallest singular values of $\hat{P}$. This operation can correctly estimate the number of user signals in the signal space.

However, as we mentioned before, each CM solution y corresponding to each ground user in the UAV-BS system should have the specific structure $w \otimes w^*$. In common wireless UAV-BS communications, the sample size is generally much larger than the number of users served, which leads to an overdetermined system. We require $n > \delta^2$; hence, the set of independent solutions y is not unique. Thus, we need the condition $y = w \otimes w^*$ to restrict the solution space. Simple SVD operations cannot guarantee this structure; it is not appropriate to compute the solutions y and hope that they have the specified structure [32]. Operations are needed to make sure that each solution can be expressed as $w \otimes w^*$. We will simply introduce the extended QZ iteration method in the ACMA to solve this structural problem.

In general, the CM solution space that satisfies Equation (21) can be written as

$$
y = c_1\beta_1 + c_2\beta_2 + \ldots + c_{\hat{d}}\beta_{\hat{d}} = w \otimes w^*, \tag{22}
$$

where $\{c_1, \ldots, c_{\hat{d}}\}$ is a set of constants, $\{\beta_1, \ldots, \beta_{\hat{d}}\}$ is the basis of the kernel of $\hat{\boldsymbol{P}}$, and $\hat{d}$ is the dimension of $Ker(\hat{\boldsymbol{P}})$. The linearly independent $\boldsymbol{y}$ corresponds to linearly independent $\boldsymbol{w}$, which now leads to a linearly independent set of constants. Let SVD$(\hat{\boldsymbol{P}}) = \boldsymbol{U}_p \boldsymbol{\Sigma}_p \boldsymbol{V}_p{}^*$, as mentioned above, be the basis of $Ker(\hat{\boldsymbol{P}})$, which can be estimated as the singular vectors corresponding to singular zero values. The dimension of $Ker(\hat{\boldsymbol{P}})$ is equal to the number of singular zero values, which is also the number of target user signals, so we have $\hat{d} = d$ (proved in Reference [32]). Then, the last d rows of $\boldsymbol{V}_p{}^*$ corresponding to singular zero values can form a basis $\{\beta_1, \ldots, \beta_d\}$. After finding the basis of this CM solution space, once we obtain $\{c_1, \ldots, c_d\}$, we will have separated the CM signal space from the non-CM ones. Notice that, here, we say "singular zero values" because this is under the ideal conditions, which means that there is no noise in the system. When additive noise is applied to the system, there will be a threshold for deciding when small singular values can be seen as "singular zero values". The discussion about the threshold will be introduced in the next section.

Let $\boldsymbol{B}_k = mat(\boldsymbol{\beta}_k), k \in \{1, \ldots, d\}$, and $\boldsymbol{Y} = mat(\boldsymbol{y})$; then, Equation (22) can be expressed as

$$\boldsymbol{Y} = c_1 \boldsymbol{B}_1 + \ldots + c_d \boldsymbol{B}_d = \boldsymbol{w}^* \cdot \boldsymbol{w}^{\mathsf{T}}. \tag{23}$$

A proper set of constants can make the result of Equation (23) rank 1 and positive semidefinite; hence, it can be factorized as $\boldsymbol{w}^* \cdot \boldsymbol{w}^{\mathsf{T}}$. Assume that we already have the weight vectors $\boldsymbol{w}_1 \ldots \boldsymbol{w}_d$; then, $\boldsymbol{\beta}_i = \boldsymbol{w}_i \otimes \boldsymbol{w}_i{}^*, i \in \{1, \ldots, d\}$ can form the basis of $Ker(\hat{\boldsymbol{P}})$, and each $\boldsymbol{Y}_i$ can be factorized as

$$\boldsymbol{Y}_i = \lambda_{i1} \boldsymbol{w}_1{}^* \boldsymbol{w}_1^{\mathsf{T}} + \lambda_{i2} \boldsymbol{w}_2{}^* \boldsymbol{w}_2^{\mathsf{T}} + \ldots + \lambda_{id} \boldsymbol{w}_d{}^* \boldsymbol{w}_d^{\mathsf{T}} = \boldsymbol{W}^* \Lambda_i \boldsymbol{W}^{\mathsf{T}}. \tag{24}$$

If we could find matrix $\boldsymbol{Q}$ and $\boldsymbol{Z}$ to make $\boldsymbol{Q}\boldsymbol{Y}_i\boldsymbol{Z} = \boldsymbol{\Lambda}_i$, then the set of constants $\{c_1, \ldots, c_d\}$ corresponding to each user signal (making the solution rank 1) could be derived from the eigenvalue matrices $\boldsymbol{\Lambda}_i$. Consequently, the UAV-BS beamforming problem can be seen as an eigenvalue problem. To reduce the calculation complexity and save the calculation cost in the UAV-BS, it is not necessary to fully diagonalize $\boldsymbol{Y}_i$, as eigenvalues can also be obtained if $\boldsymbol{Q}\boldsymbol{Y}_i\boldsymbol{Z}$ are upper-triangular matrices.

The extended QZ iteration sets the initial $\boldsymbol{Q}^{(0)} = \boldsymbol{Z}^{(0)} = \boldsymbol{I}$. For the $k-$th iteration, this method aims to make

$$\begin{array}{ll} min & ||\boldsymbol{Q}^{(k)}(\boldsymbol{Y}_1\boldsymbol{Z}^{(k-1)})||_{LF}^2 + \ldots + ||\boldsymbol{Q}^{(k)}(\boldsymbol{Y}_d\boldsymbol{Z}^{(k-1)})||_{LF}^2 \\ min & ||(\boldsymbol{Q}^{(k)}\boldsymbol{Y}_1)\boldsymbol{Z}^{(k)}||_{LF}^2 + \ldots + ||(\boldsymbol{Q}^{(k)}\boldsymbol{Y}_d)\boldsymbol{Z}^{(k)}||_{LF}^2 \end{array}, \tag{25}$$

which necessitates making the lower-triangular part the minimal norm. Methods, such as HT and SVD, can be used to construct upper-triangular $\boldsymbol{R}_i = \boldsymbol{Q}\boldsymbol{Y}_i\boldsymbol{Z}$ [32]. With the derived $\boldsymbol{R}_i$, we extract the diagonal entries of each $\boldsymbol{R}_i$ and form a coefficient matrix:

$$\boldsymbol{C} = \begin{bmatrix} (R_1)_{11} & \cdots & (R_1)_{dd} \\ \vdots & & \vdots \\ (R_d)_{11} & \cdots & (R_d)_{dd} \end{bmatrix}^{-1}. \tag{26}$$

Finally, each set of constants $\{c_1, \ldots, c_d\}$ is given by rows of $\boldsymbol{C}$. Hence, each $\boldsymbol{Y}_i = \boldsymbol{w}^* \cdot \boldsymbol{w}^{\mathsf{T}}$ can be calculated, and the beamformers $\boldsymbol{w}_i$ for target user signal recovery $\boldsymbol{s}_i^{\mathsf{T}} = \boldsymbol{w}_i^{\mathsf{T}} \hat{\boldsymbol{V}}$ can be estimated as the singular vectors corresponding to the largest singular value in each $\boldsymbol{Y}_i$. The user signal space can be successfully separated from the interference space. However, sometimes, we still need to run the Gerchberg–Saxton algorithm (GSA) several times to solve the phase invariance problem in order to make the solutions near optimal. At the $k-$th iteration, the update rule is

$$w_i^{\mathsf{T}}(k+1) = \left[\frac{(w_i^{\mathsf{T}}(k)\hat{V})_1}{|(w_i^{\mathsf{T}}(k)\hat{V})_1|}, \ldots, \frac{(w_i^{\mathsf{T}}(k)\hat{V})_n}{|(w_i^{\mathsf{T}}(k)\hat{V})_n|} \right]. \tag{27}$$

5. Performance Evaluation

In this section, we perform numerical simulations to test the efficiency of the analytical method for blind beamforming in a UAV-BS. We consider an uplink wireless communication scenario in which a UAV-BS serves d ground users in a specific urban region—for instance, a 200 m $\times$ 200 m square region. In this region, the UAV-BS hovers in the sky, while the users stay on the ground. The time-varying channel is considered according to the model in Equation (10). We assume that the UAV moves at a speed of 10 m/s [28], while the ground users maintain static. The carrier frequency is set to 28 GHz in the mmWave band. So, $f_{Doppler}$ can be easily derived. The positions of the UAV-BS and ground users are randomly distributed in the horizontal direction. For each position setting in each simulation run, we average the simulation results over 100 times.

The UAV-BS is equipped with an $m = m_x \times m_y$ element URA, while each ground user is equipped with a single antenna. In this uplink process, ground users transmit independent and randomly generated signals by using QPSK modulations with n snapshots. Each signal is assumed to have $L_i = 4$ MPCs with the existence of LoS and the three strongest NLoS components; thus, the channel in this scenario can be regarded as a mixture of LoS and NLoS parts. We set the path-loss factors $\alpha = 0.95$ and $\beta = 2.25$ in Equations (8) and (9) according to the mmWave channel measurements in Reference [37]. The small-scale factor σ_f can be set to a small value, such as 0.01. By combining the channel model mentioned in Section 2.2 and the parameters in this section, we can simulate different channels for the tests of this method. According to detailed parameter and environment settings in this section, combined with the algorithm and models mentioned above, we can have the antenna outputs matrix $\boldsymbol{X}$, signal matrix $\boldsymbol{S}$ and the beamforming matrix $\boldsymbol{H}$. Then, we can easily evaluate the algorithm performance in this specific-simulated application scenario.

In the following simulations, additive Gaussian white noise is added to the system. We first assume that $m = 4 \times 4$, and there are $\delta = 6$ signals that appear in the system; four of them are transmitted by ground users (target CM signals). By exploring the number of singular near-zero values after $SVD(\hat{\boldsymbol{P}})$, we can estimate how many ground users are located in the region served by the UAV-BS. As shown in Figure 2, we tested the efficiency of the estimation of the number of ground users under different SNR and snapshot settings.

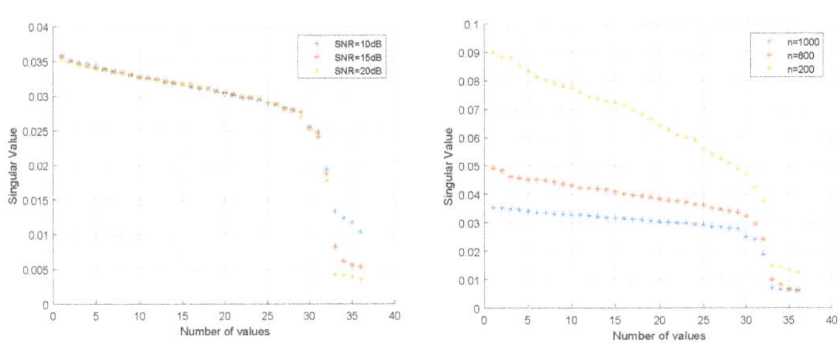

(**a**) Different SNR settings (n = 1000) (**b**) Snapshot settings (SNR = 15 dB)

Figure 2. Estimation of the number of ground users by examining small singular values when using different SNR/snapshot settings.

Due to the existence of Gaussian white noise, the singular values corresponding to the ground users are raised above zero. However, there is still a gap between the near-zero values and other values. With the increasing SNR and number of snapshots, the gap becomes more distinct. A reasonable threshold was given in Reference [32], which the large singular values should satisfy (with a probability better than 95%)

$$min(large\ singular\ value) > \frac{1}{\sqrt{n}} - \frac{\delta - 0.5}{n}. \tag{28}$$

We then tried to evaluate the performance of the analytical method in minimizing the system's MSE under different channel settings. For the LoS channel, the LoS component dominates the channel response. For the NLoS channel, all MPCs are regarded as NLoS components. The Rayleigh fading channel represents an ideal NLoS situation in which $\boldsymbol{H} \sim Rayleigh(1)$. We used the classic MMSE method, random beamforming, and the Artificial Bee Colony (ABC) method to compare with the analytical method. As we do not know the channel $\boldsymbol{H}$ in a blind situation, the MMSE receiver cannot be directly employed in practice. However, here, we assume that the MMSE receiver already knows the channel. The classic MMSE will give us a weight vector

$$\boldsymbol{W}_{MMSE} = \boldsymbol{H}^{\mathrm{H}}(\boldsymbol{H}\boldsymbol{H}^{\mathrm{H}} + \frac{\sigma^2}{P}\boldsymbol{I})^{-1}, \tag{29}$$

where σ^2 is the system noise power, and P is the system signal power; $\frac{\sigma^2}{P}$ is actually SNR^{-1} on the natural scale. The random beamformer is defined as $\boldsymbol{W}_{rand} = e^{j\boldsymbol{\Psi}}$, where each element of the random phase matrix $\boldsymbol{\Psi}$ will give us a random phase within $[0, 2\pi]$ [20]. The ABC method here is slightly different from the ABC beamforming process in Reference [20]; we transform it into an uplink process here. In this test, we set the food source number $Ns = 200$ with $iteration = 500$.

From Figure 3, we can observe that, for the LoS and NLoS channel settings, the analytical blind beamforming achieves better performance in minimizing the system's MSE, while the MMSE and ABC receivers do not perform as well. For the Rayleigh fading channel, the efficiency of the analytical method is close to that of the MMSE because, in this situation, noise is the dominant force affecting the signal recovery. The actual values depend on the randomness of the noise and the channel distortion. Nevertheless, the analytical blind beamforming is useful in most common scenarios.

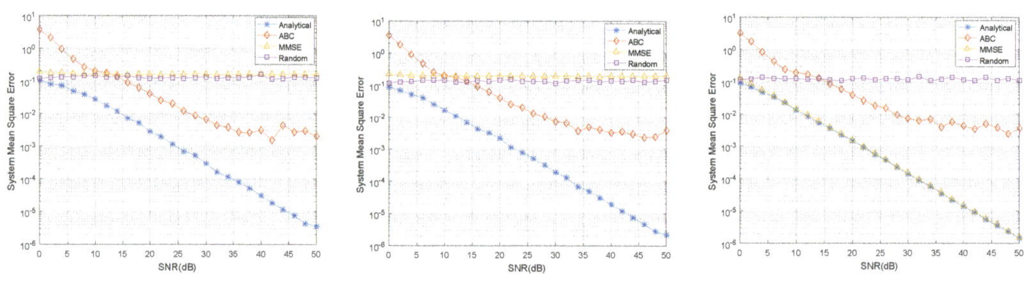

(**a**) LoS channel (**b**) NLoS channel (**c**) Rayleigh fading channel

Figure 3. The system's MSE using different beamforming methods with different channels.

We also tested the ACMA beamforming method under some sub-6-GHz bands. As shown in Figure 4, the results are very close to each other. The carrier frequency in our simulations mainly decides the values of the path loss and $f_{Doppler}$, but it does not affect the performance of the method in achieving a minimal system MSE.

Figure 5 shows the constellations of the originally sent user signals and recovered user signals. It directly shows the performance of the ACMA in beamforming for signal recovery. Compared with random beamforming, the analytical method can successfully derive both the amplitude and phase information. Due to the phase ambiguity of blind beamforming methods, sometimes, there are phase differences between the original signals and the recovered ones. Operations, such as differential encoding, can be applied to cancel the phase differences. However, we mainly focused on blind beamforming for signal recovery; thus, we have omitted the details about the encoding part.

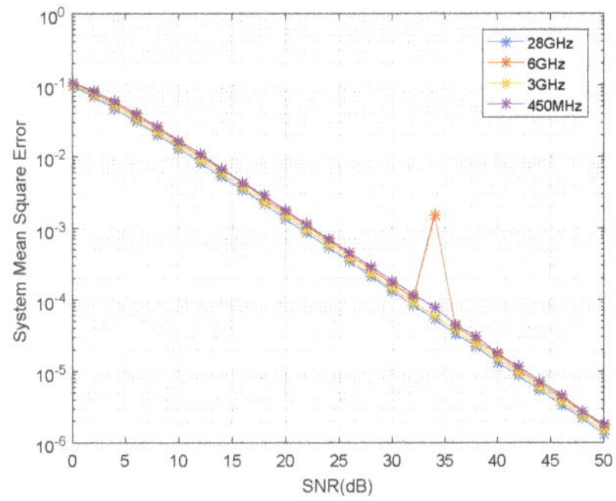

Figure 4. The MSE performance of the analytical method with different frequency bands.

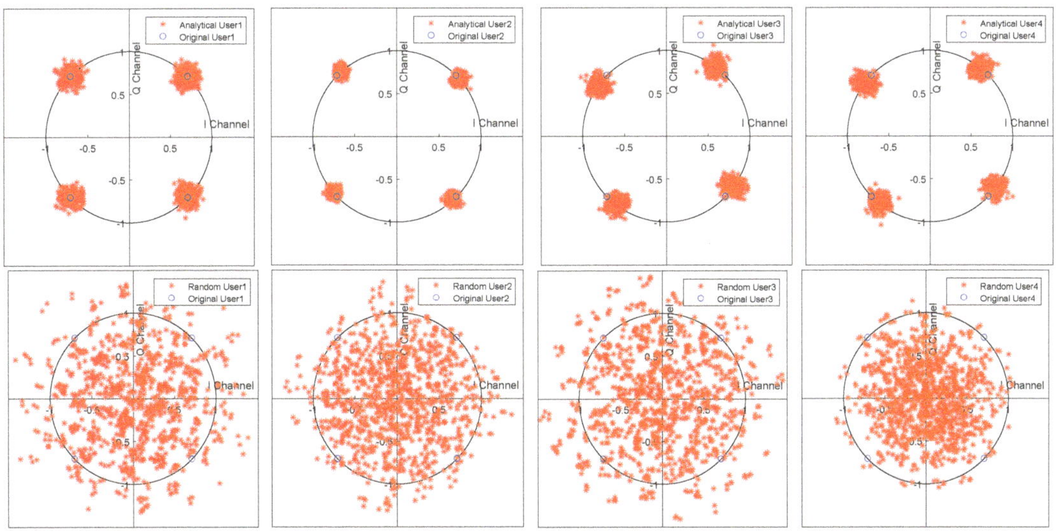

Figure 5. Constellation diagrams of the received user signals when SNR = 15 dB. The upper row shows the recovered signals of four different users using the ACMA with the original constellation. The bottom row shows the signals when using random beamforming.

In Figure 6, we show the compensated channel response for $d = 4$ different users after applying analytical beamforming. We can see that the compensated channel $\boldsymbol{wH}$ looks the same as an impulse. This, in essence, shows that the channel effects on each transmitted user signal are nearly erased at the UAV-BS's receiver. In other words, the ACMA receivers only recover user signals ($\boldsymbol{w}_i\boldsymbol{HS} = \boldsymbol{s}_i$), which can be seen as directing the receiving antennas towards different users.

Then, we evaluate the system's sum rate when applying the analytical method under different settings. The system's sum rate can be described as

$$Sumrate = \sum_{i=1}^{d} \log_2 \left(1 + \frac{P \cdot |\boldsymbol{h}_i^{\mathrm{H}} \boldsymbol{w}_i|^2}{\sigma^2}\right), \tag{30}$$

which summarizes achievable rates of d ground users. Here, the signal power P and noise power σ^2 are still calculated on the natural scale. For better illustration, we transform them into dB in the figures. We introduced approximate beamforming as an upper bound in our tests. Approximate beamforming is regarded as an ideal beamforming in which the beam gains are zeros along the non-user directions and are significant along the user directions [20].

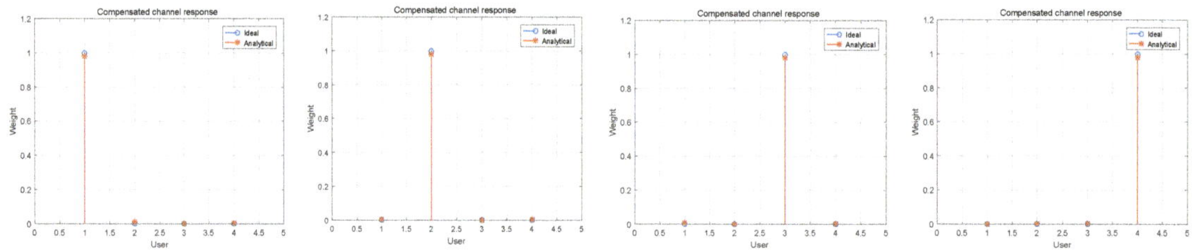

Figure 6. Compensated channel response of four different users with the ACMA in comparison with ideal beamforming when SNR = 10 dB.

In this experiment, we first evaluated the performance of these three beamforming methods with a settled noise power of $\sigma^2 = -100$ dBm and varying antenna numbers m. Then, we used a 16-element URA to evaluate the system's sum rate for varying σ^2. As shown in Figure 7, under different m and σ^2 settings, the system sum rates achieved with the analytical method are close to the approximate ones; moreover, they are much better than with the random values.

As mentioned in Section 4, the Gerchberg–Saxton algorithm (GSA) is used for the phase-retrieval problem. We proceeded with some iterations of the GSA for both analytically and randomly calculated weight vectors $\boldsymbol{w}_i$ for more accuracy. The convergence speed of GSA was tested in our simulations. As shown in Figure 8, the beamformers $\boldsymbol{w}_i$ obtained with the analytical method can achieve a very fast convergence speed, while the ones derived from random beamforming surely need more iterations to guarantee that they converge to their optimal solutions.

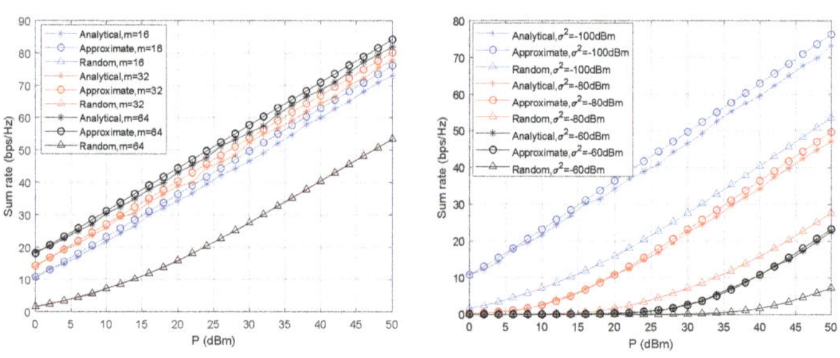

(**a**) Different m ($\sigma^2 = -100$ dBm) (**b**) Different σ^2 ($m = 16$)

Figure 7. The system's sum rate performance after analytical blind beamforming.

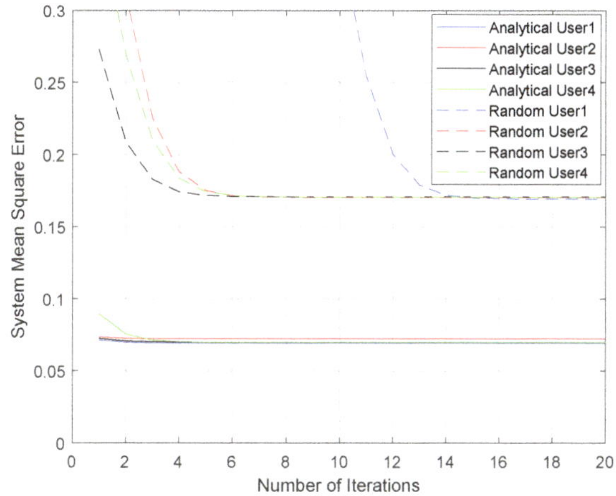

Figure 8. GSA convergence speed of four users using random initial points or points calculated with the ACMA when SNR = 20 dB.

After testing the performance of the analytical method in terms of the signal recovery accuracy and sum rate, we then tested its performance in terms of complexity. In Table 1, we compare it with classic non-blind methods (LS channel estimation/MMSE estimation + MMSE receiver) and the ABC method. As different methods depend on different parameters, sometimes, it is hard to directly compare the complexity in one united form. Thus, we compare these methods when they achieve comparable MSE results and introduce an equivalent complexity. In our settings, we need at least $m \geqslant d$ and $n \geqslant d^2 + 1$, so we can transform $\mathcal{O}(dnm)$ into $\mathcal{O}(d^4)$ [32]. When $d = 4$, we need about $L_{GSA} = 20$, $Ns = 100$, and $L_{ABC} = 300$ to make the CMA and ABC converge to their optimal results, so we obtain $\mathcal{O}(L_{GSA}) \approx \mathcal{O}(d^2)$, $\mathcal{O}(Ns) \approx \mathcal{O}(d^3)$, and $\mathcal{O}(L_{ABC}) \approx \mathcal{O}(d^4)$. The equivalent complexity of each method will depend on the actual parameter settings.

Table 1. Comparison of the complexity of the methods when achieving comparable system MSE values (notations: m: antenna number, d: user number, n: snapshots, Ns: food source number, L_{GSA}: iteration number of the GSA, L_{ABC}: iteration number of ABC).

Method	Complex Operations	Equivalent Complexity
LS+MMSE	$m^2d + 2d^2m + dmn$	$\mathcal{O}(d^4)$
MMSE+MMSE	$5m^2d + 2d^2m + dmn$	$\mathcal{O}(d^4)$
CMA	$L_{GSA}2dmn + 2m^2n$	$\mathcal{O}(d^6)$
ACMA	$9(d^2)^2n + 9m^2n$	$\mathcal{O}(d^6)$
ABC	$N_sdm + L_{ABC}(2N_s + 1)dm$	$\mathcal{O}(d^7)$

We then tested the simulation runtimes under specific parameter settings, as shown in Figure 9. We can see that the ACMA can achieve a comparable runtime with that of the classic methods with respect to different values of m. However, with an increasing user number, there is a dramatic runtime increase for the ACMA, but it is still more efficient than ABC. In fact, when the user number increases, ABC needs more food sources and iterations to obtain good results, so we set $N_s = 100$ and $L_{ABC} = 300$ for only the controlled variables.

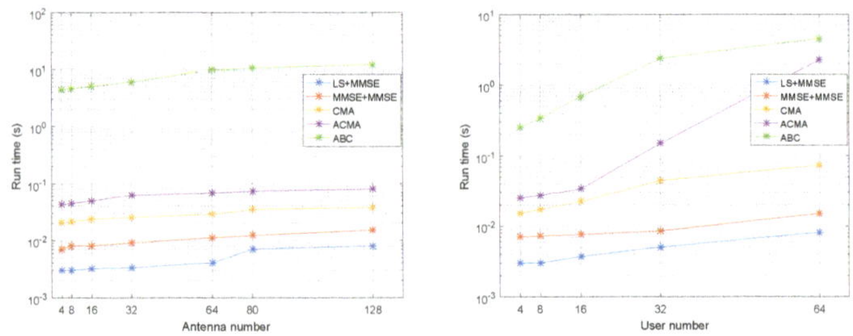

(**a**) $d = 4$, $n = 1000$, $L_{GSA} = 20$, $Ns = 100$, $L_{ABC} = 300$. (**b**) $m = d$, $n = d^2 + 1$, $L_{GSA} = 20$, $Ns = 100$, $L_{ABC} = 300$.

Figure 9. Simulation runtimes for different parameter settings.

In relation to Figure 3, there is a trade-off problem between the complexity and the system MSE. Indeed, the analytical method is more complex than the traditional MMSE method, but it achieves more accurate signal recovery beamforming when considering new features in the UAV communication channels. Accurate channel estimation is crucial for non-blind methods, but it is still difficult to acquire due to the time variance and other features [27]. From another perspective, battery endurance and energy cost are two inevitable factors in UAV-assisted networks that must still be solved, though non-blind methods usually need pilots or training sequences, which impact the bandwidth efficiency, power consumption, and battery life [38]. The computational complexity is not as difficult to deal with because chips are becoming more capable.

Through the evaluations in this section, we can observe that, when applied to a simulated UAV-BS system, the analytical blind beamforming method can achieve good MSE performance in terms of signal recovery and a reasonable system sum rate. Complex and difficult MIMO channel estimation is omitted to save costs. In other words, this method can successfully solve the MIMO UAV-BS beamforming problem without CSI. At the same time, it can achieve a reasonable computational complexity when there are not too many ground users. One possible UAV-BS application scenario can be a situation in which some specific users/target signals need higher-quality communications.

6. Conclusions

In this paper, we investigated the blind beamforming method with the ACMA at URA receivers for an MIMO UAV-BS system in mmWave bands. We firstly formulated the UAV-BS beamforming problem as a system MSE minimization problem. Then, we introduced the analytical method, ACMA, in order to derive a proper beamformer for signal recovery. Instead of relying on detailed CSI, the analytical method only explores information from antenna outputs. This method separates the user signal space from other interference spaces by using algebraic operations, such as SVD and diagonalization; then, it uses a specific structure to restrict the signal space to obtain a proper beamformer. We evaluated this method under different settings, such as different channels, different antenna settings, and different noise environments. New channel features, such as UAV mobility and time variance, were taken into consideration. The simulation results show that the analytical method exhibits good performance in terms of minimizing the system MSE, and, at the same time, it achieves a reasonable system sum rate. When the UAV-BS is not serving too many users, it achieves a complexity that is comparable with that of the MMSE method. Future research might study the overall power consumption increased by computation and transmission length. Practical experiments might be included in future work.

Author Contributions: Funding acquisition, project administration, supervision, K.F.; methodology, investigation, writing—original draft, P.L.; writing—review and editing, Y.C. All authors have read and agreed to the published version of the manuscript.

Funding: This work was supported by the National Natural Science Foundation of China (No. 61763 018), the 03 Special Project and 5G Program of the Science and Technology Department of Jiangxi Province (No. 20193ABC03A058), the China Scholarship Council (CSC, No. 201708360150), the Key Foundation of Education Committee of Jiangxi (No. GJJ170493, No. GJJ190451), the Program of Qingjiang Excellent Young Talents in Jiangxi University of Science and Technology (JXUSTQJBJ2019004), and the cultivation project of the State Key Laboratory of Green Development and High-Value Utilization of Ionic Rare-Earth Resources in Jiangxi Province (20194AFD44003).

Data Availability Statement: Data are only available upon request due to restrictions regarding, e.g., privacy and ethics. The data presented in this study are available from the corresponding author upon request. The data are not publicly available due to their relation to other ongoing research.

Conflicts of Interest: The authors declare no conflict of interest.

References

1. Samimi, M.K.; MacCartney, G.R.; Sun, S.; Rappaport, T.S. 28 GHz Millimeter-Wave Ultrawideband Small-Scale Fading Models in Wireless Channels. In Proceedings of the 2016 IEEE 83rd Vehicular Technology Conference (VTC Spring), Nanjing, China, 15–18 May 2016; pp. 1–6. [CrossRef]
2. Xiao, Z.; Xia, P.; Xia, X.G. Enabling UAV cellular with millimeter-wave communication: Potentials and approaches. *IEEE Commun. Mag.* **2016**, *54*, 66–73. [CrossRef]
3. Du, J.; Xu, W.; Deng, Y.; Nallanathan, A.; Vandendorpe, L. Energy-Saving UAV-Assisted Multiuser Communications With Massive MIMO Hybrid Beamforming. *IEEE Commun. Lett.* **2020**, *24*, 1100–1104. [CrossRef]
4. Miao, W.; Luo, C.; Min, G.; Zhao, Z. Lightweight 3-D Beamforming Design in 5G UAV Broadcasting Communications. *IEEE Trans. Broadcast.* **2020**, *66*, 515–524. [CrossRef]
5. Jiang, F.; Swindlehurst, A.L. Optimization of UAV Heading for the Ground-to-Air Uplink. *IEEE J. Sel. Areas Commun.* **2012**, *30*, 993–1005. [CrossRef]
6. Jian, O.; Yi, Z.; Min, L.; Jia, L. Optimization of beamforming and path planning for UAV-assisted wireless relay networks. *Chin. J. Aeronaut.* **2014**, *27*, 313–320. [CrossRef]
7. Bor-Yaliniz, R.I.; El-Keyi, A.; Yanikomeroglu, H. Efficient 3-D placement of an aerial base station in next generation cellular networks. In Proceedings of the 2016 IEEE International Conference on Communications (ICC), Kuala Lumpur, Malaysia, 22–27 May 2016; pp. 1–5. [CrossRef]
8. Mozaffari, M.; Saad, W.; Bennis, M.; Debbah, M. Communications and Control for Wireless Drone-Based Antenna Array. *IEEE Trans. Commun.* **2019**, *67*, 820–834. [CrossRef]
9. Zhang, Z.; Zhu, Q.; Zhang, P. Fast Beam Tracking Discontinuous Reception for D2D-Based UAV MmWave Communication. *IEEE Access* **2019**, *7*, 110487–110498. [CrossRef]
10. Rupasinghe, N.; Yapici, Y.; GÂÍÂźvenc, I.; Ghosh, M.; Kakishima, Y. Angle Feedback for NOMA Transmission in mmWave Drone Networks. *IEEE J. Sel. Top. Signal Process.* **2019**, *13*, 628–643. [CrossRef]
11. Yi, W.; Liu, Y.; Bodanese, E.; Nallanathan, A.; Karagiannidis, G.K. A Unified Spatial Framework for UAV-Aided MmWave Networks. *IEEE Trans. Commun.* **2019**, *67*, 8801–8817. [CrossRef]
12. Mei, W.; Zhang, R. Cooperative Downlink Interference Transmission and Cancellation for Cellular-Connected UAV: A Divide-and-Conquer Approach. *IEEE Trans. Commun.* **2020**, *68*, 1297–1311. [CrossRef]
13. Alzenad, M.; El-Keyi, A.; Lagum, F.; Yanikomeroglu, H. 3-D Placement of an Unmanned Aerial Vehicle Base Station (UAV-BS) for Energy-Efficient Maximal Coverage. *IEEE Wirel. Commun. Lett.* **2017**, *6*, 434–437. [CrossRef]
14. Sun, J.; Masouros, C. Deployment Strategies of Multiple Aerial BSs for User Coverage and Power Efficiency Maximization. *IEEE Trans. Commun.* **2019**, *67*, 2981–2994. [CrossRef]
15. Savkin, A.V.; Huang, H. Deployment of Unmanned Aerial Vehicle Base Stations for Optimal Quality of Coverage. *IEEE Wirel. Commun. Lett.* **2019**, *8*, 321–324. [CrossRef]
16. Liu, L.; Zhang, S.; Zhang, R. Multi-Beam UAV Communication in Cellular Uplink: Cooperative Interference Cancellation and Sum-Rate Maximization. *IEEE Trans. Wirel. Commun.* **2019**, *18*, 4679–4691. [CrossRef]
17. Li, S.; Duo, B.; Yuan, X.; Liang, Y.C.; Di Renzo, M. Reconfigurable Intelligent Surface Assisted UAV Communication: Joint Trajectory Design and Passive Beamforming. *IEEE Wirel. Commun. Lett.* **2020**, *9*, 716–720. [CrossRef]
18. Ge, L.; Dong, P.; Zhang, H.; Wang, J.B.; You, X. Joint Beamforming and Trajectory Optimization for Intelligent Reflecting Surfaces-Assisted UAV Communications. *IEEE Access* **2020**, *8*, 78702–78712. [CrossRef]
19. Li, L.; Ren, H.; Cheng, Q.; Xue, K.; Chen, W.; Debbah, M.; Han, Z. Millimeter-Wave Networking in the Sky: A Machine Learning and Mean Field Game Approach for Joint Beamforming and Beam-Steering. *IEEE Trans. Wirel. Commun.* **2020**, *19*, 6393–6408. [CrossRef]

20. Xiao, Z.; Dong, H.; Bai, L.; Wu, D.O.; Xia, X.G. Unmanned Aerial Vehicle Base Station (UAV-BS) Deployment With Millimeter-Wave Beamforming. *IEEE Internet Things J.* **2020**, *7*, 1336–1349. [CrossRef]
21. Zhang, W.; Zhang, W.; Wu, J. UAV Beam Alignment for Highly Mobile Millimeter Wave Communications. *IEEE Trans. Veh. Technol.* **2020**, *69*, 8577–8585. [CrossRef]
22. Colpaert, A.; Vinogradov, E.; Pollin, S. Aerial coverage analysis of cellular systems at LTE and mmwave frequencies using 3D city models. *Sensors* **2018**, *18*, 4311. [CrossRef]
23. Song, Q.; Zheng, F.C.; Zeng, Y.; Zhang, J. Joint Beamforming and Power Allocation for UAV-Enabled Full-Duplex Relay. *IEEE Trans. Veh. Technol.* **2019**, *68*, 1657–1671. [CrossRef]
24. Yuan, Q.; Hu, Y.; Wang, C.; Li, Y. Joint 3D Beamforming and Trajectory Design for UAV-Enabled Mobile Relaying System. *IEEE Access* **2019**, *7*, 26488–26496. [CrossRef]
25. Zhu, L.; Zhang, J.; Xiao, Z.; Cao, X.; Wu, D.O.; Xia, X.G. 3-D Beamforming for Flexible Coverage in Millimeter-Wave UAV Communications. *IEEE Wirel. Commun. Lett.* **2019**, *8*, 837–840. [CrossRef]
26. Liu, L.; Zhang, S.; Zhang, R. CoMP in the Sky: UAV Placement and Movement Optimization for Multi-User Communications. *IEEE Trans. Commun.* **2019**, *67*, 5645–5658. [CrossRef]
27. Huang, J.; Wang, C.X.; Chang, H.; Sun, J.; Gao, X. Multi-frequency multi-scenario millimeter wave MIMO channel measurements and modeling for B5G wireless communication systems. *IEEE J. Sel. Areas Commun.* **2020**, *38*, 2010–2025. [CrossRef]
28. Mao, K.; Zhu, Q.; Song, M.; Hua, B.; Zhong, W.; Ye, X. A Geometry-Based Beamforming Channel Model for UAV mmWave Communications. *Sensors* **2020**, *20*, 6957. [CrossRef]
29. Li, X.; Jin, S.; Suraweera, H.A.; Hou, J.; Gao, X. Statistical 3-D Beamforming for Large-Scale MIMO Downlink Systems Over Rician Fading Channels. *IEEE Trans. Commun.* **2016**, *64*, 1529–1543. [CrossRef]
30. Du, L.; Li, L.; Zhang, P.; Miao, D.; Liu, Z. Vector Perturbation Precoding Under Imperfect CSI and Inaccurate Power Scaling Factors. *IEEE Access* **2019**, *7*, 89162–89171. [CrossRef]
31. Rajashekar, R.; Di Renzo, M.; Hari, K.; Hanzo, L. A Beamforming-Aided Full-Diversity Scheme for Low-Altitude Air-to-Ground Communication Systems Operating With Limited Feedback. *IEEE Trans. Commun.* **2018**, *66*, 6602–6613. [CrossRef]
32. Van der Veen, A.J.; Paulraj, A. An analytical constant modulus algorithm. *IEEE Trans. Signal Process.* **1996**, *44*, 1136–1155. [CrossRef]
33. Molisch, A.F., Multiantenna Systems. In *Wireless Communications*; Wiley: Hoboken, NJ, USA, 2011; pp. 445–498. [CrossRef]
34. Mynbaev, D.K.; Scheiner, L.L., Digital Signals and Digital Transmission. In *Essentials of Modern Communications*; Wiley: Hoboken, NJ, USA, 2020; pp. 203–258. [CrossRef]
35. Li, J.; Sun, Y.; Xiao, L.; Zhou, S.; Koksal, C.E. Analog beam tracking in linear antenna arrays: Convergence, optimality, and performance. In Proceedings of the 2017 51st Asilomar Conference on Signals, Systems, and Computers, Pacific Grove, CA, USA, 29 October–1 November 2017, pp. 1193–1198. [CrossRef]
36. Gene, H.G.; Charles, F.V.L. *Matrix Computation*; The Johns Hopkins University Press: Baltimore, MD, USA, 2013; p. 747.
37. Rappaport, T.S.; MacCartney, G.R.; Samimi, M.K.; Sun, S. Wideband Millimeter-Wave Propagation Measurements and Channel Models for Future Wireless Communication System Design. *IEEE Trans. Commun.* **2015**, *63*, 3029–3056. [CrossRef]
38. Zeng, Y.; Wu, Q.; Zhang, R. Accessing From the Sky: A Tutorial on UAV Communications for 5G and Beyond. *Proc. IEEE* **2019**, *107*, 2327–2375. [CrossRef]

sensors

MDPI

Article

Algorithms for Delivery of Data by Drones in an Isolated Area Divided into Squares

Adrian Marius Deaconu [1], Razvan Udroiu [2,*] and Corina-Ştefania Nanau [1]

1 Department of Mathematics and Computer Science, Transilvania University of Brasov, 29 Eroilor Boulevard, 500036 Brasov, Romania; a.deaconu@unitbv.ro (A.M.D.); corina.nanau@unitbv.ro (C.-Ş.N.)
2 Department of Manufacturing Engineering, Transilvania University of Brasov, 29 Eroilor Boulevard, 500036 Brasov, Romania
* Correspondence: udroiu.r@unitbv.ro; Tel.: +40-268-421-318

Abstract: Drones are frequently used for the delivery of materials or other goods, and to facilitate the capture and transmission of data. Moreover, drone networks have gained significant interest in a number of scenarios, such as in quarantined or isolated areas, following technical damage due to a disaster, or in non-urbanized areas without communication infrastructure. In this context, we propose a network of drones that are able to fly on a map covered by regular polygons, with a well-established mobility schedule, to carry and transfer data. Two means exist to equidistantly cover an area with points, namely, grouping the points into equilateral triangles or squares. In this study, a network of drones that fly in an aerial area divided into squares was proposed and investigated. This network was compared with the case in which the area is divided into equilateral triangles. The cost of the square drone network was lower than that of the triangular network with the same cell length, but the efficiency factors were better for the latter. Two situations related to increasing the drone autonomy using drone charging or battery changing stations were analyzed. This study proposed a Delay Tolerant Network (DTN) to optimize the transmission of data. Multiple simulation studies based on experimental flight tests were performed using the proposed algorithm versus five traditional DTN methods. A light Wi-Fi Arduino development board was used for the data transfer between drones and stations using delivery protocols. The efficiency of data transmission using single-copy and multiple-copy algorithms was analyzed. Simulation results showed a better performance of the proposed Time-Dependent Drone (TD-Drone) Dijkstra algorithm compared with the Epidemic, Spray and Wait, PRoPHET, MaxProp, and MaxDelivery routing protocols.

Keywords: drones; network; DTN; mobility schedule; routing algorithms; data delivery

Citation: Deaconu, A.M.; Udroiu, R.; Nanau, C.-Ş. Algorithms for Delivery of Data by Drones in an Isolated Area Divided into Squares. *Sensors* **2021**, *21*, 5472. https://doi.org/10.3390/s21165472

Academic Editor: Margot Deruyck

Received: 26 July 2021
Accepted: 10 August 2021
Published: 13 August 2021

1. Introduction

Delay tolerant networks (DTNs) allow communication in environments in which frequent transmission discontinuities are present [1–4]. They have applications in numerous fields, such as space communication networks [1–3], smart cities [5], intelligent transport networks, rural networks, environmental monitoring networks, and vehicle networks [4,6]. Within DTNs, message transmission is based on the store-carry-forward paradigm [7,8]. Devices update their communication routes based on the topological changes of the network, and the mobility of the devices plays an important role [9–11].

The role of routing [12–14] in DTNs is to find the best path to send data through the network to reach the destination. The routing strategies used by DTNs are classified based on criteria such as connection type between nodes, the time at which the path for messages is established, the amount of information held by the nodes about the network, and the number of copies of a message that a node sends.

Algorithms containing different amounts of DTN information are proposed for investigation in [14]. It has been shown that the performances of these algorithms gradually increase, depending on the amount of information about the network they use. Based on the number of copies of a message, single-copy algorithms (forward based) and multiple-copy algorithms (flood based) have been investigated [15,16]. The Direct Delivery algorithm only sends the message to the destination node, and is suitable for both small- and high-mobility networks, in which the probability of meeting between the nodes is high. Comparisons between classical flood-based protocols, such as Epidemic [17], Spray and Wait [18], PRoPHET [19], and MaxProp [20], were undertaken in [15]. Maximum flow with the static approach in buffer-limited delay tolerant networks was investigated in [21,22].

In public transport networks, the connection type between nodes is based on a well-known schedule and well-defined routes. A DTN composed of pedestrians and cyclists equipped with smart devices was investigated using routing protocols in an Opportunistic Network Environment (ONE) [23]. The MaxProp protocol yielded the best results in terms of delivery probability and average latency. DTN communications in the network storage depend on the store-carry-forward mechanism. A DTN system applied in a communication network on a railway line was found to reduce the message delivery time by 20%, depending on the schedule of trains [8]. In [24], an algorithm to search for the shortest safe path on the network with a time-dependent and edge-length danger factor was proposed. This work is suitable for the optimization of heavy trucks carrying inflammable materials, poison gas, or explosive cargo, and traveling within a city. Based on the simulation in ONE software, the Scheduling-Probabilistic Routing Protocol using History of Encounters and Transitivity (PROPHET) improves the delivery rate and optimizes the delivery delay with low overhead in DTNs for IoT applications [25].

Drone networks are now used in numerous applications across domains including topography [26], aerial observation [27], delivery [28,29], agriculture [30], communications [25,31,32], atmospheric sciences [33–35], and rescue missions [36].

Drone flight path planning can be categorized as off-line planning, on-line planning, or cooperative planning [37,38]. The relatively short flight distance of drones due to their limited battery energy [39] can be extended using drone networks.

Flight simulations using a topological map of hexagons, in which each hexagon contains one drone, have shown the potential for application in indoor rescue missions [36]. In [40], coverage path planning with UAVs was studied, addressing simple geometric flight patterns, such as back-and-forth and spiral, and more complex grid-based solutions considering full and partial information about the area of interest. The area was divided into squares and a drone was flown from a square to another. Parcel delivery missions using a drone were simulated based on heuristic flight path planning (HFPP) and other routing algorithms [28]. The results showed that HFPP delivers up to 33% more data packets compared with Encounter-Based Routing and Epidemic routing protocols.

Vehicular ad hoc networks (VANETs), which are a type of mobile ad hoc network (MANET), can be used in an intelligent transport system. VANETs allow the mobile vehicles to establish three main categories of communication: vehicle to vehicle, vehicle to infrastructure, and infrastructure to infrastructure. A specific application of VANET applied to a drone network allows messages to be sent via wireless links [32]. A bio-inspired coordination protocol for a drone flying ad-hoc network (FANET) used for agriculture applications [30] has been used in an ad hoc simulator for a preliminary analysis of the feasibility of drone network design. The performance of the branch and bound search-based mode selection (BBS-MS) for drone-based air-to-ground wireless networks was investigated in [31].

In this study, we investigated six routing algorithms in a network of drones with a mobility schedule to ensure communication between isolated areas or in areas with technical damage. We then qualitatively compared the algorithms. The major contributions of this study are as follows:

- The proposed drone network is independent of the Internet, and is of the DTN type.
- A topological map as a collection of regular polygons (squares) was proposed.

- A time-dependent variant of Dijkstra's algorithm, which determines the fastest route by taking into account the time when the message reaches the node and the time allocated for data transfer, was developed for the proposed network.
- Five classical algorithms for DTN networks were adapted and tested for the proposed network.
- Simulations based on flight tests were performed to analyze the efficiency of the data transmission. The results obtained by the single-copy and multiple-copy algorithms were compared, in the case of buffer limited capacity.
- Practical application of this work is to facilitate the transmission of information in regions quarantined due to an infectious outbreak, such as COVID-19 pandemic, in regions with technical damage due to a disaster, and in non-urbanized areas without electricity access or communication infrastructure.

The remainder of the paper is organized as follows. In Section 2, a drone network architecture based on the mission profile, and drone sensing and communication, are proposed. Then, the algorithms and protocols for the delivery of data using drones are proposed, described, and analyzed. In Section 3, simulation results are summarized and discussed. Finally, conclusions and possible future work are presented in Section 4.

2. Materials and Methods

2.1. Drone Network Architecture and Communication

A network map is essential for both drone flight control and simultaneous localization and mission tasks. As noted in [36], three forms of map generation exist: metric, topological, and hybrid maps. A metric map is represented as a grid, geometric, or feature map. Topological maps are represented by graphs comprised of nodes and edges, where nodes represent places, and the edges represent the paths between the nodes [36]. A hybrid map consists of small metric map locations in nodes. These nodes are connected by edges, which are the paths between the metric maps [36]. In this work, a topological map was proposed.

The 2D network surface is intended to be covered with equidistantly distributed points. These points form regular polygons covering the 2D surface. Below, we prove that the only possible means of covering the 2D surface in this manner is by using equilateral triangles or squares. First, we prove the following lemma:

Lemma 1. *There are 3 ways to cover a 2D surface using regular polygons: equilateral triangles, squares, or regular hexagons.*

Proof of Lemma 1. The sum of the degrees of the angles of a polygon with n vertices ($n \geq 3$) is $180° \cdot (n-2)$. Each angle of a regular polygon has $[360° \cdot (n-2)]/n$ degrees. If a point of the surface is a vertex of a polygon, then it is a vertex for m polygons ($m \geq 3$) around this point (Figure 1). Thus, at such a point we have:

$$m \cdot \frac{180° \cdot (n-2)}{n} = 360° \Leftrightarrow \frac{2 \cdot n}{n-2} = m \tag{1}$$

□

It results that $2 \cdot n/(n-2) = 2 + 4/(n-2)$ is an integer value greater or equal to 3. This implies that $n-2$ is a divisor of 4 and, because $n \geq 3$, it follows that $n-2$ is one of the values 1, 2, or 4, which is equivalent to the fact that n is 3, 4, or 6, and m is 6, 4 or, respectively, 3. This means that the 2D surface can only be covered with equilateral triangles, squares, or regular hexagons.

Theorem 1. *There are two ways to equidistantly cover a 2D surface with points.*

Proof of Theorem 1. Using the result from Lemma 1 and the fact that the points are considered equidistant on the 2D surface, it follows that there are only two ways to cover the 2D surface: with equilateral triangles or with squares, because using regular hexagons is impossible (the vertices of the hexagons are not equidistantly positioned). □

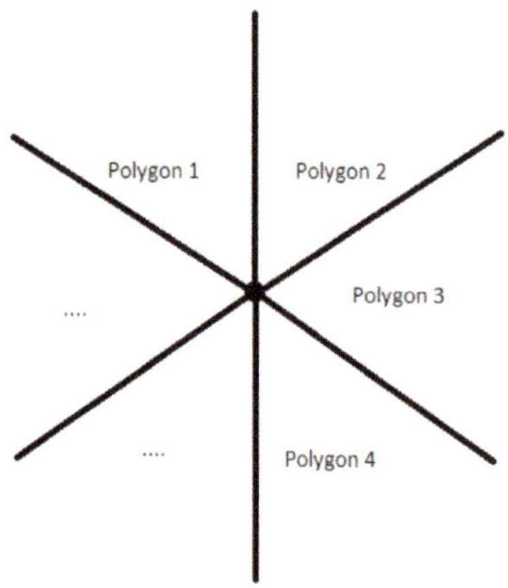

Figure 1. Regular polygons sharing the same vertex.

A network to cover a surface with hexagon cells containing three equilateral triangles was proposed in [41]. In the current paper, a network is proposed to cover a similar surface with square cells containing two operational squares (Figure 2). Each operational cell is covered by a drone. A battery charging/changing dock, which is shared by two drones, is placed in the center of each cell. The docks are able to automatically change/recharge drones, without manual intervention, allowing fully autonomous drone management. The mission profile of each drone consists of several phases or steps: engines start, take off, climb at the cruise altitude, cruise, hovering and data exchange, descent, landing, and engines shut-down. The cruise phase of the flight mission profile consists of four segments (Figure 3). Charging or changing the drone battery is a necessary step at the end of the flight mission.

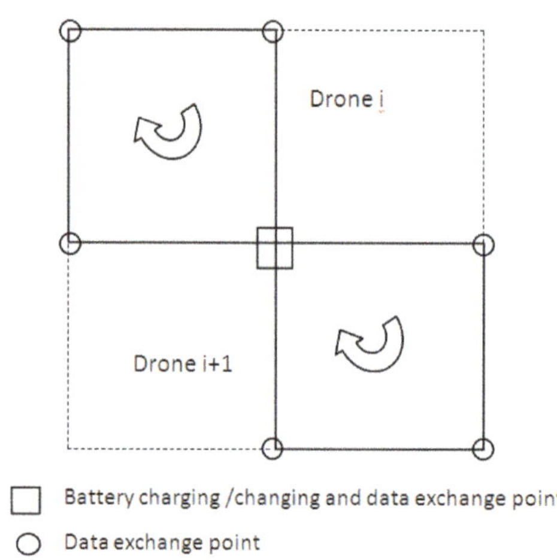

Figure 2. Square cell with two drones.

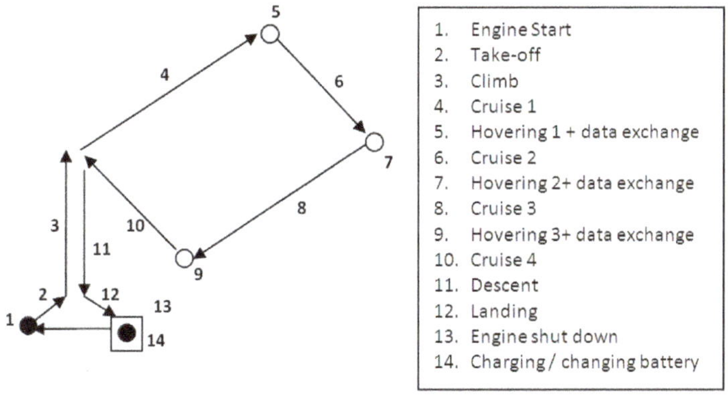

Figure 3. Square shape flight mission profile of the drone.

A drone with a quadcopter configuration [42], i.e., a DJI Mavic 2 Pro (DJI, Shenzhen, China) with a size of 214 × 91 × 84 mm (length × width × height), and takeoff weight of 905 g, was considered in this study. The performance characteristics of this drone are presented in Table 1 [43]. All of the drones of the proposed squares network operate in a pre-programmed manner. The drone network communicates with data exchange points via wireless links.

Table 1. Drone performance [43].

Parameter	Value
Max ascent/descent speed	4 m/s; 3 m/s
Max flight time (no wind)	31 min (at a consistent 25 km/h)
Max flight distance (no wind)	18 km (at a consistent 50 km/h)
Drone battery	3850 mAh, 1800 mA, 3.83 V

The endurance of drones can be improved using various methods, such as changing the battery [44,45], charging the battery via wires [46,47], wireless recharging [48], solar cells [49,50], laser-beam in-flight recharging [49,51], and tethered drones [49]. An average time of 90 min is needed to fully charge an empty battery. An automated means of charging a battery via a wire can be performed using a charging platform installed on the ground and a drone retrofit-kit mounted on the drone [46,47]. Thus, the landing gear of the drone is connected by touch with the charging platform after the drone lands, and charging starts automatically. The main disadvantage of this system is that the drone is locked on the ground during the charging of the battery.

An automatic battery changing and recharging system was investigated in [44]. The battery is automatically recharged after it is changed at the station. The electricity needed at each station is able to be provided by a solar panel that charges a battery located at the station. The changing time depends on the efficiency of the changing mechanism, and varies between 15 s [44] and 60 s [45]. In the current study, a maximum changing time before the drone is ready to take off of 60 s was considered.

Four flight tests were performed based on the square-shaped flight mission in the Brasov area of Romania. The flight tests were performed at a temperature of 2 °C, humidity of 69%, and wind speed of 3.5 km/h. A Samsung S9 smartphone device, on which DJI Go 4 app software (DJI, Shenzhen, China) was installed, connected to a DJI remote controller, was used to program the flight mission segments and to remotely control the drone (Figure 4). A flight altitude of 30 m was chosen [52] and the cruise flight distance of each drone was 12,000 m.

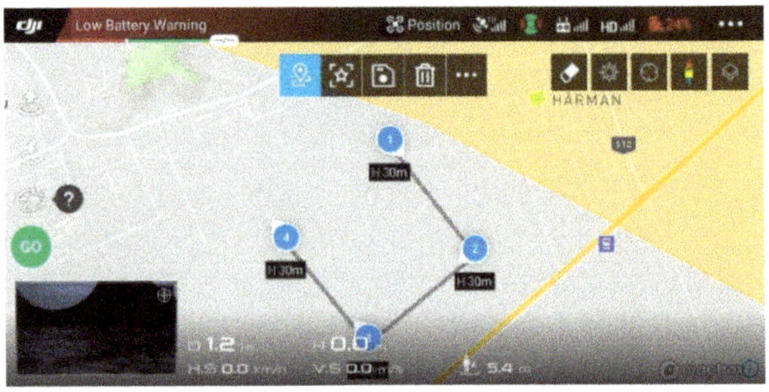

Figure 4. Square-shaped flight mission profile.

The theoretical flight times calculated based on the drone specifications (max. ascent, descent speeds, and cruise speed) are not realistic for simulations. The main factors that influence the experimental flight times are acceleration and deceleration of the drone, and wind speed. The mean flight time for each flight segment was calculated based on the flight tests (Table 2). Moreover, the average calculated cruise speed of the drone obtained from the flight tests was 12.93 m/s.

Table 2. Flight test results.

Mission Phase	Mean Flight Time	Standard Deviation
Take off + climb (30 m)	8.24 s	0.193
Cruise segment (3000 m)	232 s	0.187
Descent + landing (30 m)	12.12 s	0.085
Transfer data (3 points)	120 s	-
Total flight on square cell	1308 s	0.651

The percentage of the remaining drone battery obtained at the end of the flight mission was 16% for the rectangular cell. The charging time for the drone battery was 79 min. A safety multiplier of 1.1 was applied to obtain the considered charging time, and the resulting time was used in the simulations. Thus, 87 charging minutes were considered for the rectangular cell flight. The average values of the flight times for each segment were used as input parameters for the simulation of the drone networks in the DTN algorithms.

Eight high-resolution and two infrared sensors were used on the DJI Mavic 2 Pro. These sensors enabled omnidirectional obstacle sensing, to determine the relative speed and distance between the drone and the object, and to ensure good stability in forward and hovering flight. Left, right, up, down, forward, and backward obstacle sensing were used.

A NodeMCU Lua Wi-Fi, V3, ESP-12E, CP2102 Wi-Fi Arduino development board (Espressif Systems, Shanghai, China) was used for the data transfer between the drones and the stations. The main characteristics of the NodeMCU are shown in Table 3. The data were stored on a micro-SD card. The Wi-Fi board and micro-SD card module are very lightweight and have very low power consumption. The experimental layout, consisting of the NodeMCU Lua Wi-Fi board and the micro-SD card module mounted on a breadboard, is shown in Figure 5a. Arduino code was used to program the Wi-Fi boards.

Table 3. Wi-Fi NodeMCU main characteristics [53].

Parameter	Value
ESP8266 chip	26 MHz, 4 MB flash, 160 KB RAM
Dimensions (L × W)	48 mm × 25 mm
Operating temperature	−40 °C to + 125 °C
Weight	8 g

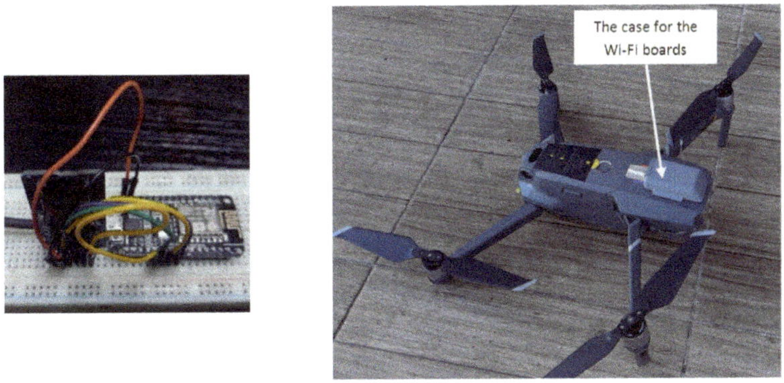

(**a**) (**b**)

Figure 5. (**a**)Testing stand of the NodeMCU Lua Wi-Fi board and the micro-SD card module mounted on a testing breadboard. (**b**)The drone with the 3D-printed case and the Wi-Fi boards mounted on it.

A protective mounting case (Figure 5b) for the Wi-Fi boards was designed using the SolidWorks version 2016 software (Dassault Systèmes, Waltham, MA, USA), and then 3D printed using material extrusion technology from PLA (BCN3D Technologies, Barcelona, Spain) on a BCN3D Sigma R19 printer (BCN3D Technologies, Barcelona, Spain).

The Wi-Fi board, micro-SD module, SD card, connection wires, and power cable weighed a total of 21.8 g, thus representing an increase of 2.4% in the total weight of the drone.

2.2. Algorithms and Protocols for Delivery of Data Using Drones

The DTN was modeled with a graph having fixed and mobile nodes. There are no connections between the fixed nodes, so there is no possibility of direct data transmission because the considered distance between the nearest two nodes is 3000 m. Network connections are provided by mobile nodes (drones), but their condition is not always the same; they have periods when they are active and periods when they are inactive. This means that there is not always an end-to-end path available between any two nodes in the graph.

We tested five well-known routing algorithms for DTNs (Epidemic, Spray and Wait, PRoPHET, MaxProp, and MaxDelivery [54]), and a newly proposed TD-Drone Dijkstra approach, on the square-shaped network shown in Figure 6. We performed tests on the network by choosing random sources and random destinations that could be located on any node (marked as a gray circle or a gray rectangle). We considered the drones worked each day from 7:00 a.m. to 6:00 p.m. and the messages may leave a source node between 7:00 a.m. and 5:00 p.m. We considered 1000 messages randomly sent within this interval of time.

Epidemic is the basic form of a flood-based routing protocol: when two nodes meet, they identify the packages that the other node has and transfer the packages that it does not have. At the end of the process, the two nodes have the same content in the buffer. This process is repeated each time two nodes come into contact. When a node has a copy of a message, it waits to meet the destination. In this case, the resource consumption is high but, in a high mobility network, the delay of message transmission is small. In the current network configuration, the algorithm produces poor results due to the low number of contacts between nodes.

Spray and Wait is an algorithm with two phases, one for sending messages (spray) and one to wait for the contact with the destination node (wait). This algorithm circulates in two variants—standard and binary—depending on the number of spread copies of the message. It acts in the same manner as Epidemic, with an important difference: the number of spread copies is constant. The spray phase of the standard approach consists of spraying

L copies of the message by the source node itself. The spray phase of the binary approach consists of spraying half of the number of copies to a meeting node. In this case, not only does the source spray messages, but also every node that has more than one copy. The nodes that have only one copy enter the wait phase. This algorithm has the disadvantage that nodes must keep track of other nodes' movement, but the advantage is that the level of flooding is limited.

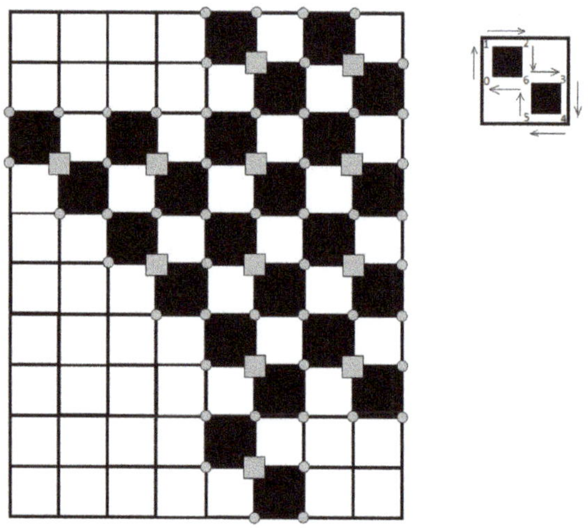

Figure 6. Squares network considered for experiments.

PRoPHET is similar to Epidemic, with the exception that it uses information from the buffer of the other node to update its predictability vector. Each node calculates the predictability of the message delivery and sends the message onward only if the contact node has higher predictability than its own. The disadvantage of this approach is the relationship between the overhead ratio and the number of nodes—as the number of nodes increases, the overhead ratio increases [15]. This protocol is known for the complexity of its forwarding strategy. Thus, it consumes a significant quantity of resources to process and store historical values. This approach is feasible for networks with high computation and infrastructure capabilities.

MaxProp is an algorithm based on prioritizing packet transmission and discarding. The packets in the queue are divided into two categories: those below the "n" hop threshold (up to that point), and those above this threshold. Newer packages that have not traveled too much are considered a priority and the guarantee that they will reach their destination is considered to be high. This algorithm also requires high computation and infrastructure capabilities. This protocol has low performance when nodes have small buffer sizes because of the adaptive threshold calculation, but gives better performance with a larger buffer size. It has a so-called slow start problem, because, in the case of a big network, it may take a very long time before each node receives the delivery predictability of other nodes because of the disconnecting nature of the networks, as shown in [20].

MaxDelivery [54] is an algorithm based on prioritizing message delivery using an appropriate buffer management strategy that consists of a forwarding, dropping, and buffer-cleaning mechanism.

Next, we propose our time-dependent Dijkstra algorithm to find a route between two nodes in our drone network. Dijkstra's algorithm is used to find the shortest path connecting two nodes in a network [55]. Our routing problem can be modeled using a time-dependent oriented network [56] defined as follows.

Definition 1. *A triple G = (V, A, f) is called a time-dependent oriented network, where V is a set of vertices,* $A \subseteq V \times V$ *is a set of arcs, and f is the time dependency function defined on each arc,* $f: A \times T \rightarrow T$, $T \subseteq R+$ *is called the time set and when moving on the arc a = (u, v)* $\in$ *A from node u to node v, f(a, t)* $\in$ *T is the moment of arrival at node v if node u is left at the moment t* $\in$ *T. Of course, f(a, t) > t, for each arc a* $\in$ *A and for any moment t* $\in$ *T.*

In our problem, V is the equidistantly distributed set of 2D points, A is the set of connections between these points ensured by drones, and T = {0, 1, 2, ... } is the set of counting seconds in a day. The value of f(a, t) must be computed as quickly as possible. Thus, for each arc a = (u, v), the arrival moments of drones at node v are maintained in order to enable a binary search to be undertaken when f(a, t) is calculated at arc a and time t. The arrival moments of drones for all arcs are pre-calculated (once before starting to use the algorithm) because an exact schedule of drones is known based on each drone's starting second in a day, its travel on each arc, its data transfer at nodes, and its wireless charge/battery change time.

For a given source node s $\in$ V, the function dists: $V \times T \rightarrow N$ is introduced, where dists(u, ts) is the distance in seconds from s to u starting from the moment ts, i.e., it is the minimum number of seconds needed to go from node s to node u $\in$ V on the arcs of A if source s is left at moment ts. Of course, dists(s, ts) = 0 for every ts $\in$ T. If node u is not reachable (accessible) from s, then dists(u, ts) = $+\infty$. For a given destination node d, our problem is to determine dists(d, ts) at a given time ts. The pseudo-code for the time-dependent Dijkstra algorithm is presented in Figure 7.

```
Input: G = (V,E), f, s, d, ts
Output: p, dists(d)
for each node u from V\{s} do
    dists(u) = +∞;
    settled(u) = false;
end for;
dists(s) = 0; settled(s) = false;
p(s) = s;
Q.insert(s, 0);
while not Q is empty do
    found = false;
    while not Q is empty do
        The last node u from Q is popped out;
        if not settled(u) then
            found = true;
            break;
        end if;
    end while;
    if not found then
        break;
    end if;
    settled(u) = true;
    if u = d then
        break;
    end if;
    for every arc a = (u, v) of E do
        newdist = dists(u) + f(a, ts + dists(u));
        if newdist < dists(v) then
            dists(v) = newdist;
            Q.insert(v, newdist);
            p(v) = u;
        end if;
    end for;
end while;
```

Figure 7. Pseudo-code of TD-Drone Dijkstra algorithm.

Q is a priority queue. This means that, at any moment, the nodes from Q are sorted in ascending order according to their distance in seconds from s. At the end of the algorithm, if destination d was reached, the values kept in the so-called predecessor vector p are used to determine the route from s to d. If a node v was reached, then this was done using the arc (p(v), v), where node p(v) is called the predecessor of v. After the time-dependent Dijkstra algorithm is executed using the staring moment ts, the route from s to d (if it exists, i.e., $dists(d) < +\infty$) is determined using the algorithm presented in Figure 8. Because the nodes of the route are found by the above algorithm in inverse order (from d to s), the route must be reversed at the end.

```
Input: dists(d), p, s, d
Output: a route from s to d (if exists)
if dists(d) < +∞ then
    v = d;
    while not p(v) = s do
        Add v to the route;
        v = p(v);
    end while;
    Add s to the route;
    Reverse the route;
else
    There is no route from s to d;
end if;
```

Figure 8. Pseudo-code for the construction of the route after the execution of the TD-Drone Dijkstra algorithm.

For each data file that has to be delivered, a "json" file is attached that stores all of the information needed to transfer the file from the source to the destination: delivery type (Dijkstra, Epidemic etc.); file information (name, size, time-to-leave); route information (node codes: stations and drones); etc. The json files that store message information for the algorithms Epidemic, Spray and Wait, PRoPHET, MaxProp, and MaxDelivery implemented in the ONE environment are similar to those considered in [41]. Each route starts and ends with a station id. Each station id (except for the destination) is followed by a drone id, and each drone id is followed by a station id. When a drone arrives at a station, a transfer is initiated between the drone and the station. The drone transfers to the station all of the files that have the station's id in the attached json files. After the drone transfer to the station is completed, the station transfers to the drone all of the files that have the drone's id in the json file.

The range of the Wi-Fi boards was tested. The connections between the boards and file transfer were performed at a distance of up to 85 m with no obstacles in between. In Figure 9, the console output of the Wi-Fi board is presented. The following steps are executed: setup (upload speed is set to 115,200, MAC address is obtained), the connection between two Wi-Fi boards is established, the file transfer is performed, and, finally, the Wi-Fi boards are disconnected.

In our model, when transferring files, the drone hovers over the station at a height of 30 m, which is significantly less than the maximum distance obtained in range tests. The transfer speed, including writing on and reading from the SD card, was also tested at a distance of 30 m. An average of 5.81 Mbps was obtained, which means that a file of 10 MB was transferred in 13.77 s.

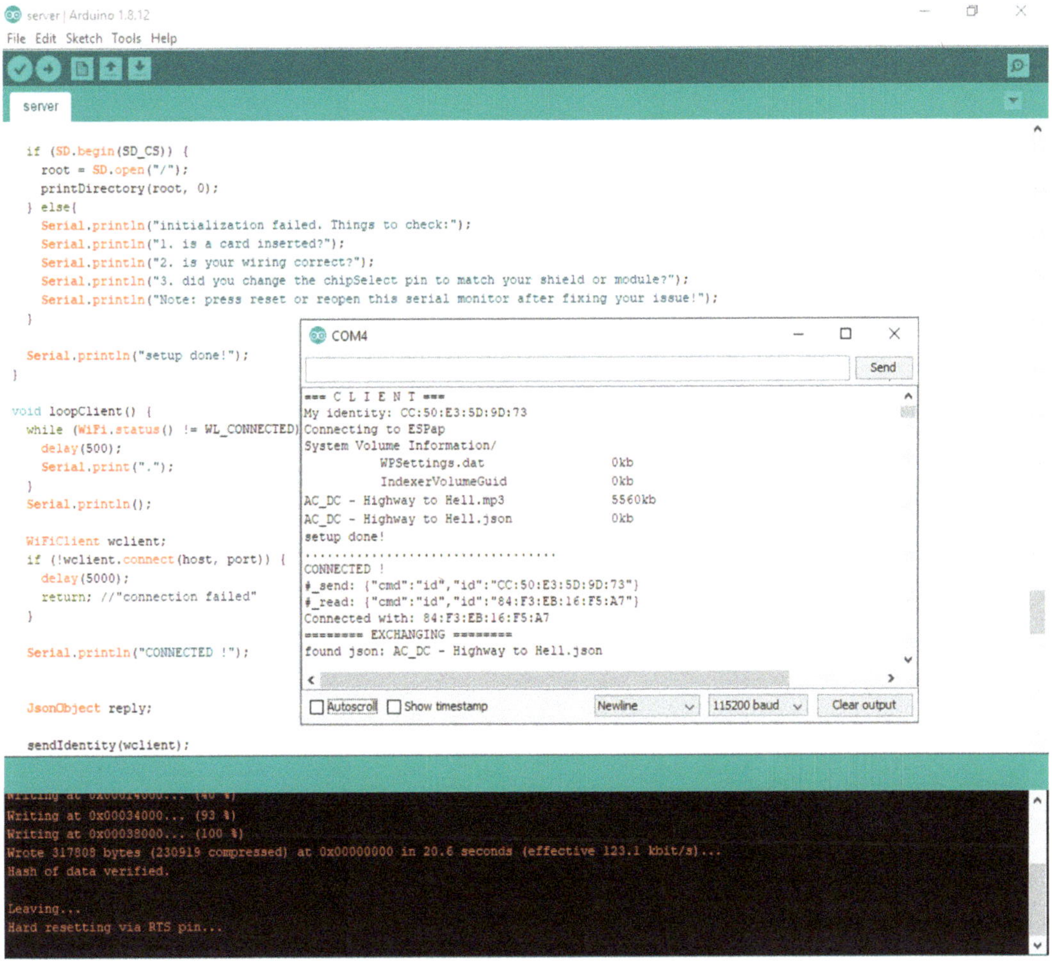

Figure 9. The console output of the Wi-Fi board.

3. Simulation Results and Discussion

The proposed network of drones can be applied in various scenarios, such as in remote quarantined or isolated areas, following technical damage due to a disaster (e.g., an earthquake), or in non-urbanized areas without electricity access or communication infrastructure. For instance, a remote quarantined zone (e.g., due to the COVID-19 pandemic), in which buildings are separated by a safe distance, is considered. In this scenario, each building may be a house in which patients are isolated, a warehouse storing food or drugs, a laboratory in which medical tests are performed for patients, or a location at which doctors are working (isolated from patients). Data packages (medical images, tests, results, prescriptions from doctors, etc.) must be sent between these buildings. The communication between these buildings can be achieved by drones organized in a square-shaped network.

To validate the proposed method, simulations were performed using the drone network maps as a collection of squares. Most of the simulations were performed using the Java-based simulator ONE and its provided facilities, using an ASUS ROG GL752VW-T4015D laptop with Intel® Core™ i7-6700HQ 2.60GHz processor (Asus, Taipei City, Taiwan), and 8 GB of RAM. The routing protocols used for the simulation within the ONE simulator were Epidemic, Spray and Wait, PRoPHET, MaxProp, and MaxDelivery.

The following main steps were used for implementation of the ONE scenarios. The first step consists of defining the map (Figure 10) in wkt file format, in which the coordinates of all the points on the map, including points that establish the route of each drone, are defined. The initializations of the algorithms that define the mobility of drones consist of establishing the initial positions of drones and the recharging/changing points, associating each drone with a recharger/changing point, establishing the stationary points for data transfer, and defining the route of each drone. The final step is the establishment of the simulation parameters, as shown in Table 4. The time parameters, such as the travel autonomy time, the hovering time for the transfer points, and the parking time at the charging or changing points, were established based on the experimental flight tests of the DJI Mavic 2 Pro drone.

Table 4. Simulation parameters for the square-shaped flight mission.

Parameter	Value
Number of drones for cruise	24
Number of fixed transfer points	63
Number of charging/ changing battery points	12
Average cruise speed of a drone	46.55 km/h (12.93 m/s)
Flight height of drones	30 m
Operating time of the drone in one day	11 h
Data transmission speed	2 Mbps
Drone buffer space	2 Gb
Message size	500 kb–1 Mb
Message time to live	10 h
Source and destination of messages	any station
No. of route simulations	1000

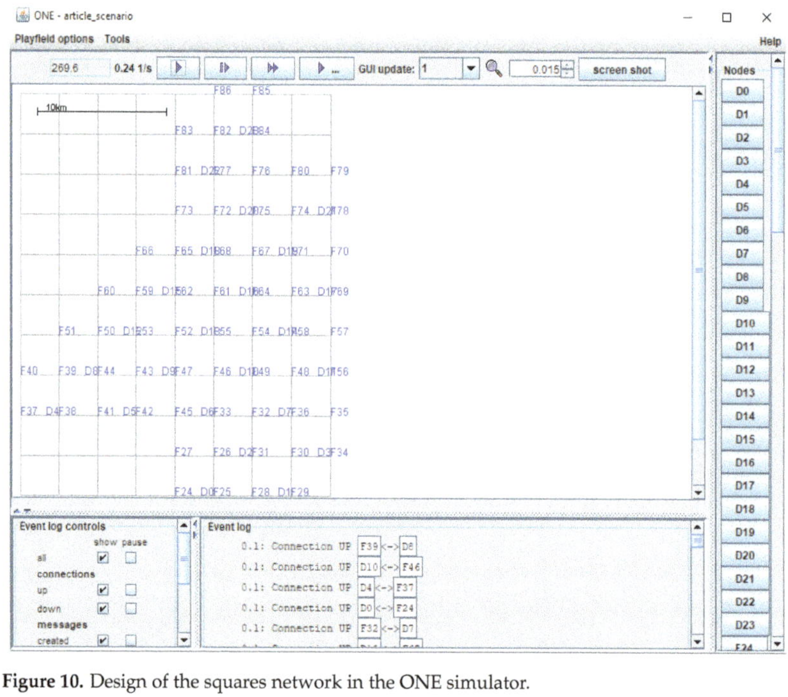

Figure 10. Design of the squares network in the ONE simulator.

The proposed time-dependent Dijkstra variant was implemented in Visual C++ 2017 programming language. The application has about 1100 lines of C++ source code.

The application written in C++ was executed for each of the two considered situations: squares with battery charging, and squares with battery changing. For each case, the same 1000 route simulations considered in the ONE experiments were executed.

The delivery rate and latency metrics were used to measure the performance of all six routing protocols analyzed in this paper. The delivery rate is determined as a ratio between the number of delivered messages and the number of created messages. The latency is the average time needed for a message to reach the destination starting from the source (departure node).

Detailed results of the comparison between changing the battery and charging the battery cases are presented in Table 5. The results obtained in this paper by simulation were compared with those obtained in [41].

Table 5. Efficiency factors in drone network.

Algorithm	Delivery Rate				Latency (hours)			
	Battery Changing		Battery Charging		Battery Changing		Battery Charging	
	Squares	Triangular	Squares	Triangular	Squares	Triangular	Squares	Triangular
Epidemic	0.166	0.209	0.135	0.146	0.81	0.72	2.28	2.13
Spray and Wait	0.211	0.179	0.141	0.156	0.52	0.56	1.75	1.92
PRoPHET	0.594	0.762	0.143	0.319	0.61	0.52	2.28	2.49
MaxProp	0.646	0.743	0.135	0.261	0.52	0.47	1.72	1.90
MaxDelivery	0.203	0.271	0.139	0.160	1.08	0.71	2.11	1.80
TD-Drone Dijkstra	0.954	0.973	0.540	0.664	0.43	0.45	1.69	1.48

In the case of the squares network and battery changing, the values of the delivery rate were within the range of 0.166 to 0.646 for the routing protocols Epidemic, Spray and Wait, PRoPHET, MaxProp, and MaxDelivery. The best delivery rate was 0.954 for the TD-Drone Dijkstra protocol (Figure 11). The worst result in terms of latency was obtained for MaxDelivery, and TD-Drone Dijkstra's average latency was the best, as expected (Figure 12).

The delivery rate in the case of drone battery charging was between 0.135 and 0.143 for the routing protocols Epidemic, Spray and Wait, PRoPHET, MaxProp, and MaxDelivery. The maximum delivery rate was 0.54 and was obtained for the TD-Drone Dijkstra protocol. These low values were obtained because of the large delay due to the battery charge. The worst results for latency were obtained for PRoPHET and Epidemic, and TD-Drone Dijkstra's latency was also the best.

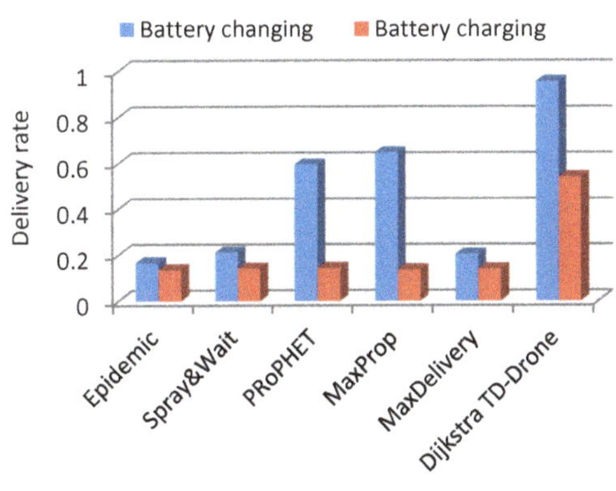

Figure 11. Delivery rate in the squares drone network.

The comparison between average latencies was performed on the routes where the delivery was successful for all of the algorithms (66 routes for the case of battery change and 64 routes in the case of battery recharge).

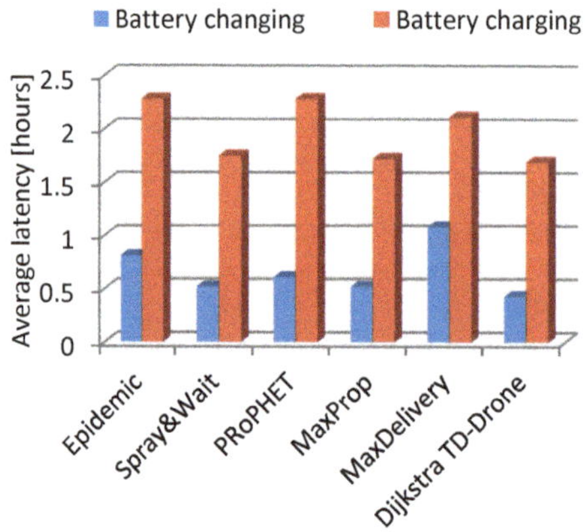

Figure 12. Average latency in the squares drone network.

The best delivery rate in all cases was given by the TD-Drone Dijkstra algorithm because its buffer load was the lowest (the data was loaded only in the stations and drones belonging to the calculated route). The latency in the case of Dijkstra's algorithm was high because it can deliver most of the packages. Generally, the other algorithms can deliver only on the shorter distances due to buffer restrictions. It is known that Dijkstra's algorithm results in the shortest path, and, therefore, the best latency.

The delivery rate was better for the triangles drone network (Figures 13 and 14), with the exception of the Spray and Wait algorithm in the battery changing case. For PRoPHET, Max Prop, and TD-Drone Dijkstra in the battery charging case, the latency was significantly better.

Latency in the case of changing the battery in the triangles drone network was generally better (Figure 15), with the exception of Spray and Wait, and in the case of Max Delivery a significantly better latency was obtained. In the case of battery charging (Figure 16), latency was better in the triangles drone network for three algorithms (Epidemic, Max Delivery, and TD-Drone Dijkstra), and, for the other three algorithms, the latency was better in the squares drone network.

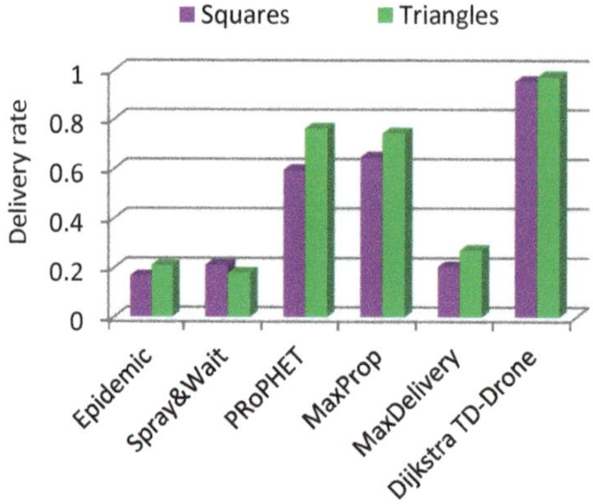

Figure 13. Delivery rate, battery changing, squares vs. triangles.

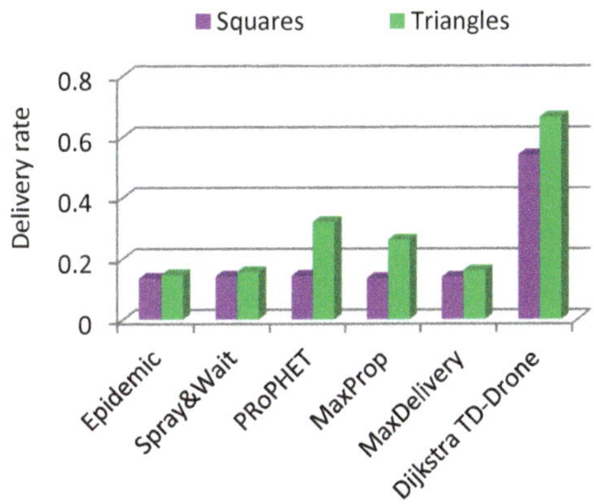

Figure 14. Delivery rate, battery charging, squares vs. triangles.

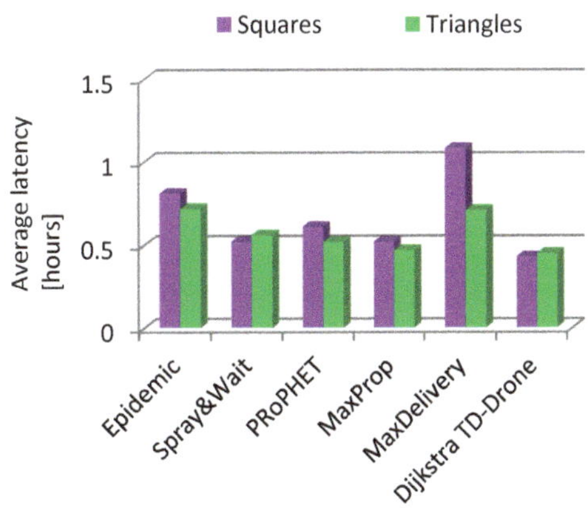

Figure 15. Average latency: battery changing, squares vs. triangles.

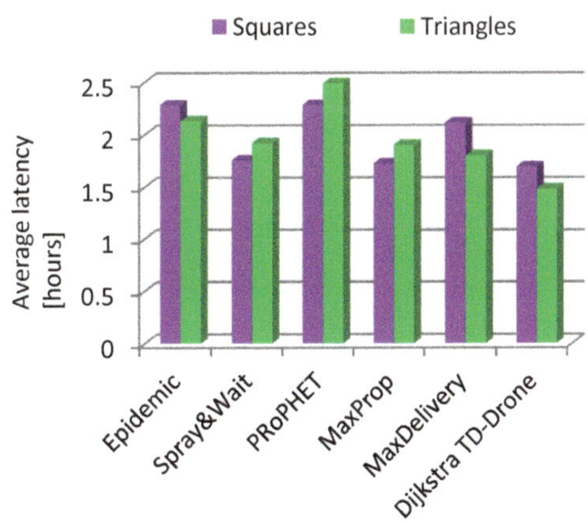

Figure 16. Average latency: battery charging, squares vs. triangles.

In essence, the delivery rate was considerably better for the triangles drone networks, and latency was generally better for the triangles case. The significant disadvantage of the triangles drone network is that double the number of drones is needed to cover approximately the same surface, and double the number of loading/changing stations is required, although the total number of fixed communication stations is similar (63 vs. 65).

Epidemic, Spray and Wait, PRoPHET, and MaxProp are classic algorithms used for DTN. However, in our case, as shown, the time-dependent Dijkstra algorithm adaptation can be successfully used because since the flight timetables are known. There are numerous advantages of the TD-Drone Dijkstra algorithm: an exact and optimum route is a priori calculated, ensuring the fastest time of delivery from departure to destination if the route exists; a message is not unnecessarily sent in the network if no route exists from departure to destination; multiple copies of the messages are not unnecessarily spread through drone and station buffers, resulting in unnecessary overloading of the buffers; and, finally, the rate of delivery success is maximized. The drawback of the TD-Drone Dijkstra algorithm is that the route is calculated using the information about the operating drones and stations at the moment of route calculation and, if a drone or a station from the route is down on this route, the message does not reach the destination. All the routes passing through the station or drone that is down are compromised until the fault is detected. Moreover, this problem reappears when the drone/station is fixed until the moment this information is updated. However, the chance of this problem occurring is low and, if it appears, it may be fixed in time following the repair of the drone/station or when the current status of the network is updated. Using the example of the Spray and Wait or Epidemic algorithms presented in this paper, any message has the chance to reach the destination even if drones or stations are down because a copy of the message is spread in the network.

4. Conclusions and Future Work

This paper presents a novel method of communication in quarantined or isolated areas, or areas with technical damage, using networks of drones that fly based on a well-established mission plan and schedule, on 2D surfaces covered by squares. Two situations of drone battery management—charging and battery changing stations—were investigated.

A network of square cells with two drones in each cell was proposed to cover a geographical area. The drone network was simulated based on input data from experimental flight tests of a quadcopter using six routing algorithms.

A TD-Drone Dijkstra algorithm (single-copy algorithm) and multiple-copies algorithms were proposed to simulate a Delay Tolerant Network of drones. Results showed a delivery rate ranging from 0.166 to 0.954 in the drone network with battery changing, and from 0.135 to 0.540 in the drone network with battery charging. The best latency of 0.43 h for a drone network with battery changing was obtained using the TD-Drone Dijkstra algorithm. Thus, the best results were obtained for the TD-Drone Dijkstra algorithm, which was able to deliver most of the data packages in the shortest time.

The traditional DTN algorithms, such as Epidemic, Spray and Wait, and MaxProp, produced lower results due to the small number of contacts between nodes, and a low number of message exchanges. The fastest communication was established for the drone squares network with battery changing. It was found that the battery change scenario led to an increase in the delivery rate of 76% compared to the battery charge scenario.

It was found that the battery change scenario led to an increase in the delivery rate of ~200% compared to the battery charge scenario.

The fastest communication was found for the drone triangular network with battery charging, and for the drone square network with battery changing. However, the drone square network is considerably cheaper than the drone triangular network. Thus, if cost is not an issue, the triangle network of drones may be implemented because better performances can be achieved. However, if there is a budget constraint, then the square network type is more suitable.

In future work, we will aim to apply this network of drones to parcel delivery during emergencies in remote quarantined zones.

Author Contributions: Conceptualization, A.M.D. and R.U.; methodology, R.U. and A.M.D.; software, A.M.D. and C.-Ş.N.; validation, R.U. and A.M.D.; formal analysis, A.M.D. and R.U.; investigation, R.U. and A.M.D.; resources, R.U., C.-Ş.N. and A.M.D.; data curation, C.-Ş.N. and A.M.D.; writing—original draft preparation, R.U., A.M.D. and C.-Ş.N.; writing—review and editing, R.U. and A.M.D.; visualization, A.M.D. and R.U.; supervision, A.M.D. and R.U.; project administration, R.U. and A.M.D.; funding acquisition, A.M.D., C.-Ş.N. and R.U. All authors have read and agreed to the published version of the manuscript.

Funding: This research was funded by University Transilvania of Brașov.

Institutional Review Board Statement: Not applicable.

Informed Consent Statement: Not applicable.

Data Availability Statement: Not applicable.

Acknowledgments: The authors acknowledge of Transilvania University of Braşov for providing the infrastructure used in this work.

Conflicts of Interest: The authors declare no conflict of interest.

References

1. Alessi, N.; Caini, C.; de Cola, T.; Martin, S.; Mayer, J.P. DTN Performance in Complex Deep-Space Networks. In Proceedings of the ASMS 2018, Berlin, Germany, 10–12 September 2018; pp. 1–7.
2. Cerf, V.; Hooke, A.; Torgerson, L.; Durst, R.; Scott, K.; Fall, K.; Weiss, H. Delay-Tolerant Networking Architecture. Internet RFC 4838. 2007. Available online: www.rfc-editor.org/rfc/rfc4838.txt (accessed on 25 June 2020).
3. Bertolazzi, M.; Caini, C. Mars to Earth Data Downloading: A Directory Synchronization Approach. *Future Internet* **2019**, *11*, 173. [CrossRef]
4. Tornell, S.M.; Calafate, C.T.; Cano, J.C.; Manzoni, P. DTN Protocols for Vehicular Networks: An Application Oriented Overview. *IEEE Commun. Surv. Tutor.* **2015**, *17*, 868–887. [CrossRef]
5. Zguira, Y.; Rivano, H.; Meddeb, A. Internet of Bikes: A DTN Protocol with Data Aggregation for Urban Data Collection. *Sensors* **2018**, *18*, 2819. [CrossRef] [PubMed]
6. Urquiza-Aguiar, L.; Tripp-Barba, C.; Aguilar Igartua, M. A Geographical Heuristic Routing Protocol for VANETs. *Sensors* **2016**, *16*, 1567. [CrossRef]
7. Warthman, F. *Delay-and Disruption-Tolerant Networks (DTNs)*; Version 3.2.; Warthman Associates: Palo Alto, CA, USA, 2015.
8. Wong, K.S.; Wan, T.C. Current State of Multicast Routing Protocols for Disruption Tolerant Networks: Survey and Open Issues. *Electronics* **2019**, *8*, 162. [CrossRef]
9. Tikhonov, E.; Schneps-Schneppe, D.; Namio, D. Delay Tolerant Network Potential in a Railway Network. In Proceedings of the 2020 IEEE 26th Conference of Open Innovations Association (FRUCT), Yaroslavl, Russia, 20–24 April 2020; pp. 438–448.
10. Natkaniec, M. Ad hoc mobile wireless networks: Principles, protocols, and applications. *IEEE Commun. Mag.* **2009**, *47*, 12–14. [CrossRef]
11. Picu, A.; Spyropoulos, T. Forecasting DTN performance under heterogeneous mobility: The case of limited replication. In Proceedings of the 9th Annual IEEE Communications Society Conference on Sensor, Mesh and Ad Hoc Communications and Networks (SECON), Seoul, Korea, 18–21 June 2012; pp. 569–577.
12. Mengjuan, L.; Yan, Y.; Zhiguang, Q. *A Survey of Routing Protocols and Simulations in Delay-Tolerant Networks, Wireless Algorithms, Systems, and Applications, Vol. 6843 of Lecture Notes in Computer Science*; Springer: Berlin/Heidelberg, Germany, 2011; pp. 243–253.
13. Nikam, D.D.; Singh, H.P.; Kushwah, R. Conditional Shortest Path Routing in Delay Tolerant Networks. *Int. J. Adv. Res. Sci. Eng.* **2017**, *6*, 1.
14. Jain, S.; Fall, K.; Patra, R. Routing in a Delay Tolerant Network. In Proceedings of the 2004 conference on Applications, Technologies, Architectures, and Protocols for Computer Communications, Portland, OR, USA, 30 August–3 September 2004; Volume 34, pp. 145–158. [CrossRef]
15. Kawakib, K.A.; Mohd, H.O.; Suhaidi, H. Survey and Comparison of Operating Concept for Routing Protocols in DTN. *J. Comput. Sci.* **2016**, *12*, 141–152.
16. Spyropoulos, T.; Psounis, K.; Raghavendra, C.S. Single-copy routing in intermittently connected mobile networks. In Proceedings of the 1st Annual IEEE Communications Society Conference on Sensor and Ad Hoc Communications and Networks, Santa Clara, CA, USA, 4–7 October 2004; pp. 235–244.
17. Vahdat, A.; Becker, D. *Epidemic Routing for Partially-Connected Ad Hoc Networks*; Technical Report CS-2000–06; Duke University: Duhram, NC, USA, 2000.
18. Raghavendra, C.; Spyropoulos, T.; Psounis, K. Spray and Wait: An Efficient Routing Scheme for Intermittently Connected Mobile Networks. In Proceedings of the ACM SIGCOMM Workshop on Delay-Tolerant Networking (WDTN), Philadelphia, PA, USA, 26 August 2005; pp. 252–259.

19. Lindgren, A.; Doria, A.; Schelen, O. Probabilistic routing in intermittently connected networks. *ACM SIGMOBILE Mob. Comput. Commun.* **2003**, *7*, 19–20. [CrossRef]
20. Burgess, J.; Gallagher, B.; Jensen, D.; Levine, B.N. MaxProp: Routing for Vehicle-Based Disruption-Tolerant Networks. In Proceedings of the IEEE INFOCOM 2006, Barcelona, Spain, 23–29 April 2006.
21. Nanau, C.Ş. Maximum flow in buffer-limited delay tolerant networks. The static approach. *Bull. Transilv. Univ. Bras. Ser. III Math. Inform. Phys.* **2020**, *13*, 363–372. [CrossRef]
22. Schiopu, C.; Ciurea, E. Maximum flows in planar dynamic networks with lower bounds. *Fundam. Inform.* **2018**, *163*, 189–204. [CrossRef]
23. Spaho, E. Energy consumption analysis of different routing protocols in a Delay Tolerant Network. *J. Ambient Intell. Humaniz. Comput.* **2020**, *11*, 3833–3839. [CrossRef]
24. Jigang, W.; Jin, S.; Ji, H.; Srikanthan, T. Algorithm for time-dependent shortest safe path on transportation networks. *Procedia Comput. Sci.* **2011**, *4*, 958–966. [CrossRef]
25. Mao, Y.; Zhou, C.; Ling, Y.; Lloret, J. An Optimized Probabilistic Delay Tolerant Network (DTN) Routing Protocol Based on Scheduling Mechanism for Internet of Things (IoT). *Sensors* **2019**, *19*, 243. [CrossRef]
26. Chițonu, G.C.; Rolfsen, C.N.; Deaconu, O. Accessibility to Transylvania's cultural heritage through BIM-Heritage and community involvement. *IOP Conf. Ser. Mater. Sci. Eng.* **2020**, *789*, 012010. [CrossRef]
27. Hassanalian, M.; Abdelkefi, A. Classifications, applications, and design challenges of drones: A review. *Prog. Aerosp. Sci.* **2017**, *91*, 99–131. [CrossRef]
28. Iranmanesh, S.; Raad, R. A Novel Data Forwarding Strategy for a Drone Delay Tolerant Network with Range Extension. *Electronics* **2019**, *8*, 659. [CrossRef]
29. Kellermann, R.; Biehle, T.; Fischer, L. Drones for parcel and passenger transportation: A literature review. *Transp. Res. Interdiscip. Perspect.* **2020**, *4*, 100088. [CrossRef]
30. Rango, F.; De Potrino, G.; Tropea, M.; Santamaria, A.F.; Fazio, P. Scalable and ligthway bio-inspired coordination protocol for FANET in precision agriculture applications. *Comput. Electr. Eng.* **2019**, *74*, 305–318. [CrossRef]
31. Hu, T.; Wang, Y.; Ma, B.; Zhang, J. Orbit Angular Momentum MIMO with Mode Selection for UAV-Assisted A2G Networks. *Sensors* **2020**, *20*, 2289. [CrossRef] [PubMed]
32. Nazib, R.A.; Moh, S. Routing Protocols for Unmanned Aerial Vehicle-Aided Vehicular Ad Hoc Networks: A Survey. *IEEE Access* **2020**, *8*, 77535–77560. [CrossRef]
33. Lee, T.R.; Buban, M.; Dumas, E.; Baker, C.B. On the Use of Rotary-Wing Aircraft to Sample Near-Surface Thermodynamic Fields: Results from Recent Field Campaigns. *Sensors* **2019**, *19*, 10. [CrossRef] [PubMed]
34. Alaoui-Sosse, S.; Durand, P.; Medina, P.; Pastor, P.; Lothon, M.; Cernov, I. OVLI-TA: An Unmanned Aerial System for Measuring Profiles and Turbulence in the Atmospheric Boundary Layer. *Sensors* **2019**, *19*, 581. [CrossRef]
35. Nolan, P.J.; Pinto, J.; González-Rocha, J.; Jensen, A.; Vezzi, C.N.; Bailey, S.C.C.; De Boer, G.; Diehl, C.; Laurence, R., III; Powers, C.W.; et al. Coordinated Unmanned Aircraft System (UAS) and Ground-Based Weather Measurements to Predict Lagrangian Coherent Structures (LCSs). *Sensors* **2018**, *18*, 4448. [CrossRef]
36. Da Rosa, R.; Aurelio Wehrmeister, M.; Brito, T.; Lima, J.L.; Pereira, A.I.P.N. Honeycomb Map: A Bioinspired Topological Map for Indoor Search and Rescue Unmanned Aerial Vehicles. *Sensors* **2020**, *20*, 907. [CrossRef] [PubMed]
37. Yang, P.; Tang, K.; Lozano, J.A.; Cao, X. Path planning for single unmanned aerial vehicle by separately evolving waypoints. *IEEE Trans. Robot.* **2015**, *31*, 1130–1146. [CrossRef]
38. Zhang, S.; Zhang, H.; Di, B.; Song, L. Cellular controlled cooperative unmanned aerial vehicle networks with sense-and-send protocol. *IEEE Int. Things J.* **2018**, *6*, 1–13.
39. Kima, S.J.; Lima, G.J.; Chob, J. Drone flight scheduling under uncertainty on battery duration and air temperature. *Comput. Ind. Eng.* **2018**, *1171*, 291–302. [CrossRef]
40. Cabreira, T.M.; Brisolara, L.B.; Ferreira, P.R., Jr. Survey on Coverage Path Planning with Unmanned Aerial Vehicles. *Drones* **2019**, *3*, 4. [CrossRef]
41. Udroiu, R.; Deaconu, A.M.; Nanau, C.-Ş. Data Delivery in a Disaster or Quarantined Area Divided into Triangles Using DTN-Based Algorithms for Unmanned Aerial Vehicles. *Sensors* **2021**, *21*, 3572. [CrossRef] [PubMed]
42. Udroiu, R.; Blaj, M. Conceptual design of a VTOL remotely piloted aircraft for emergency missions. *Sci. Res. Educ. Air Force* **2016**, *18*, 207–214. [CrossRef]
43. DJI. DJI Mavic 2 Pro Specification. Available online: www.dji.com (accessed on 15 June 2020).
44. Herath, H.M.C.W.B.; Herath, H.M.S.; Sumangala, S.W.; de Silva, O.; Chathuranga, D.; Lalitharatne, T.D. Design and Development of an Automated Battery Swapping and Charging Station for Multirotor Aerial Vehicles. In Proceedings of the 17th International Conference on Control, Automation and Systems (ICCAS 2017), Jeju, Korea, 18–21 October 2017; pp. 356–361.
45. Lee, D.; Zhou, J.; Lin, W.T. Autonomous battery swapping system for quadcopter. In Proceedings of the International Conference on Unmanned Aircraft Systems (ICUAS), Denver, CO, USA, 9–12 June 2015; pp. 118–124.
46. High Power Drone Charging Pad and Infrastructure. Available online: https://skycharge.de/ (accessed on 26 October 2020).
47. Costea, I.M.; Plesca, V. Automatic battery charging system for electric powered drones. In Proceedings of the 2018 IEEE 24th International Symposium for Design and Technology in Electronic Packaging (SIITME), Iasi, Romania, 25–28 October 2018; pp. 377–381.

48. Mostafa, T.M.; Muharam, A.; Hattori, R. Wireless Battery Charging System for Drones via Capacitive Power Transfer. In Proceedings of the IEEE PELS Workshop on Emerging Technologies: Wireless Power Transfer (WoW), Chongqing, China, 20–22 May 2017. [CrossRef]
49. Boukoberine, M.N.; Zhou, Z.; Benbouzid, M. Power Supply Architectures for Drones-A Review. In Proceedings of the 45th Annual Conference of the IEEE Industrial Electronics Society (IECON 2019), Lisbon, Portugal, 14–17 October 2019; pp. 5826–5831.
50. Lin, C.-F.; Lin, T.-J.; Liao, W.-S.; Lan, H.; Lin, J.-Y.; Chiu, C.-H.; Danner, A. Solar Power Can Substantially Prolong Maximum Achievable Airtime of Quadcopter Drones. *Adv. Sci.* **2020**, *7*, 20. [CrossRef]
51. Achtelik, M.C.; Stumpf, J.; Gurdan, D.; Doth, K.-M. Design of a flexible high performance quadcopter platform breaking the MAV endurance record with laser power beaming. In Proceedings of the IEEE/RSJ International Conference on Intelligent Robots and Systems, San Francisco, CA, USA, 25–30 September 2011. [CrossRef]
52. Taddia, Y.; Corbau, C.; Zambello, E.; Pellegrinelli, A. UAVs for Structure-From-Motion Coastal Monitoring: A Case Study to Assess the Evolution of Embryo Dunes over a Two-Year Time Frame in the Po River Delta, Italy. *Sensors* **2019**, *19*, 1717. [CrossRef]
53. WiFi NodeMCU ESP8266. Available online: www.espressif.com/en/products/socs/esp8266 (accessed on 25 June 2020).
54. Nănău, C.Ş. MaxDelivery: A new approach to a DTN Buffer Management. In Proceedings of the 21ST IEEE International Symposium on a World of Wireless, Mobile and Multimedia Networks (IEEE WOWMOM 2020), Cork, Ireland, 31 August–3 September 2020; pp. 60–61.
55. Lewis, R. Algorithms for Finding Shortest Paths in Networks with Vertex Transfer Penalties. *Algorithms* **2020**, *13*, 269. [CrossRef]
56. Wang, Y.; Yuan, Y.; Ma, Y.; Wang, G. Time-Dependent Graphs: Definitions, Applications, and Algorithms. *Data Sci. Eng.* **2019**, *4*, 352–366. [CrossRef]

sensors

MDPI

Article

Machine Learning for the Dynamic Positioning of UAVs for Extended Connectivity

Francisco Oliveira [1,2] and Miguel Luís [2,3,*] and Susana Sargento [1,2]

1 Department of Electronics, Telecommunications and Informatics (DETI), University of Aveiro, 3810-193 Aveiro, Portugal; fbpo@ua.pt (F.O.); susana@ua.pt (S.S.)
2 Instituto de Telecomunicações, 3810-193 Aveiro, Portugal
3 ISEL-Instituto Superior de Engenharia de Lisboa, Instituto Politécnico de Lisboa, 1959-007 Lisboa, Portugal
* Correspondence: nmal@av.it.pt

Abstract: Unmanned Aerial Vehicle (UAV) networks are an emerging technology, useful not only for the military, but also for public and civil purposes. Their versatility provides advantages in situations where an existing network cannot support all requirements of its users, either because of an exceptionally big number of users, or because of the failure of one or more ground base stations. Networks of UAVs can reinforce these cellular networks where needed, redirecting the traffic to available ground stations. Using machine learning algorithms to predict overloaded traffic areas, we propose a UAV positioning algorithm responsible for determining suitable positions for the UAVs, with the objective of a more balanced redistribution of traffic, to avoid saturated base stations and decrease the number of users without a connection. The tests performed with real data of user connections through base stations show that, in less restrictive network conditions, the algorithm to dynamically place the UAVs performs significantly better than in more restrictive conditions, reducing significantly the number of users without a connection. We also conclude that the accuracy of the prediction is a very important factor, not only in the reduction of users without a connection, but also on the number of UAVs deployed.

Keywords: unmanned aerial vehicle; UAV positioning; machine learning; wireless communications

Citation: Oliveira, F.; Luís, M.; Sargento, S. Machine Learning for the Dynamic Positioning of UAVs for Extended Connectivity. *Sensors* **2021**, *21*, 4618. https://doi.org/10.3390/s21134618

Academic Editors: Margot Deruyck and Omprakash Kaiwartya

Received: 10 May 2021
Accepted: 1 July 2021
Published: 5 July 2021

1. Introduction

Unmanned Aerial Vehicle (UAV) networks are receiving increasing attention mostly due to their potential in a number of innovative services and applications, which are useful not only for military, but also for public and civil purposes. They are useful, mainly due to their versatility and dynamic nature, as their nodes and links can change, along with their positions, according to their needs. For this reason, they are a good candidate to reinforce existing networks when they fail or are overloaded [1].

The beauty of UAV networks lies in the fact that the position of the network nodes, the UAVs, can be controlled, which can be dynamically changed to optimize the network performance, and according to the users' needs and their mobility. To find the best placement for the UAVs, the UAV network should be able to extend the base network with the limited capacity, by either moving the UAVs according to the needs, in real-time, or predicting ahead of time where the users will be and how they will move. The first method has the advantage of allocating exactly what is needed for the current situation, but has the disadvantage of the delays associated with UAVs positioning. The second method does not have any delays, as it can mobilize the UAVs ahead of time and optimize their position according to predicted movements; however, the accuracy of the prediction will influence how well the UAV network can help the existing network. Moreover, the volatile nature of the UAVs network, and the fact that UAVs are energy-restricted nodes, require that the placement needs to be determined with a low-complexity approach and be adapted over time.

Considering the previous requirements, this paper proposes an algorithm to calculate the best positions for UAVs in the network, in order to provide a connection to as many users as possible, satisfying their services and requirements. The algorithm predicts, through machine learning approaches the users' position, and according to their needs, it predicts the position of the UAVs through machine learning, according to the predicted requirements and location.

Several machine learning approaches are tested to determine the users' positions. Random Forest and Gradient Boosting have the best results, while Lasso, Ridge and ElasticNet are tied at the last place. In terms of the time to train and predict, which is a very important metric, Random Forest has a training time almost five times higher and a prediction time 36 times higher, than Gradient Boosting.

The proposed algorithm for UAVs positioning is tested with real data of the users' location and their connection through base stations. The test comprises the influence of several parameters, such as the base station bandwidth and UAVs network range, to understand the optimal conditions for the use of a UAV network to extend an already existing network of ground stations, and in which situations should the algorithm be used. The results show that the algorithm performs better in less restrictive network conditions, and that the accuracy of the prediction is indeed an important factor to improve the users connectivity.

The main contributions of this paper are summarized as in the following:

- Test a set of machine learning approaches for user mobility prediction with real sets of users and their location;
- Proposal of a UAV positioning algorithm able to identify overloaded areas and areas that can receive more user's traffic, considering the users' positioning, users' mobility and each users' needs;
- Build of a bridge of UAVs that is able to transport the traffic through underloaded areas, in a dynamic approach and updated over time;
- Test the approach with the real set of users and with new metrics to understand the usefulness of UAVs to help in the transport of the users' traffic.

The rest of this paper is organized as follows. Section 2 presents the related work. Section 3 describes the problem and our proposed approach. Section 4 explains the methodology for choosing the most suitable machine learning algorithm for the users' prediction. Section 5 describes the proposed algorithm, while the results are presented in Section 6. Finally, Section 7 concludes this paper and presents the future work.

2. Related Work

In UAV networks, the coverage optimization is always a challenge that needs to be addressed. The network coverage optimization problem has been confronted by a number of different research works, resulting in multiple possible solutions. Galkin et al. [2] investigated how different UAV network parameters, such as density and height above ground, as well as environmental parameters, such as the building density and building heights, can influence the coverage probability. Kuhlman et al. [3] proposed an automated, physics-aware planner that made use of an information value map for path planning with a Markov Decision Process-based approach. Lyu et al. [4], in a scenario with no ground base stations, proposed a solution where UAV-mounted mobile base stations provide wireless connectivity to a number of ground terminals. The algorithm places a priority on ground terminals on the boundary, to maximize coverage efficiency. Reina et al. [5] presented a solution based on a multi-sub-population genetic algorithm and compares it to single-population genetic algorithms and other meta-heuristic optimization algorithms, and found that the multi-sub-population genetic algorithm achieves better performance than the other options. Sabino et al. [6] also proposed a genetic algorithm, in this case, a Multi-Objective Evolutionary Algorithm, to optimize the UAV node placement considering minimization of the number of UAVs, assuring the necessary bandwidth and a limited number of UAVs. Khan et al. [7] proposed a system where coverage of an area is given to a UAV that can handle the specific quality of service requirements. Each UAV's capability to provide the required quality of service is evaluated through a reputation-based auction mechanism. They also address the real-time monitoring framework challenge present in this solution by using a permission blockchain architecture considering Support Vector Machine.

Other works also take into account the energy management of the UAVs as an important component to the network coverage optimization. Peng et al. [8] proposed an algorithm based on an Echo State Network to predict trajectories of user equipments, and a Kuhn–Munkres-based algorithm to find the most energy-efficient trajectories for the UAVs. Liu et al. [9] used Deep Reinforcement Learning for both the control of the UAVs and for controlling energy consumption, communications coverage and connectivity. The work in [10] proposed a framework to derive the localization error of terrestrial nodes in urban areas when using UAVs as anchor nodes. Such framework includes height-dependent UAV to ground channel characteristics and a highly detailed UAV energy consumption model. Following a different problem in [11] the authors derive a model to minimize the total UAV energy consumption while satisfying the communication throughput requirement of each ground user. The model includes both propulsion energy and communication-related energy. The work in [12] presents a detailed survey on the applicability of ML techniques for UAV-based communications.

However, all these works consider a static network of UAVs. In a real network, users are moving, and the needs of UAVs also move with the users. This paper proposes an approach to predict this movement, and move the UAVs in advance to their best locations.

3. Problem Statement

In a network, traffic is not uniformly distributed across all base stations, with people naturally migrating more to some areas and less to others. This will cause some base stations to have more users than they can handle, while other base stations are largely unused, causing connection problems to some users, even though the network as a whole has more than enough resources to accommodate all users.

By identifying where the network will be overloaded, we will redistribute the traffic of overloaded areas to places that have a lesser strain on the network, in an attempt to make the distribution of users throughout the city as uniform as needed. This is the purpose of the UAVs in the network: to divert the traffic to underloaded areas, so that all users can be connected with their services and requirements. The UAVs will then be placed in the best areas to redistribute the traffic in the network, as illustrated in Figure 1.

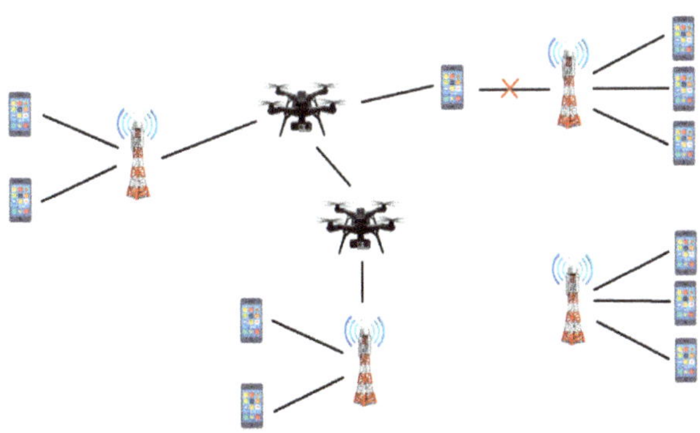

Figure 1. UAV assisted communication network.

To identify the overloaded areas in the network preemptively, we will use a machine learning-based prediction algorithm. Several machine learning techniques are tested, and the one with the best performance will be chosen. To be able to train it, we need a suitable dataset that has records of a variety of users with information about the position, connection type and quality, with fairly frequent updates, like once every 15 min. Furthermore, the data should also have detailed information about the positioning of the ground base stations, which service they provide, their capacity and the area they can provide connection to. Finally, this data should contemplate a location like a big city, with 50 to 100 km^2.

4. Position Prediction Approach

The real data used in the user position prediction results from a large-scale research initiative called Mobile Data Challenge (MDC) [13,14]. The dataset was collected with the Lausanne Data Collection Campaign (LDCC), which was responsible for providing close to 200 volunteers with data collection software on their smartphones. The data collection started in October 2009 and finished in March 2011.

Using this dataset, we extracted the information necessary for training, which consisted of the position of the users over time. We focused all the data on a circular area in the city of Lausanne with a 5 km radius. This area was then divided into rectangles of approximately 200 m horizontally and 140 m vertically.We also divided the data according to its date: month, weekday and hour.

The end result is a four-dimensional matrix in which the first dimension comprises the coordinates that represent the rectangular section, the second dimension is the month, the third is the weekday and the fourth is the hour, each cell containing a number that represents the quantity of users. Figure 2 shows an example of the matrix for an area in May, where each line represents a different weekday, by the numbers 0 to 6, and each column represents a different hour, from 0 h to 23 h.

	0	1	2	3	4	5	6	7	8	9	10	11	12	13	14	15	16	17	18	19	20	21	22	23
0	0	0	0	0	0	0	2	0	0	28	51	16	14	64	31	70	83	8	0	0	24	0	0	0
1	0	0	0	0	0	0	0	4	0	0	1	0	0	0	2	27	43	44	32	0	0	0	0	0
2	0	0	0	0	0	0	0	1	1	15	0	2	0	0	0	3	0	29	26	1	2	0	0	0
3	0	0	0	1	0	0	2	0	0	5	12	15	1	13	2	2	7	17	17	0	1	0	0	0
4	0	0	0	0	0	0	1	0	0	0	0	0	0	0	0	0	73	22	9	0	2	0	0	0
5	0	0	0	0	0	0	0	0	6	0	0	16	10	9	0	6	87	5	0	0	0	0	0	0
6	0	0	0	0	0	0	0	0	0	13	7	0	29	8	51	99	0	0	0	8	0	0	5	0

Figure 2. Example of the number of users in the matrix format.

4.1. Feature Selection

To perform feature selection, we use a method called Boruta [15], which compares the features to a randomized version of themselves, called shadow feature, only deeming them as useful if the features can perform better than their shadows. Performing better than the shadow becomes a very definite threshold for this method, that does not need any human input. Besides that, the algorithm also runs several times, to understand how each feature performs favorably and how many times it does not. Since the result for each run is a "useful" or "not useful" criterion, the results follow a binomial distribution and the features are positioned on it. Boruta then divides the binomial distribution into three different areas which are the features to drop, features to keep tentatively and features to keep. The features to drop and features to keep areas are defined by being the tails of the distribution (the extreme portions of the distribution), where each tail represents 0.5% of the distribution. Features in the features to drop area should be dropped as they have little value as predictors. Features to keep should be kept, as they are very useful for the prediction. The features to keep tentatively are the features that the Boruta method could not guarantee that they are useful, and should be used according to the discretion of the person building the model. In the case of this dissertation, those features will be kept, since the number of features is not very high.

The Boruta algorithm is executed using a random forest regressor as the estimator, used to evaluate if a shadow feature is better or not than the feature, for 20 iterations. The results can be seen in Table 1.

Table 1. Boruta results.

Feature	Decision
Month	Drop
Weekday	Keep tentatively
Hour of the day	Keep
Latitude	Keep
Longitude	Keep
Last known value	Keep

Features in the keep tentatively category are not guaranteed to be useful, and they are left for human responsibility. In this case, the feature "Weekday" was kept.

4.2. Machine Learning Algorithms and Hyper-Parameters

The machine learning methods used in the users' mobility prediction are shown in Table 2. They are separated into three different categories, which are Linear, Ensemble and Deep Neural Network. Ensemble learners use simpler machine learning algorithms, like the decision trees in this case, and improve them either by averaging the values of several algorithms, or by continuously adjusting bias values on the course of several iterations of the same algorithm.

Table 2. Machine learning in the users' mobility prediction.

Method	Type
Lasso	Linear
Ridge	Linear
Elastic Net	Linear
Random Forest	Ensemble
Gradient Boosting	Ensemble
Neural Network	Deep Neural Network

For all the machine learning algorithms except neural networks, the sklearn python library is used. This library provides implementations of the algorithms with default values on most algorithms that were not changed, unless it is specifically mentioned in the remaining manuscript.

4.2.1. Lasso, Ridge and Elastic Net

The sklearn python library offers implementations of these three algorithms with built-in cross-validation. This means that the algorithm uses the cross-validation capabilities to select the best hyper-parameters, in order to build the best model according to the available data. therefore, the choice of the hyper-parameters was left entirely to the responsibility of the algorithm.

4.2.2. Random Forest Regressor

The random forest uses a number of decision trees with various sub-samples of the dataset; it then uses the average of them all to improve the predictive accuracy and to control over-fitting. To decide on the value for the maximum depth and the number of estimators, an empirical research was done. The algorithm was trained using various values of both max depth and number of estimators which can be seen in Figures 3 and 4, respectively.

For the max depth, the optimal value seems to be 21, even though the differences in the performance start to be very small after 15. It is worthwhile to note that a higher depth also implies a bigger training time; therefore, if a small training time is needed, a smaller value of depth, like 15, might also be acceptable, as the difference in performance does not make a significant impact.

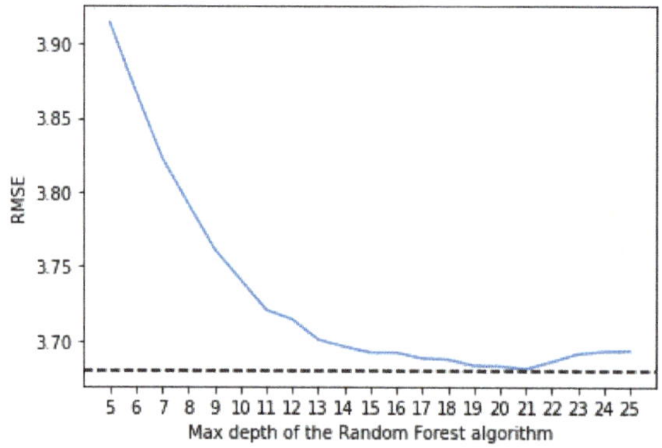

Figure 3. Comparing Root-Mean-Square Error (RMSE) values with different values of max depth for the Random Forest algorithm.

Considering the number of estimators, the results are more irregular, but the number which offers the best performance is 650, at the cost of a significantly bigger training time when compared to, for example, using 150 estimators (4 h versus 52 min).

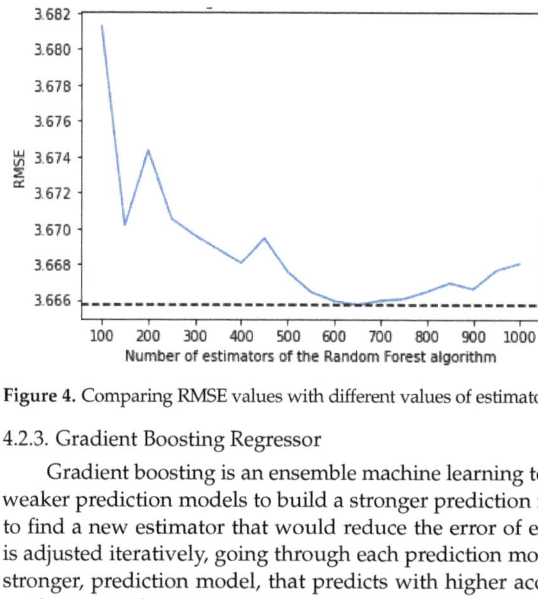

Figure 4. Comparing RMSE values with different values of estimators for the Random Forest algorithm.

4.2.3. Gradient Boosting Regressor

Gradient boosting is an ensemble machine learning technique that uses a number of weaker prediction models to build a stronger prediction model. For each model, it tries to find a new estimator that would reduce the error of each prediction. This estimator is adjusted iteratively, going through each prediction model. The final result is a single, stronger, prediction model, that predicts with higher accuracy than any of the weaker prediction models that it uses. Like the Random Forest algorithm, some parameters will also be tested empirically, to research their impact on the performance of the model.

Figure 5 shows how the three available loss algorithms perform, where "ls" stands for least squares regression, "lad" stands for least absolute deviation and "huber" is a combination of the two. While the "huber" function performs notably worse, "lad" and "ls" have more similar performances, but "ls" achieve better performance. Figure 6 shows that the best value for the max depth is 8, while Figure 7 shows that 150 is the number of estimators that have the best performance.

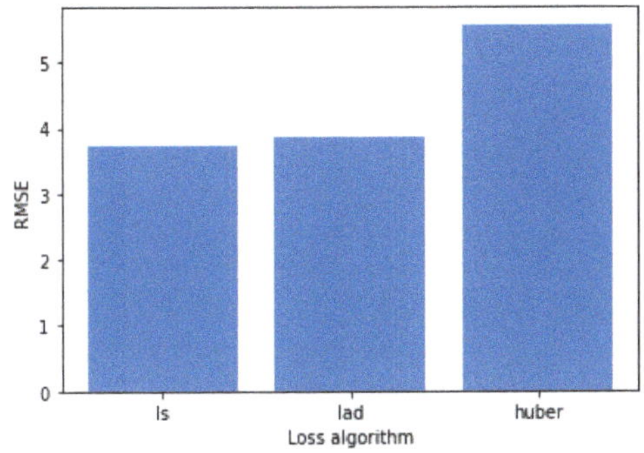

Figure 5. Comparing RMSE values with different loss algorithms for the Gradient Boosting algorithm.

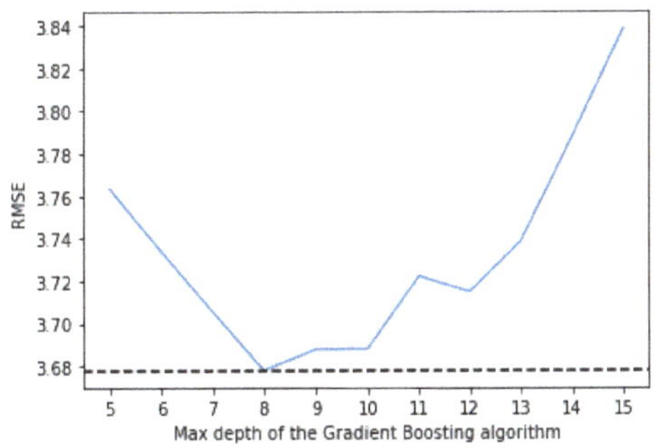

Figure 6. Comparing RMSE values with different values of max depth for the Gradient Boosting algorithm.

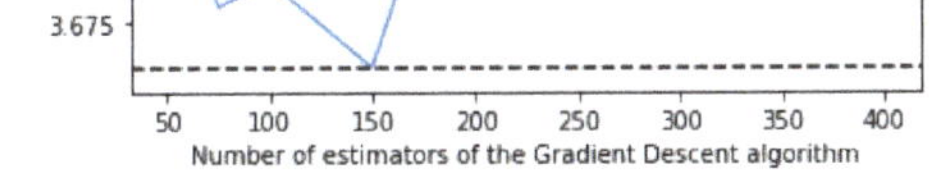

Figure 7. Comparing RMSE values with different values of estimators for the Gradient Boosting algorithm.

4.2.4. Neural Networks

The neural network uses nodes, which is where the computation happens, divided into layers. Besides the input layer and the output layer, a neural network may have several layers, each one with several nodes. To decide both the number of layers and the number of nodes in each layer, several different models were tried and evaluated.

In Figure 8 we can observe the impact that the number of layers has on the RMSE values. While it does not have a large variation, using four layers seems to consistently offer the best results.

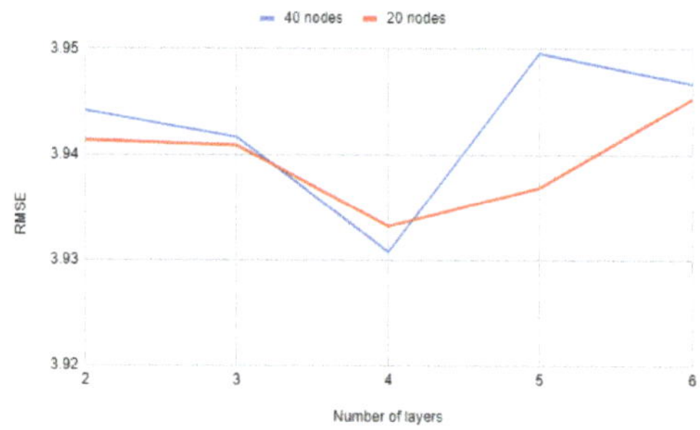

Figure 8. Comparing RMSE values with different number of layers and nodes per layer for the neural network.

In Figure 9 we use four layers for each model, but varying the number of nodes from 10 to 80. Here in this figure, the variation is also very small, but using 40 nodes gives the best RMSE score.

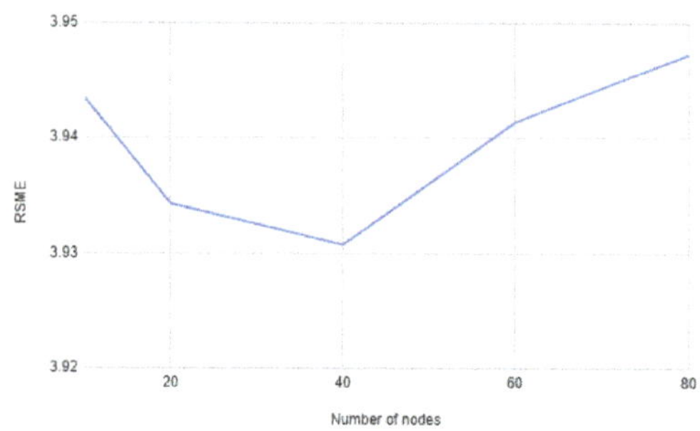

Figure 9. Comparing RMSE values with different number of nodes per layer for the neural network.

Another aspect that impacts the performance of a neural network is the batch size. Figure 10 shows the variation in performance according to the batch size. The variation is highly irregular, not converging to any value, as shown in Figure 10. Since it is not practical to test every possible value for the batch size, the best option is to try to find a local minimum that offers good performance when compared to the other local minima. Therefore, a batch size of 2500 offers the best performance value.

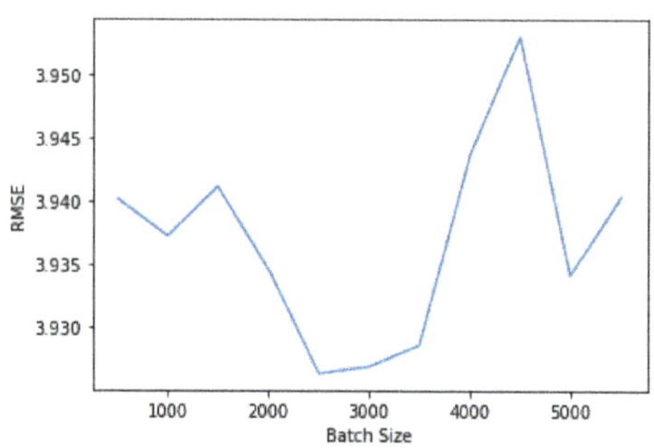

Figure 10. Comparing RMSE values with different number of batch size for the neural network.

With these results, we can come to the conclusion that the best neural network model for this data is one with four layers, with 40 nodes each, and with a batch size of 2500. This model is the one used in the performance tests in the following section.

4.3. Performance of the Machine Learning Algorithms

To compare the performance of the machine learning algorithms, we use metrics such as Mean Absolute Error (MAE), Root-Mean-Square Error (RMSE), R-squared (R^2) as well as training and prediction times. We used 80% of the dataset for training, and the remaining 20% for testing, i.e., 13 months for training and 3 months for validation, with the data being shuffled previously.

In Figures 11–13 all the algorithms are compared according to their performance with the RMSE, R^2 and MAE metrics, respectively. Random forest and Gradient Boosting have the best results, while Lasso, Ridge and ElasticNet are tied at the last place. All algorithms have RMSE values significantly higher than the MAE values, indicating a high variation on the error values with a significant portion of them being higher than the average. Random forest and Gradient Boosting have the best results, while Lasso, Ridge and ElasticNet are tied at the last place. All algorithms have RMSE values significantly higher than the MAE values, indicating a high variation on the error values with a significant portion of them being higher than the average.

Another important metric is the time required for training and predicting. Table 3 shows these values when training with approximately 5 million samples of data. Random Forest, that performs slightly above Gradient Boosting, has a training time almost 5 times higher and a prediction time 36 times higher. If prediction speed is a priority, then Gradient Boosting might be the best option, rather than Random Forest.

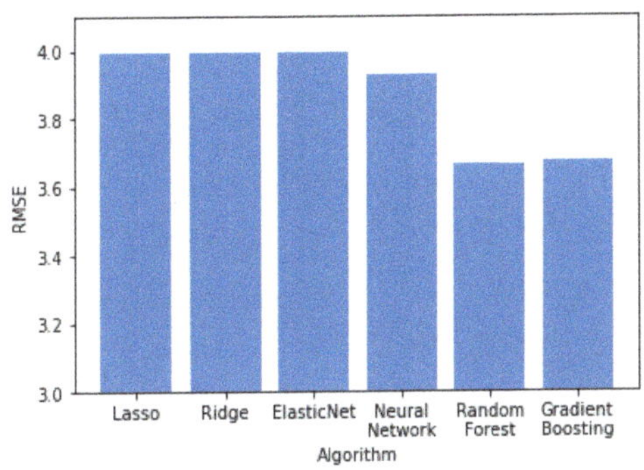

Figure 11. Performance of the algorithms according to their RMSE value.

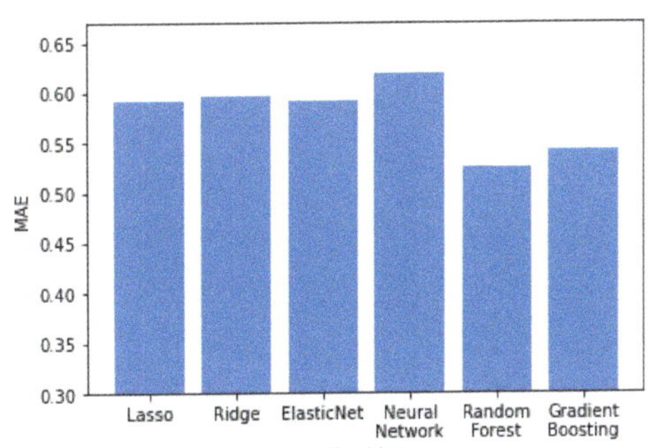

Figure 12. Performance of the algorithms according to their MAE value.

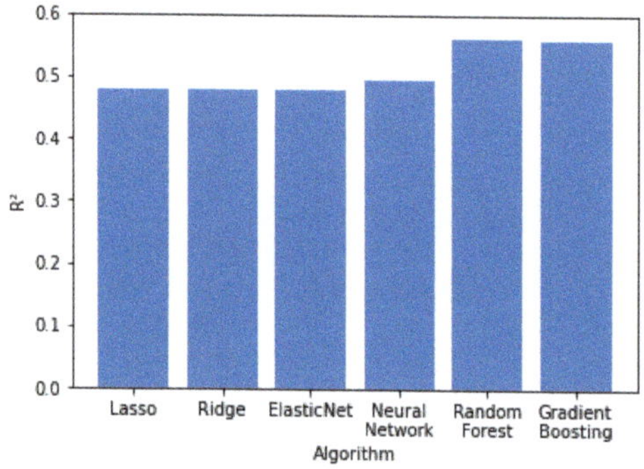

Figure 13. Performance of the algorithms according to their R^2 value.

Table 3. Training and prediction times for all tested algorithms.

Algorithm	Training Time (s)	Prediction Time (s)
Random Forest	14,242	179
Gradient Boosting	2842	5
Neural Network	323	23
Lasso	28	0.02
Ridge	4	0.02
ElasticNet	27	0.02

4.4. Prediction Results

Through the prediction approach, the Random Forest predictor was used to generate a user distribution map. Figure 14 shows the real values for the 15–16 h interval, while Figure 15 shows the predicted values for the same interval. While the prediction shows user activity in the same regions than the real values do, it still misses some of the regions with low user values. For example, in Figure 14 there is a clear trace of users along a road in the top left region, which is not present in the predicted values. This is due to the low amount of users and data.

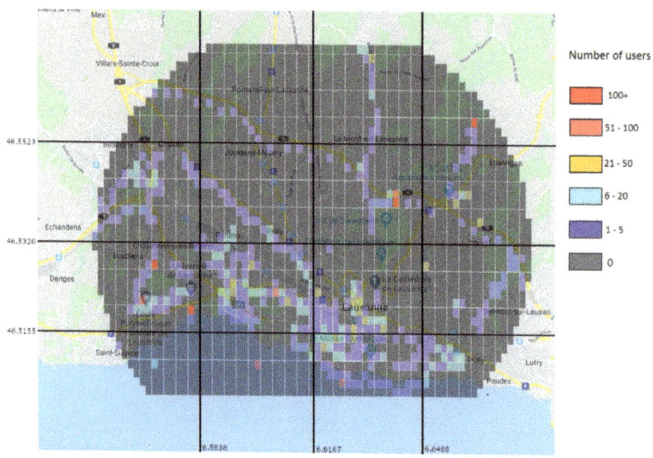

Figure 14. Real values of the number of users on an April Monday at 15–16 h.

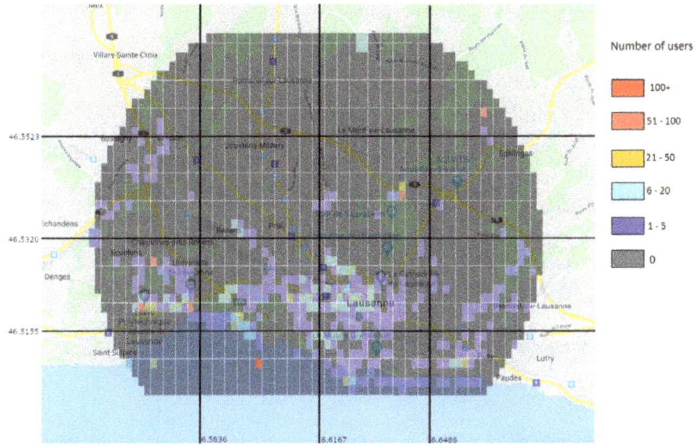

Figure 15. Prediction of the number of users on an April Monday at 15–16 h.

5. UAV Positioning for Network Traffic Redistribution

5.1. UAVs as Traffic Redistributors

Aerial UAVs are highly mobile and dynamic, which make them excellent to be used as traffic redistributors: they can follow the users as they move throughout the day. In order for them to redistribute traffic, they need to build a bridge of UAVs from the overloaded area to the target area. UAVs will not serve as base stations, but only as extensions of the already existing base stations. With this purpose, there will be UAVs with either one of these two functions: connecting to the users or providing interconnection between the target base stations and the UAVs responsible for connecting to the users. UAVs that connect to the users will be in the areas that are overloaded, moving with the users' movement prediction, and will be the ones responsible for providing connection to those users. The UAVs that provide interconnection will transport the traffic towards the target base station in a less populated area. The number of UAVs will depend on the distance between the overloaded area and the target area, the range of the UAVs, the users' needs and the links capacity according to the distance.

Each UAV is limited in its capabilities. The number of users each one can serve is limited by its bandwidth, transmission speed and services required by the users. While the bandwidth and transmission speed are fixed for each UAV, even though transmission speed might vary according to the number of interconnecting UAVs, the services required by the users will most likely differ from user to user. This variation will be taken into account when calculating the capacity of each UAV, and also the number of UAVs needed to satisfy the traffic needs of each area.

5.2. UAV Positioning Algorithm

The UAV positioning algorithm is responsible for identifying overloaded areas and areas that can receive more user's traffic: through the mobility prediction of the users and the location of the base stations, decide to which areas traffic will be redirected and, therefore, decide the best placement of the UAVs.

The algorithm proposed in this paper has three main steps. After predicting the number of users in each area:

1. It identifies overloaded areas and iterates through all overloaded areas;
2. It finds the nearest areas to offload traffic; and
3. It builds a bridge of UAVs that transports the traffic to those areas.

5.2.1. Identifying Overloaded Areas

To identify overloaded areas, the average user bandwidth requirements as well as the available bandwidth per base station are required to be known. With these two values, we can determine the maximum number of users per base station and use it to determine, for each base station, the number of users they can provide a connection to. Consequently, we can also calculate and predict, for each area, the number of users without access to enough bandwidth, if any. Algorithm 1 is proposed to run the previous steps.

Algorithm 1: Number of predicted users without a connection.

```
Output: Number of users in excess in each area (extra_users_area) and the number of users
        connected to each base station (bs_users)
for area in All_Areas do
    users = get_users_in_area(area) ;     // Using prediction algorithm to get the
     number of users
    while users > 0 and not_all_base_stations_are_full(area) do
        for base_station in base_stations_in_this_area do
            if users == 0 then
                exit loop;
            end
            if base_station_has_bandwidth(base_station) then
                users -= 1;
                bs_users[base_station] = 1;
            end
        end
    end
    extra_users_area[area] = users;
end
```

The algorithm assigns a base station to each user that is predicted to be inside it, until all base stations are saturated or until all users are distributed. Overloaded areas can now be predicted and used to distribute the UAVs.

5.2.2. Finding the Nearest Areas to Offload Traffic

After identifying an area with predicted users in excess, we need to offload their traffic into other areas that can handle the traffic increase. For this purpose, we propose Algorithm 2 that finds a number of available areas closer to the overloaded area that requires the traffic offload. This number can be changed to define how many areas the algorithm will return to forward the traffic to, to reduce the probability of overloading any areas with this redistribution.

Algorithm 2: Areas available to receive more traffic.

```
Input  : Overloaded areas and amount of extra users per area
Output: Areas available to receive offloaded traffic (list_of_available_areas)
for area in All_Areas do
    distance = get_distance_between(overloaded_area, area);
    available_user_spots = get_available_spots(area);
    if available_user_spots > users_spots_available_required then
        list_of_available_areas.add( (distance, area) );
    end
end
sort_by_distance(list_of_available_areas);
return get_first_n_elements_of_list(n, list_of_available_areas);
```

The loop iterates through all areas, from now on denominated as target areas, calculates the distance from the current area to the target area, and checks if it can receive the necessary traffic. If it can, then we add a new tuple to the list created, containing the distance calculated before and the identifier of the target area. When the loop finishes, we sort the list by the distance in order to select only the n closest target areas, and return them in a list format. Since each area may have multiple base stations covering them and two different areas may be covered by the same base station, it is possible that the areas returned will not be able to support all the traffic required, which will require preventive measures when using the output of Algorithm 2.

With this list, we can then proceed to redistribute the users through the areas, with the help of the aerial UAVs.

5.2.3. UAVs Distribution

To redistribute traffic, UAVs are used as a path to other areas. The number of UAVs used should be minimized when possible, which requires the use of UAVs that were already placed before to serve as or complement paths to new areas. To avoid overloading any UAVs, each UAV has a counter on how many users it supports (considering the users' requirements), so if it is needed, a reinforcement UAV is deployed. For this purpose, for every UAV that has been placed, there is a list of all other UAVs with a connection, not only directly, but also indirectly through other UAVs. Therefore, whenever Algorithm 2 returns a new area aiming to redirect the traffic, first it always checks if an available UAV path to the area already exists. If that is not the case, then we search for the shortest available path to find a UAV that has a connection closest to the target area. When this UAV is found, the amount of UAVs needed to reach the target area is then calculated according to the distance to the target area and the range of each UAV. UAVs are then placed in a straight line in intervals of length equivalent to their range. Finally, a list is created, which registers the position of every UAV, and the UAVs connected to.

5.2.4. Uniting All Parts

With all key parts defined before, the entire system will now be integrated. In this integration, it is important to assure that the number of users connected to each base station will not surpass the ones that it can receive. The procedure to perform the integration and place the required UAVs to cover all predicted users in the network is described in Algorithm 3.

The algorithm contains a loop through all areas to verify if there are any users left without a connection, and if there are attempts to distribute them through other available areas. The list with the target areas is found using Algorithm 2, and then the algorithm iterates through all target areas, to check if a UAV route is already available, i.e., a path made of UAVs previously built. If this is not the case, the algorithm finds the UAV in the network closest to the target area (line 9) and, from it, it builds a path of UAVs to the target area. After a UAV path is established, the algorithm begins to distribute the user's traffic through the base stations in the target area (line 15 to 26), and ends when all user's traffic has been redirected to that target area, or if all base stations are already at their maximum capacity (considering each user's traffic).

After the loop through all target areas, the distribution may not be able to provide connection to all predicted users in the current area, caused by two or more target areas sharing base stations, therefore providing less bandwidth availability than initially assumed. When this occurs, the algorithm performs the process again for the current area, finding other target areas and distributing the traffic through them, until there are no users left without a connection or all base stations are at their maximum capacity. The overall workflow of the proposed system, which encompasses all the algorithms presented before, is illustrated in Figure 16.

Algorithm 3: Location of the UAVs for traffic redistribution.

```
Input  :Number of users in excess in each area (extra_users_area) and predicted number
        of users per base station (bs_users)
Output:UAVs distributed across the areas
for area in All_Areas do
    while extra_users_area > 0 do
        available_areas = get_available_areas(n) ;              // using Algorithm 2
        add_UAV(area);
        for available_area in available_areas do
            users_to_distribute = get_amount_of_users_to_distribute(n, extra_users_area);
            if UAV_route_not_available(area, available_area) then
                closest_UAV, number_of_UAVs = get_closest_UAV(area, available_area);
                for i in range(1, number_of_UAVs) do
                    UAV_position = calculate_UAV_position(closest_UAV, available_area,
                     i);
                    add_UAV(UAV_position);
                end
            end
            while users_to_distribute > 0 and not_all_base_stations_are_full(available_area) do
                for base_station in base_stations_in_this_area do
                    if users_to_distribute == 0 then
                        exit loop;
                    end
                    if bs_has_available_bandwidth(bs) then
                        extra_users[area] -= 1;
                        users_to_distribute -= 1;
                        bs_users[base_station] += 1;
                    end
                end
            end
        end
    end
end
```

Finally, it is important to mention all the signaling required to make such a system to work. First, and to have an up-to-date ML model, the system needs to be constantly collecting the number of users per cell. This information can be done using tracking systems or the user's information that can be gathered by the cellular stations. Such information must be gathered in a central entity that will be responsible for the use of the selected ML model. For every prediction stage, the updated positions must be transmitted to the UAVs in order to cope with overcrowded cells, which may include the deployment of new UAVs, the removal of unnecessary UAVs or simply the re-positioning of the UAVs in use. Resuming, in a real-time environment we have a system where the user's positioning is gathered at a central entity (just like the system that gathered the MDC dataset), which is then used to periodically train the model. In that case, the UAV placement algorithm will be determining the best positions of the UAVs with historic and current users' location and mobility, and will update their positions upon new placement results.

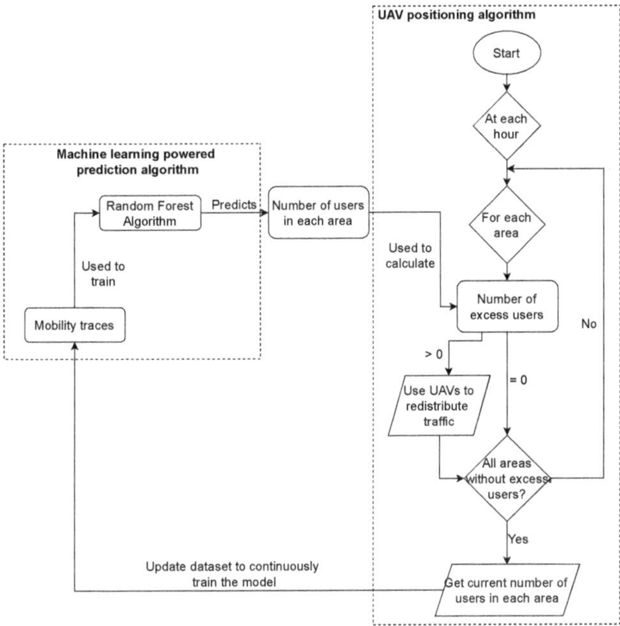

Figure 16. Overall workflow of UAV positioning system.

6. Performance Assessment

This section evaluates the performance of the algorithms by redistributing traffic through those UAVs, and checking how many users are left without a connection. We will now consider the real position of the users to check if the UAVs are placed in the best positions, that is, if the users' mobility and location prediction were performed with good performance. The redistribution takes into account the positions of the base stations and UAVs, and tries to redistribute the traffic according to the UAVs distribution as a result of the UAV positioning algorithm.

The first step is to distribute the users through the base stations. After this process, the number of users in excess and users connected to each base station are known, and therefore, we should distribute the users in excess to other available base stations using the UAVs placed before.

To evaluate the performance of the proposed approach, a new metric is created, henceforth called *score*. To calculate this score, we will use the number of UAVs as well as the number of users left without a connection. This process will be explained further in the following section.

6.1. Score Formula

The proposed score formula takes into consideration the number of UAVs used and how many users were left without a connection. To calculate the final score, we start with 0 points, which is the best possible score. Then, we will add the number of users in excess and UAVs. However, one user in excess does not necessarily mean one more point to the score, because we can give different weights to the extra users or the UAVs depending on what we want to prioritize. For example, if the priority is to use the least amount of UAVs, then each UAV may add two or three points to the total score while each extra user only adds half or one point. On the other hand, if the priority is that no user is left without a connection, then each extra user may add two or three points while each UAV only adds half or one.

The results will be shown with different weights for the extra users (w_u) and number of UAVs (w_d), in order to reach the best conclusion regardless of the priority. The different score formulas are the following:

- Balanced—both inputs have the same weight, serving as a good baseline when there is not any special focus on either reducing the number of users without a connection or the number of UAVs to be placed $(w_u = w_d)$;
- User focused—a bigger weight is given to the number of users without a connection, while a smaller weight is given to the number of UAVs $(w_u = 4\, w_d)$;
- UAV focused—the bigger weight is used for the number of UAVs $(w_u = 0.25\, w_d)$.

Thus, the score is given by

$$Score = \frac{(w_d \cdot n_d + w_u \cdot n_u)}{\Delta_w}, \tag{1}$$

where n_d represents the number of UAVs used, n_u the amount of users with coverage by the extra users, and Δ_w a normalization factor.

To reflect about the usefulness of the UAV positioning algorithm, the score is also calculated before applying the algorithm, where the number of UAVs will inevitably be 0, so only the number of extra users will be used for the calculation. We then calculate the difference between the beginning and the end of the redistribution, and measure the percentage by which the score changed. When a bigger weight has been given to the number of UAVs, the score may actually be higher after applying the algorithm, and the percentage will be negative.

6.2. Results

To better understand the strengths and weaknesses of the proposed approach, tests were performed changing the base station available bandwidth, the user bandwidth and the ranges of the UAVs. For each set of parameters, the score was calculated through the three strategies presented before.

The placement algorithm was executed five different times, corresponding to 5 h, from 14h to 19h, and the score results were averaged. The data used in the performance assessment was taken from the dataset used to train the prediction algorithm. The prediction is executed hourly, using the Random Forest method, as well as the UAV positioning algorithm, calculating the score for every hour, and averaging it in the end. It is assumed that the users require a fixed amount of bandwidth for the duration of the hour, but this value may change in the subsequent hour.

The bandwidth required for each user is assumed to be randomly selected with a given probability. For example, for the standard example of an average user bandwidth of 50, the possible values are 25, 50 and 75, all with a probability of 33.3%. The bandwidth has no explicit unit, but they all use the same unit and are therefore comparable to one another.

The results will show the score obtained or the percentage decrease when comparing the network before and after applying the algorithm. Each mark corresponds to the mean of 5 runs and its 95% confidence interval (the confidence intervals are small, and may not be visible in some of the points).

6.2.1. Base Station Bandwidth

This sub-section evaluates the impact of the base station bandwidth. For this set of results, the average user bandwidth is 50 and all UAVs are considered to have a range of 240 m.

The bandwidth available in each base station is a very strong factor in how high the score is, with the biggest score values coming from lower values of the base station bandwidth, as observed in Figure 17. This is due to a higher number of UAVs placed, since more user's traffic will have to be redirected; it is also due to a lower tolerance for mistakes in the prediction, since there will be a lower number of base stations that can receive traffic from other areas.

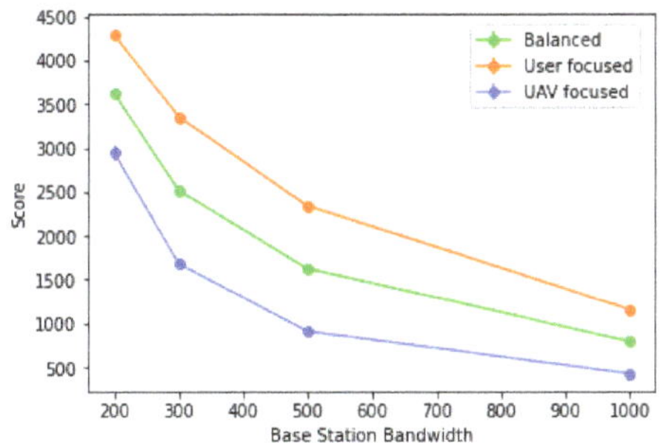

Figure 17. Score of the algorithm with different values of base station bandwidth.

The algorithm is also more efficient for larger values of base station bandwidth, with higher values also having a stronger percentage decrease in the score, as observed in Figure 18. However, for all types of score, the trend is the same: the actual values are very different depending on the score type. In the "user focused" type, the score always decreases, never going into negative values, where the same cannot be said for the "balanced" type, where the percentage goes negative towards the lowest values of the base station bandwidth. Finally, for the "UAV focused" type, the values tend to be worse, going into negative values when the base station bandwidth values reach 500 or less. The efficiency of the algorithm using this score is also lower overall, when compared to the other two. This suggests a sharp increase in the number of UAVs placed when the base station bandwidth decreases, caused by more users needing traffic redistribution and also more routes created, since each base station will also accept less traffic.

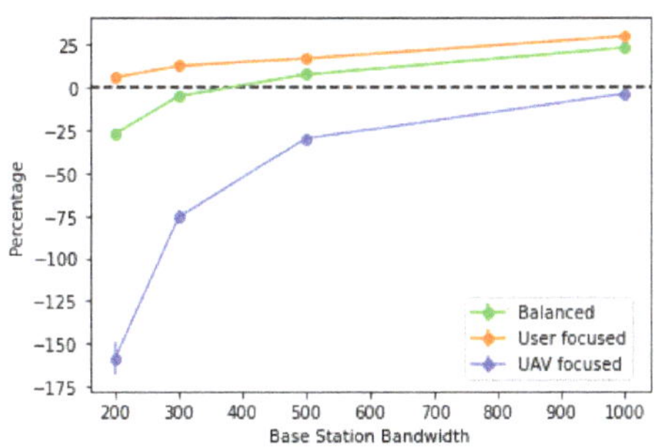

Figure 18. Percentage decrease of the score with different values of base station bandwidth.

6.2.2. Average User Bandwidth

On the user bandwidth evaluation, the bandwidth of the base station was set to 1000, and the average user bandwidth varied between 33 (each user with a value of 25 with 67 % probability, and 50 with 33% probability), 50 (with 25, 50 and 75 all with equal probability), and 75 (each user consuming either 50 with 33% probability or 75 with 67% probability). All UAVs are considered to have a range of 240 m.

The variation of the average user bandwidth has the opposite impact when compared to the base station bandwidth, as it just increases or decreases the number of users that fit into each base station. Naturally, as the average user bandwidth decreases, so does the score, since more users will be able to fit in the same number of base stations, as shown in Figure 19. With an increase in the average user bandwidth, less users will fit into each base station, increasing the number of users without a connection, and consequently, the number of UAVs placed. With higher values of average user bandwidth, less base stations will be available and less users will be hosted, therefore decreasing the tolerance for prediction errors.

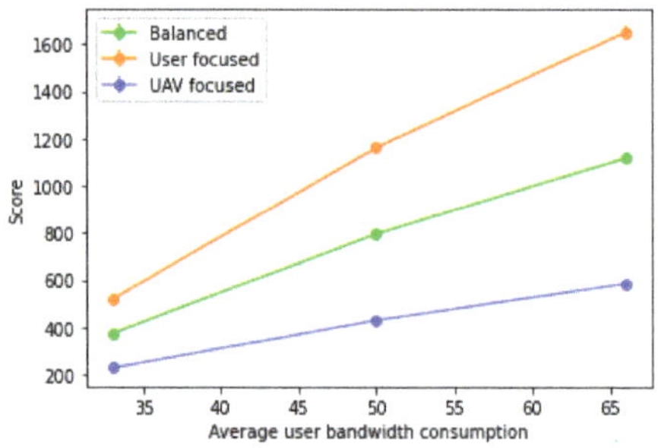

Figure 19. Score of the algorithm with different values of average user bandwidth.

Again, when comparing the decrease percentage (Figure 20), the results are very similar to what happens with the base station bandwidth variations. In this case, as the average user bandwidth consumption increases, the decrease percentage of the score gets lower. Moreover, it is noticeable that, with the "UAV focused" score, the percentage decrease is substantially lower than the other two types, which are very similar overall. This suggests that the main factor for the lower performance is the high number of UAVs placed, since with more users without a connection, and less availability on base stations to receive traffic, more UAV paths have to be built to redistribute the traffic.

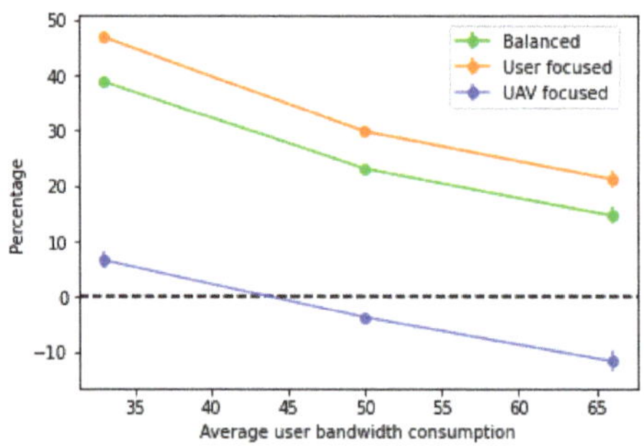

Figure 20. Percentage decrease of the score with different values of average user bandwidth.

6.2.3. UAV Coverage Range

This test evaluates the influence of the UAV coverage range, where the range varied between 100, 240 and 500 m. The base station bandwidth was set to 1000 and the average user bandwidth is 50 with 25, 50 and 75 as possible values, all with equal probability.

The UAV coverage range will determine two key aspects: the number of UAVs needed to reach a remote base station, which will influence the number of UAVs needed in total; and also how tolerant to prediction errors the positioning of the UAVs is, which will influence the number of users left without a connection. With this in mind, we can expect that scores will decrease as the ranges increase, as shown in Figure 21.

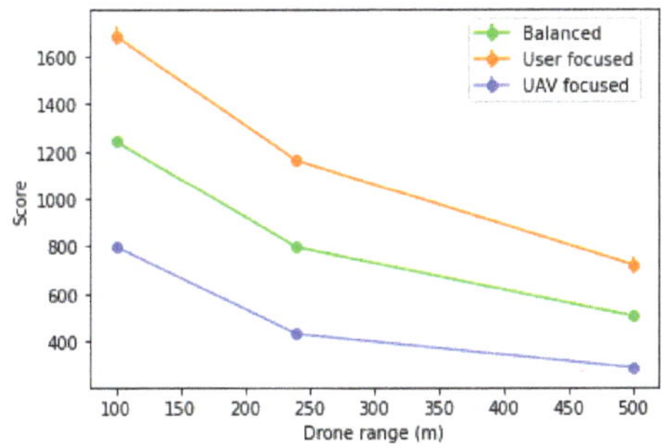

Figure 21. Score of the algorithm with different values of UAV coverage range.

In the decreased percentage of the score, illustrated in Figure 22, similar results are observed. The decreased percentage of the score increases with the range. The UAV coverage range highly influences the performance, since the larger it is, the more tolerant the network is to bad UAV positioning originated through inaccurate prediction. While it does not change the amount of users' traffic each base station can receive or the amount of traffic that exists, it changes how well the network can adapt to the current situation and user distribution, and also the number of UAVs placed.

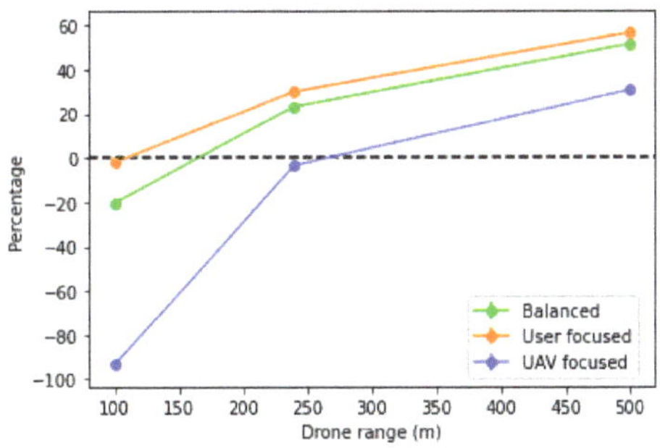

Figure 22. Percentage decrease of the score with different values of UAV coverage range with a base station bandwidth of 1000.

6.2.4. Predicted Users Vs Real Users

To further illustrate the situation and the conditions in which the algorithm operates, Figures 23 and 24 show the positions and number of users without a connection in various situations. Figure 23 shows them as the machine learning algorithm predicted, and Figure 24 shows the real number of users without a connection after the traffic distribution, and the positions of the UAVs. The most important aspect to take from the figures is that the prediction algorithm returns less users without a connection than in the reality, which will inevitably lead to a significant number of users without a connection after applying the UAV positioning algorithm.

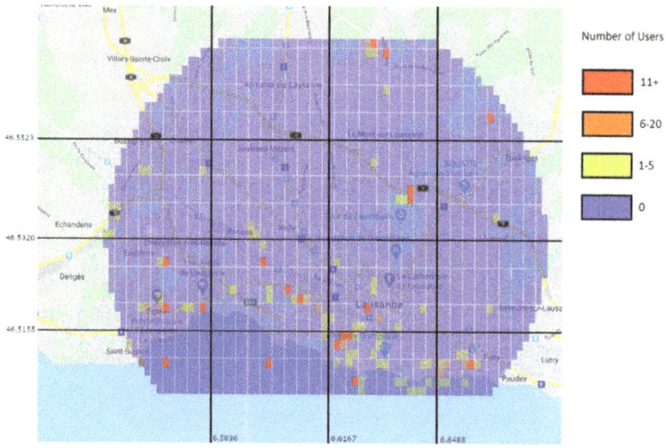

Figure 23. Prediction of users left without a connection (color marks).

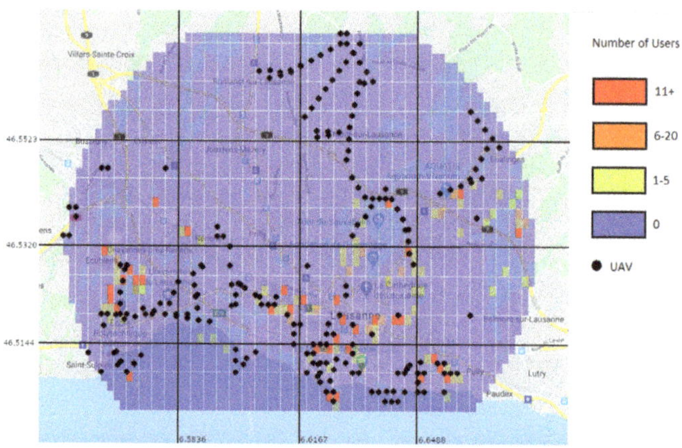

Figure 24. Users left without a connection (color marks) and positioning of the UAVs after the traffic redistribution (black marks).

As illustrated in Figure 24, there is still a significant number of areas that are not directly covered by a UAV, meaning that the prediction algorithm failed to correctly determine the number of extra users in those areas. While this can be improved by the use of a dataset with a large number of users, there are also other options that can lessen the impact of a less accurate prediction. One option is to reduce the threshold for the number of users needed until an area can be considered overloaded. Doing this will increase the number of UAVs placed, but also reduce the impact the prediction algorithm will have when it underestimates the number of users in an overloaded area. Another option is to allow UAVs to make some moving decisions after they are placed. This means that UAVs will be able to move to neighbouring areas if the movement would be beneficial overall.

Furthermore, we also investigated how many users each UAV was connected to, on average, which is illustrated in Figure 25. Using real values will generally achieve a much larger number of users per UAV, since the amount of users that benefit from the UAV placement will be significantly larger than the extra amount of UAVs placed. However, the same does not apply for more restrictive network parameters, as the number of extra UAVs placed starts to be higher than the amount of users it benefits, causing a decline in the number of users per UAV.

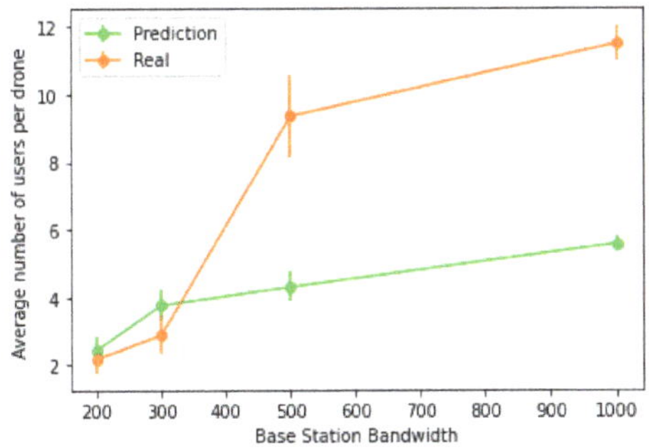

Figure 25. Average number of users per UAV when using real values versus predicted values.

7. Conclusions

This paper presented an approach to predictively place and move aerial UAVs in a communication network to extend and balance the users' traffic. It investigated the performance of multiple machine learning algorithms to predict the users' location and movement. The performance results showed that the random forest and gradient boosting presented the best performance, with random forest having a bigger prediction and training time. Moreover, this paper designed and developed a UAV positioning algorithm that returns the positions of the UAVs in a UAV network, with the objective of extending an existing network and balancing the traffic load. This algorithm was tested with real and predicted data. We concluded that the prediction algorithm slightly underestimates the number of users in most areas, causing the UAV positioning algorithm to position less UAVs than it should, and in turn increasing the amount of users left without a connection. The UAV positioning algorithm performs favorably for less restrictive network conditions.

Future work will focus on using the prediction algorithm on areas with larger number of users, and study the impact that the signal propagation, interference, noise, fading and other impairments of the wireless medium have on the UAV network created by the UAV positioning algorithm. We will also consider the UAV energy and power consumption in the UAV placement algorithm, as well as the cost of its replacement.

Author Contributions: Conceptualization, F.O.; methodology, F.O., M.L. and S.S.; investigation, F.O.; writing—original draft preparation, F.O., M.L. and S.S.; writing—review and editing, M.L. and S.S.; supervision, M.L. and S.S.; funding acquisition, S.S. All authors have read and agreed to the published version of the manuscript.

Funding: This work was funded by FCT/MEC through national funds under the project PTDC/EEI-ROB/28799/2017 FRIENDS: Fleet of dRones for radIological inspEction, commuNication anD reScue, and by the European Regional Development Fund – ERDF, included in the "Urban Innovative Actions" programme, through project Aveiro STEAM City (UIA03-084).

Institutional Review Board Statement: Not applicable.

Informed Consent Statement: Not applicable.

Data Availability Statement: Restrictions apply to the availability of these data. Data was obtained from Idiap Research Institute and are available at https://www.idiap.ch/en/dataset/mdc. with the permission of Idiap Research Institute.

Conflicts of Interest: The authors declare no conflict of interest.

References

1. Kumar, K.; Kumar, S.; Kaiwartya, O.; Kashyap, P.K.; Lloret, J.; Song, H. Drone assisted Flying Ad-Hoc Networks: Mobility and Service oriented modeling using Neuro-fuzzy. *Ad Hoc Netw.* **2020**, *106*, 102242. [CrossRef]
2. Galkin, B.; Kibilda, J.; DaSilva, L.A. Coverage Analysis for Low-Altitude UAV Networks in Urban Environments. In Proceedings of the GLOBECOM 2017–2017 IEEE Global Communications Conference, Singapore, 4–8 December 2017; pp. 1–6. [CrossRef]
3. Kuhlman, M.J.; Švec, P.; Kaipa, K.N.; Sofge, D.; Gupta, S.K. Physics-aware informative coverage planning for autonomous vehicles. In Proceedings of the 2014 IEEE International Conference on Robotics and Automation (ICRA), Hong Kong, China, 31 May–7 June 2014; pp. 4741–4746. [CrossRef]
4. Lyu, J.; Zeng, Y.; Zhang, R.; Lim, T.J. Placement Optimization of UAV-Mounted Mobile Base Stations. *IEEE Commun. Lett.* **2017**, *21*, 604–607. [CrossRef]
5. Reina, D.G.; Tawfik, H.; Toral, S.L. Multi-subpopulation Evolutionary Algorithms for Coverage Deployment of UAV-networks. *Ad Hoc Netw.* **2017**, *68*, 16–32. [CrossRef]
6. Sabino, S.; Horta, N.; Grilo, A. Centralized Unmanned Aerial Vehicle Mesh Network Placement Scheme: A Multi-Objective Evolutionary Algorithm Approach. *Sensors* **2018**, *18*, 4387. [CrossRef] [PubMed]
7. Khan, A.S.; Chen, G.; Rahulamathavan, Y.; Zheng, G.; Assadhan, B.; Lambotharan, S. Trusted UAV Network Coverage Using Blockchain, Machine Learning, and Auction Mechanisms. *IEEE Access* **2020**, *8*, 118219–118234. [CrossRef]
8. Peng, H.; Chen, C.; Lai, C.; Wang, L.; Han, Z. A Predictive On-Demand Placement of UAV Base Stations Using Echo State Network. In Proceedings of the 2019 IEEE/CIC International Conference on Communications in China (ICCC), Changchun, China, 11–13 August 2019; pp. 36–41. [CrossRef]
9. Liu, C.H.; Chen, Z.; Tang, J.; Xu, J.; Piao, C. Energy-Efficient UAV Control for Effective and Fair Communication Coverage: A Deep Reinforcement Learning Approach. *IEEE J. Sel. Areas Commun.* **2018**, *36*, 2059–2070. [CrossRef]
10. Sallouha, H.; Azari, M.M.; Pollin, S. Energy-Constrained UAV Trajectory Design for Ground Node Localization. In Proceedings of the 2018 IEEE Global Communications Conference (GLOBECOM), Abu Dhabi, United Arab Emirates, 9–13 December 2018; pp. 1–7. [CrossRef]
11. Zeng, Y.; Xu, J.; Zhang, R. Energy Minimization for Wireless Communication With Rotary-Wing UAV. *IEEE Trans. Wirel. Commun.* **2019**, *18*, 2329–2345. [CrossRef]
12. Bithas, P.S.; Michailidis, E.T.; Nomikos, N.; Vouyioukas, D.; Kanatas, A.G. A Survey on Machine-Learning Techniques for UAV-Based Communications. *Sensors* **2019**, *19*, 5170. [CrossRef] [PubMed]
13. Laurila, J.K.; Gatica-Perez, D.; Aad, I.; Blom, J.; Bornet, O.; Do, T.; Dousse, O.; Eberle, J.; Miettinen, M. The mobile data challenge: Big data for mobile computing research. In Proceedings of the Mobile Data Challenge Workshop (MDC) in Conjunction with International Conference on Pervasive Computing, Newcastle, UK, 18–22 June 2012.
14. Kiukkonen, N.; Blom, J.; Dousse, O.; Gatica-Perez, D.; Laurila, J. Towards Rich Mobile Phone Datasets: Lausanne Data Collection Campaign. In Proceedings of the 7th ACM International Conference on Pervasive Services (ICPS), Berlin, Germany, 13–16 July 2010.
15. Kursa, M.B.; Jankowski, A.; Rudnicki, W.R. Boruta—A System for Feature Selection. *Fundam. Inform.* **2010**, *101*, 271–285. [CrossRef]

sensors

MDPI

Review

Closing Connectivity Gap: An Overview of Mobile Coverage Solutions for Not-Spots in Rural Zones

Diego Fernando Cabrera-Castellanos [1,*], Alejandro Aragón-Zavala [1] and Gerardo Castañón-Ávila [2]

1 School of Engineering and Science, Tecnologico de Monterrey, Ave. Epigmenio González 500, Querétaro 76130, Mexico; aaragon@tec.mx
2 School of Engineering and Science, Tecnologico de Monterrey, Ave. Eugenio Garza Sada 2501, Monterrey 64849, Mexico; gerardo.castanon@tec.mx
* Correspondence: A00829475@itesm.mx; Tel.: +52-442-789-6898

Abstract: Access to broadband communications in different parts of the world has become a priority for some governments and regulatory authorities around the world in recent years. Building new digital roads and pursuing a connected society includes looking for easier access to the internet. In general, not all areas where people congregate are fully covered, especially in rural zones, thus restricting access to data communications and inducing inequality. In the present review article, we have comprehensively surveyed the use of three platforms to deliver broadband services to such remote and low-income areas, and they are proposed as follows: unmanned aerial vehicles (UAV), altitude platforms (AP), and low-Earth orbit (LEO) satellites. These novel strategies support the connected and accessible world hypothesis. Hence, UAVs are considered a noteworthy solution since their efficient maneuverability can solve rural coverage issues or not-spots.

Keywords: aerial communication; FANET; not-spots; stratospheric communication platform; UAV; UAV-assisted network; 5G

Citation: Cabrera-Castellanos, D.F.; Aragón-Zavala, A.; Castanón-Ávila, G. Closing Connectivity Gap: An Overview of Mobile Coverage Solutions for Not-Spots in Rural Zones. *Sensors* **2021**, *21*, 8037. https://doi.org/10.3390/s21238037

Academic Editor: Margot Deruyck

Received: 28 September 2021
Accepted: 23 November 2021
Published: 1 December 2021

1. Introduction

Coverage indicators are essential for perceiving the reliability of the network in a determined area. Specifically, each country defines the best practices to determine the covered zones for their boundaries and, therefore, the appropriated thresholds associated with frequency bands. Commonly, most mobile operators offer coverage on the main urban area [1], limiting the countryside to lower bandwidth, thus reducing connection speeds [2]. Nevertheless, the interest in providing more connectivity in rural zones has grown in the last decade since economic development will be an immediate fact.

Extensive terminology has arisen to address the coverage holes, wherein a few or even any operator guarantee its services. The Ofcom—Office of Communication of the United Kingdom—names them as *Not-Spots*. The prior entity has the intention to reach the coverage index until 95% by 2022 [3]. Several British operators (O_2, Vodafone, EE, and Three) have implemented a sharing strategy, allowing a mutual infrastructure approach and, therefore, improving the competition in the countryside. This layout—or National Roaming—grants customers in rural areas the possibility of connecting to the strongest available signal, regardless of the chosen operator for these clients [3].

1.1. Motivation

The inequality in accessing Information and Communications Technologies (ICT) resources and the lack of opportunities for reaching development are the most significant drawbacks in developing countries, even though mobile devices accounted for 87% of broadband connections there [4]. Latin America is not so far from that situation. However, most governments have changed their ways of supporting more connectivity opportunities in the last decade.

Within the call for promoting a prosperous society, which can curb inequality and poverty, the United Nations (UN) has considered the access to fixed-broadband internet under the Goal 9 outline—Industry, Innovation, and Infrastructure—a valuable resource for population growth. By 2018, 96.5% of the entire world population can access at least 2G mobile networks where LTE covers 81.8% of the population [5]. In full swing of the Internet Era, not all villages can leverage granted-by-connectivity opportunities because of thef high cost of access, which unearths the at-risk population group's unfairness.

Considering the ongoing demands of communication infrastructure, the UN Sustainable Goal 9 aims to significantly increase ICT resources access by 2020, in addition to struggling to hook up LDC (Least Developed Countries) with affordable technology [6]. The COVID-19 pandemic has triggered comprehensive research and investment in digitalization, namely economy and education boosting, since teleworking, video conferencing systems, and remote education have been crucial parts during pandemic and post-pandemic times.

In order to assess the connectivity situation around the world, the GSM Association (GSMA) provides the GSMA Mobile Connectivity Index (MCI), which measures the performance of 170 countries based on four key enablers of mobile internet adoption—infrastructure, affordability, consumer readiness, and content and services—where the current data include 2019 [7]. The prior institution has released The State of Mobile Internet Connectivity 2020 Report , which analyzes the critical connectivity trends from 2014 to 2019 in terms of mobile internet use [4].

The coverage has not been sufficiently wide to provide the same standards compared to Europe. For instance, in [4], it is possible to check that Europe and Central Asia and North America were more 70% connected than compared to 54% in Latin America and the Caribbean. Despite these observations, it is crucial to note that the offered services have grown in the last region since its MCI overcomes a 61 score by 2019 in contrast with that obtained five years earlier: 51 [7].

Although MCI appears to be the most significant, this is not the only affair to highlight at the moment in terms of analyzing connectivity for particular contexts, such as the countryside. Therefore, it is necessary to map out the earlier metric with each country's rural population density, discovering the most important limitations that prevent people from adopting mobile internet. Table 1 depicts both MCI and Rural Population Density (RPD)—in percentage units from the total—in order to analyze the gap among fifteen Latin countries.

Table 1. Contrast between MCI and RPD of 15 Latin America countries [7].

Country	MCI	RPD
Argentina	67.2	8
Bahamas	68.7	17
Brazil	63.5	13
Chile	73.2	12
Colombia	63.7	19
Costa Rica	63.3	20
Dominican Republic	59.8	18
Ecuador	65.3	36
El Salvador	55.4	27
Haiti	32.8	44
Mexico	67.6	20
Panama	65.3	32
Peru	66.6	22
Uruguay	76.7	5
Venezuela	57.4	12

Identifying the locations where coverage is under specific boundaries appears to be suitable for sketching out the Not-Spots' presence in these contexts. Therefore, Figure 1 charts the correlation between the total population and coverage density, segregated by mobile network generation from 2G to novel 5G, in two specific countries of the target region: Colombia (Figure 1a) and Mexico (Figure 1b). This is aimed at recognizing the coverage gap inside the mentioned countries. Moreover, it likely identifies regions in which the population cannot access voice and data services.

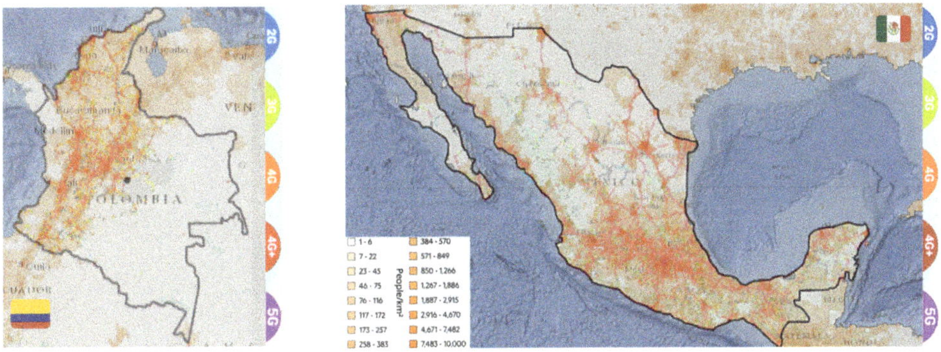

(**a**) Colombia (**b**) Mexico

Figure 1. Correlation between population and current mobile network coverage in both study cases.

1.2. Paper Outline

We have reviewed several strategies that pursue new connectivity standards by expanding network coverage, especially for developing countries, compared with developed countries such as European countries. These approaches aim to list the possible technologies that will improve connectivity in rural zones after studying the researched options in the alternative deployment of networks for optimizing those regions.

After stating the motivations and the principal purposes roughly, we outlined the article as follows: Section 2 presents a perspective of the network environment (outdoor and indoor), highlighting the solutions that engage emergent services such as the Internet of Things (IoT). Section 3 sketches out the possible researched technologies in order to enhance coverage in rural zones and achieve high Quality of Service (QoS) and the network's throughput. Sections 4 and 5 set forth the discussion and conclusions about the assessed solutions in prior sections, in addition to bearing in mind new research opportunities in this field. To this end, Figure 2 shows the overall organization of the cited references in this review article.

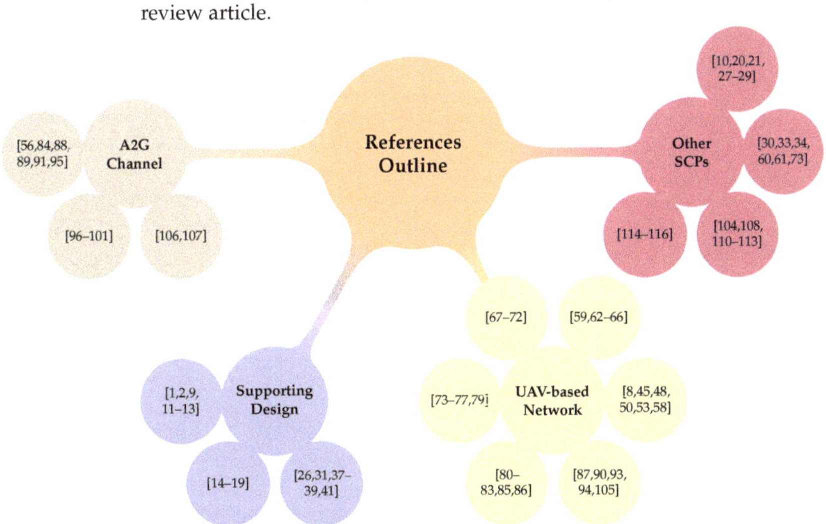

Figure 2. Some Used references in our survey.

2. The Rural Paradigm Shift

Under the perspective of granting better connectivity standards in the countryside, it is adequate to set forth the differences among several best-fitted technologies in order to find an optimal solution. The first approach is a suitable onset to focus on mobile network connections optimized for rural populations and self-steady links for IoT terminals, whether involving new communication tendencies such as Device-to-Device communications (D2D) or even 5G [8].

Outdoor and indoor environments require the above aims to lift specific responses within rural population needs. The outdoor schemes consider current traffic estimation of the mobile network by algorithmic focusing since it may provide a proper breakdown for determining the cells' coverage capacity [9]. With respect to indoor environments, achieving an extended coverage based on ad hoc Networks by lower frequency bands involving repeaters would be suitable [10]. Nonetheless, other approaches will be analyzed in Sections 2.1 and 2.2 that cope with issues for both cases.

2.1. Outdoor Perspective

Gatwaza et al. in [9] highlighted that traffic is an outstanding factor to dimension current mobile networks. In isolated zones, the challenge lies in finding out how to fix the maximum coverage per single base station with respect to complex topography and highly dispersed population distribution [11]. Information on geographical distribution is quite relevant for internet service providers (ISP) since it allows the estimation of areas that deserve specialized deployment toward determining the under-requirements of system capacity [12].

The coverage parameter defines the network's scope, resulting in the expected enhancement for lower-connectivity regions. Consequently, the channel's propagation parameters, such as Path Loss Exponent and Losses, are essential for coverage and quality analysis. For instance, CDMA and AMPS cells may overlay the targeted geographical areas to carry information among remote Base Stations (BS) appropriately [13]. Other alternatives include the use of the WiMAX—IEEE 802.16—set of standards [14] and TV White Spaces (TVWS) [15] to enable a ubiquitous network.

At the onset of 21st century, the developed countries evaluated options to achieve better QoS in rural zones. One of them was implementing high-quality in-car mobile services without the implementation of new cell sites. Thus, there was a possibility to raise roadway coverage areas by using antenna arrays set over constant on-way cars. This advance might have allowed the minimization of cost with respect to the non-installation of more BSs. Furthermore, it would provide improvements due to its implementation over dynamic CDMA signals, eradicating AMPS services [13].

With the massification of novel technologies, e.g., 5G and IoT for urban zones, the idea includes analyzing other low-deployment cost options, such as FTTx. Araujo et al. pointed out in [16] that services on FTTC (Curb) would be 70% cheaper than 5G implementation and 20% less expensive than FTTH (Home). Although the main idea is boosting countrysides as potential high opportunity zones, not all operators expect to invest in high-cost infrastructure for low-density populations because its rollout may cost 80% higher than in urban zones [17].

So far, several approaches have arisen to reach the desired coverage index. Knowing that 5G services are not considered for the countryside yet, IoT services are limited to highly reliable networks. More quantity of unfolded BSs and more coverage index may be reached, increasing efficiency [16]. The BS coverage area is greater than 0.5 Km, and having enough overlap with adjacent cells will ensure the quality of roaming at the maximum allowable distance among them [18].

In mobile networks, the handover parameter is triggered when a user equipment (UE) detects a better signal strength of the neighboring cells [1], but it can also be regarded as non-convergent in the case of rural zones. Thus, identifying the BS coverage area at network planning is a relevant part of the design process. 3G services may be the first technology to be implemented in the countryside since it is possible to monitor the network parameters—such as coverage and cell capacity—by implementing appropriate Signal-to-Noise rates (SNR) and QoS index. It is important to recall that the rural connectivity gap is proportionally greater for low-income households [19].

After reviewing some references, we found that *Stratospheric Communication Platforms* (SPC) have been trending in the last decade for outdoor solutions [20,21]. The Loon Project searched for possibilities in building a new layer for the connectivity ecosystem in the stratosphere based on weather balloons with distributed self-optimization [22]. The Loon LLC group tackled the challenge of extending internet access worldwide based on this approach until the project was closed down in 2021 [23,24]. Another intended sample was Facebook Aquila; however, it collapsed in 2018 [25].

Another kind of alternative to cover rural populations includes the use of LEO satellites. Moreover, LEO and SCPs significantly enable coverage increase and do not require new terrestrial towers. Therefore, these options can offer highly reliable data rate services while demanding simple but special maintenance attention with respect to its tracing [26]. Figure 3 states a feasible implementation of the reviewed solutions in the countryside for outdoor areas, aiming to develop new tendencies considered in Section 2.3.

Figure 3. Some solutions for outdoor networks issues.

Figure 3 entails the use of alternative-to-terrestrial wireless connectivity aimed at rural coverage. To promote the solutions mentioned previously, the SCPs bring several attributes in offering a low-cost and widespread array of services in the countryside. Among them, in [27], the author listed the advantages of deploying these airships—either a constellation or a singular aerial unit—which may be summarized into three categories: seamless countrywide coverage, power consumption trade-offs, and a higher speed of transmission.

Since SCPs enable an extension of coverage, the unobstructed free-of-shadowing LOS roaming service would boost roaming services [27], leveraging the autonomous and discrete features of these devices. For instance, Espinoza et all. led a feasibility study in [28] for two Peruvian rural areas to promote connectivity, aiming at the underserved/unserved communities despite the rough landscape of either the Andes range or the Amazon rainforest. The authors used stratospheric balloons to serve the growing customer and speed demand for up to eight years. After that, an LTE-based network complemented the established structure.

Nonetheless, the fully fledged satellite-based networks carried on a cooperative relay scheme to accomplish the ongoing high demands of reliable, seamless, and high-rate transmissions. Therefore, the space-air-ground integrated network [29]—named SAGIN—combines LEO satellites, SCPs, and terrestrial BS to achieve the future requirements of 6G communications by addressing the provided-by-satellite coverage probabilities enhanced with terrestrial gateways as the authors in acknowledged in [30] in addition to real-time dynamic adjustment of HAP-based networks.

2.2. Indoor Case

Indoor-improving techniques outlined the strategies that enhance user experiences inside closed spaces. Therefore, there is more interference resulting from physical obstacles. This case requires evaluating the best estimation of indoor coverage provided, looking for optimal system planning. The feasibility in implementing algorithmic solutions based on UE location estimation appears to be challenging since their location accuracies depend on the integrated sensors in devices used by authors in [31].

Satellite-based networks and other high-altitude platforms suffer excess losses because the slant path intersects several obstructions than compared to terrestrials. Nevertheless, using repeaters at lower frequency bands—despite the bandwidth limitation—can fulfill the requirements demanded from users [10]. These devices are low-cost and readily available, hence boosting signal propagation and simultaneously enhancing indoor coverage. Figure 4 shows a potential indoor-improvement deployment for a satellite-based backhaul.

Figure 4. Indoor solution for a satellite-based network.

In addition to satellite-based backhaul solutions, other approaches outperform the repeater's throughput, particularly for IoT devices and MTC at a glimpse, and Section 2.3 foresees the need reinforce this limitation. On one hand, the authors in [32,33] analyzed the development of indoor coverage—focused in suburban and rural areas—by deploying LTE-M, NB-IoT, LoRa, and GPRS solutions to cover a greater zone. On the other, there is a reduction in indoor time dispersion by dynamical estimation of the channel parameters in full swing machine learning techniques [34].

Regarding the first approach, Lauridsen et. al contrasted the technologies in a particular Northern Denmark region area by assessing their supported Maximum Coupling Loss [32,33]. The results accomplished 99% of indoor devices coverage for both LTE-M and NB-IoT solutions as long as these devices experienced 10 dB additional loss at its maximum level.

In the case of deep indoor devices—namely basement or underground located—NB-IoT performed the best, with 8% of outage probability, followed by LoRa (13%) and GPRS (60%) considering 30 dB additional losses. Despite the outstanding outcomes to shed light on optimal indoor coverage, the authors suggest a throughput evaluation of pico-cells or Wi-Fi access points to enhance connectivity inside closed spaces [32].

2.3. New Services

A few years ago, trending services such as IoT and 5G were considered challenging to implement in rural areas, especially for Latin America because there were no considerations to grant a reliable and high-traffic supported backhaul network. Nevertheless, these paradigms would hook up dispersed nodes located in remote zones nowadays, with staggering downlink/uplink rates, aiming to accomplish the requirements for MTC and Narrowband Internet of Things (NB-IoT) [35].

IoT promises to be a suitable technology for upgrading the countryside—a stable network may be guaranteed—following the massive number of connected things and the heterogeneous nature of IoT devices. On the other hand, there is the incursion of MTC application domains, such as agriculture management, transportation, logistics improvement, and crop automation, being one of the fastest-growing telecommunications technologies, especially in urban contexts [36]. LTE-based MTC addresses advantages in increasing capacity, traffic response, and spectral efficiency [37].

Diverse strategies have arisen from assessing the most appropriate technologies to furnish high-speed broadband and to reach desired standards such as service speed and setting up at 30 Mbps in European rural areas. Ioannou et al. in [38] stated that FTTdp (Distributed Point) solution using the G.fast standard is a cost-effective alternative to VDSL, which is the current widespread technology in Europe granting connectivity in the countryside. The authors acknowledge that FTTdP G.fast readily enables bandwidth upgrade, but the model is not cost-efficient in terms of investing in geographically sparse populations [16,17].

Consequently, LTE Fixed Wireless Access Networks (LTE FWA) could be an available, attainable solution, bearing in mind extensive LTE infrastructure in a significant rural part of the world. Regardless of whether newly emerged 5G standards are desirable for implementation, we can upgrade LTE FWA through the LTE-NR model, which creates a tight interaction between LTE and the new radio system. The also known model of E-UTRA-NR Dual connectivity—or EN-DC—allows benefits in aspects of user throughput in both low and high traffic load conditions [39].

Foreseeing the inclusion of the services mentioned above, the design of internet access solutions should be engaged with the three main factors as the authors outlined in [40]:

- Affordability —for avoiding undue hardships by employing reliable networks;
- Social shareability—to gain access through selfless (shared) connections;
- Geographical network coverage—where networks allow the user's mobility by themselves.

Simultaneously, the requirements on ubiquitous coverage will not follow the one-size-fits-all standard to pursue a more connected rural society [41]. Figure 5 summarizes the information granted by the GSMA's reports [42,43], which attempt to state the main driven innovations through an improved roll out in three foremost aspects:

- BS infrastructure —far-flung from the traditional macrocells model;
- Backhaul planning—avoiding the higher cost of urban deployment;
- Energy—mixing up with renewable sources;
- Blue Sky solutions—although those remain at the proof-of-concept stage.

These innovations will move beyond the traditional business model—such as *CapEx*—and local governments should create new regulation principles to harvest investment in network infrastructure.

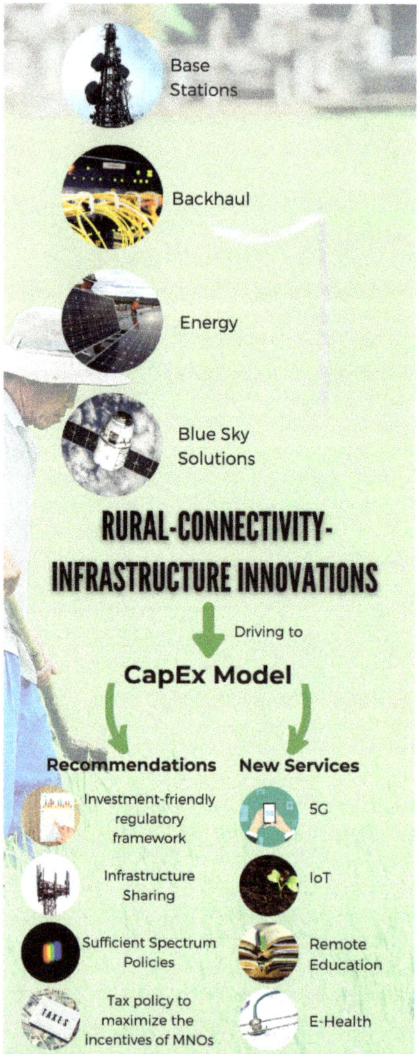

Figure 5. Innovations for rural connectivity.

3. Potential Solutions

There are several challenges to face in rural areas in terms of reliable and enhanced mobile networks. This need triggers the state-of-art study of diverse network models for the countryside in order to introduce ubiquitous solutions in which connectivity is available at anytime and at any location with respect to the population's demands in a fully connected society.

By the first attempts to overcome likely hardships, such as insufficient population for deploying infrastructure, adaptive solutions struggle with the current unfolding of Mobile Network Operators (MNO). The new platforms or devices—that enhance coverage and other rural Key Performance Indicators (KPIs)—leverage practical alternatives for outdoor environments.

There have been studies that cater to rural coverage by using the TVWS-spectrum sharing approach that utilizes free UHF band channels from analog switch-offs at a specific time and space location [15]. Indeed, the primary user (PU) exclusively uses frequency resources on bands 470 MHz and 710 MHz.

On the other hand, S. Hasan et al. [44] aimed to recover GSM whitespace—or the non-actively used and licensed GSM spectrum—to support dynamic spectrum sharing, hence achieving a suitable QoS would not be attached to low throughput and high latency. Regardless, other kinds of solutions have arisen so far that aim for a fully connected countryside.

In the following subsections, several trustworthy approaches will be set forth for diverse rural outdoor solutions, such as unmanned aerial vehicles (Section 3.1), low altitude platforms, and high altitude platforms, and satellites (Section 3.2). Then, Figure 3 graphically summarizes the solutions as mentioned above to cope with rural not-spots.

3.1. UAV-Assisted Networks

Nowadays, unmanned aircraft have commercial uses and have enabled new research interest and innovation toward improving connectivity. The smaller the airship, the better the performance in bestowing coverage, especially for isolated areas. In this case, the drone industry has addressed several civil instances and applications within an affordable and straightforward aim: leveraging UAVs' maneuverability to readily provide connectivity as an off-the-shelf alternative within the current MNO infrastructures.

Historically, the first purpose for Unmanned Aerial System (UAS) was for military and surveillance fields. During the second half of the 20th century, Warfighter's internet yielded a reliable and readily deployable UAV-based ad hoc network to boost backbone communications [45]. This exploited UAS approach resulted in higher throughput standards. Therefore, a network-centric UAS operation concept arose beyond military and political boundaries and was consequently adopted for civil and economic interests. In a nutshell, the unmanned airships outpaced soldiery endurance.

Since the use of drones has been expedited, the need for regulating them has arisen as well in terms of complying with safety standards, even though they reach lower altitudes than other larger forms of aircraft. Therefore, the Global Unmanned aircraft system Traffic Management Association—or GUTMA—appears to foster trustworthy, secure, and efficient integration of UAS into global airspace, addressing drone stakeholders practices—defined as UTM stakeholders—by close cooperation and continuous flight information management [46].

To foster a collaborative and innovative community for UTM stakeholders, GUTMA encourages governments to adopt operation-centric, safe, fair, and secure deployment of UTM solutions. Moreover, with respect to allowing the full integration of UTM services with the current network infrastructure, the first step should predict the digitalization needs of UAS technology trends [47]. Once these are set forth, Table 2 compiles some of the key specifications for the UAV-assisted network in line with the deployment scenario, namely in urban, suburban, and rural contexts.

Table 2. Context-based Specifications for UAV Networks.

Scenario	Network Parameters					Context		
	LHT (m)	UHT ()m)	BMP	LOS	NLOS	Use Case	Network Configuration	Flight Time (min)
UMa-AV	22.5	100		X		HD/M2H	5G	TBD
UMi-AV	TBD	TBD			X	M2H	LTE/LTE+	15–45
RMa-AV	10	40	X	X		L2M/LD		60–180

The gathered information in Table 2 divides the network features into two correlated fields: target scenarios and use case context. Concerning the first field, Muruganathan et al. approached the stakeholder populations in [48] and their LTE network's technical deployment in environments such as urban-macro with aerial vehicles (UMa-AV), urban-micro with aerial vehicles (UMi-AV), and rural-macro with aerial vehicles (RMa-AV). The last scenario exhibits a better mobility performance (BMP) than the others. The second field considers zone density, emphasizing the highest (HD), the medium-to-high (M2H), the low-to-medium (L2M), and the lowest (LD) densities [49].

An analysis of coverage issues should extend the operational scope through defined network architecture to successfully deploy aerial communications. A likely first option launches the UAV model by hooking up one or several Ground-BS (GBS) and using the drone as a relay node in the network. Secondly, a swarm of drones seems suitable for covering a vast extension of nodes or rural-dispersed nodes, creating a solid construction of flying ad hoc network (FANET) networking. The last strategy outpaces the challenging issues that mobile ad hoc Networks (MANET) tackled in terms of communication range since a ground node can indirectly communicate with other hops through several aerial relay nodes such as UAVs [50].

The concept of FANETs has arisen in the literature in order to top off a particular form of Vehicular Ad Hoc Network (VANET) communications and addressing scalable, reliable, real-time peer-to-peer mobile ad hoc networking between aerial and ground nodes [51]. Table 3 relates some UAV-based communication surveys where the authors have thoroughly reviewed UAS modeling strategy in fields such as civil, security, and traffic management, among others.

Table 3. Some surveys of UAV-based communications.

Publication	Brief Summary	Approaches Fields
Mozaffari et al. [52]	A group of potential benefits and applications of UAV-based communications in enhancing coverage, capacity, and reliability of wireless networks.	• The key UAV challenges include 3D deployment, performance analysis, channel modeling, and energy efficiency. • A comprehensive overview of potential applications, chief research directions, and challenging open problems, among others.
Li et al. [53]	A noteworthy integration of 5G technologies with UAV communications networks upon an emerging space-air-ground integrated network architecture.	• Space-air-ground integrated network envisions for beyond-5G communications. • 5G techniques for physical and network layer of UAV scheme and joint communication, computing, and caching.
Fotouhi et al. [54]	A development summary promotes the smooth integration between UAVs and cellular networks without a one-size-fits-all but affordable model.	• The authors surveyed interference issues and potential solutions on UVA-based flying relays and BS approaches. • The article sets forth the new regulations and protocols to grant cyber-physical security in both aerial nodes and UEs.
Shakhatreh et al. [55]	An exhibition of the next large revolution in civil applications by introducing UAV technologies to state feasible research trends and future insights.	• Addressed civil applications: road traffic's real-time monitoring, wireless coverage, remote sensing, search and rescue, surveillance, and civil infrastructure, among others. • Discussed key challenges: charging, collision avoidance, security, and networking.
Khawaja et al. [56]	Modeling Air-to-Ground (A2G) propagation channels in designing and evaluating stages of UAV communication and links attempts to improve AG channel measurement campaigns.	• AG wireless propagation channel research includes payload communications and control and non-payload (CNPC) networks. • The AG channel study tackles limitations such as large and small scale fading.
Hayat et al. [57]	Aerial network missions should vary according to the civil application aims.	• Search and rescue coverage • Network coverage • Delivery and transportation • Construction

The approaches, as mentioned earlier and among others, are comprehensively explained in Tables 4 and 5. The first acknowledges the literature of UAV-based networks between twenty and five years ago, which states the strategies that the cited authors assessed for expanding MANET coverage primarily by algorithmic solutions. The second leads our survey to the outstanding aim: to gauge the promising models for rural communications, raising the current cellular infrastructure, or even adopting a new topology for ubiquitous coverage.

Table 4. Phases of UAV-based network models.

Phase	Approaches	Strategies Models	Advantages/Findings
Early: <2011	Military Services	Airborne Communication Nodes to form a backbone network for Warfighter's internet [45].	. Allowing connection for separated forces . Reliable and easily deployed
		The biologically inspired metaphor algorithm of bird flocking for UAV nodes' placement and motion, adapting their mobility [58].	. Especially useful for rugged and mountainous terrains with heavy signal attenuation. . Achieving a stable connection and load balancing.
		Dynamically placing UAVs considered as relays nodes to provide full connectivity in a disconnected ground MANET through heuristic and algorithmic approaches [59].	. Location tracking that allows an optimal interaction between ground nodes and UAVs without introducing new MANET protocols. . Cost reduction based on finding the minimum number of needed UAVs.
	Integrated Architecture	Two-level Satellite empowered architecture (HAPs/UAVs + Satellite) to improve limited coverage, guaranteeing superior bandwidth access [60,61].	. Allowing interconnection with remote locations. . Enhancing hot-spot coverage with low latency rates. . Mitigation of shadowing impairments through a HAP/UAV repeaters configuration.
		Implementation of UAV-HALE (UAV-High Altitude Long Endurance) platform as a base station with an adaptive antenna array [62].	. Covering rural low-densely populated areas and isolated-by-relief regions. . Support the telecommunication system in emergencies. . Assist hot-spots traffic with a lower cost solution . Provide higher QoS, increasing capacity, and keeping lower computational complexity.
		An algorithmic solution to state and hedonic coalition formation consisting of a determined number of UAVs continuously collecting packets from task arrays [63].	. Performance improvement based on the self-organization of air nodes and tasks into independent coalitions. . UAVs can assess the decision to act as collectors or relays (to enhance wireless transmission). . Suitable model to tackle several aims as surveillance or wireless monitoring.
		Evaluation of A2G links coverage using UAVs at altitudes up to 500 m performing as a radio relay platform in low RF environments [64,65].	. Support over 90% coverage of the ground receivers within 10 dB of LOS Path Loss. . Excellent connectivity for low flying UAV in limited urban areas considering SWAP, even for building-blocked receivers. . For higher altitudes, the coverage becomes homogeneous in rural zones.
	MANETs Upgrade	UAV-assisted MANET model, which is rooted in four connectivity regards: global message (successful propagation to all nodes), worst-case (dividing up a close network), bisection (division cost), and k-connectivity (failed nodes threshold before a disconnection) [66,67].	. The aerial nodes can generate, receive, and forward data packets or improve network connectivity and availability. . The model will achieve better QoS and coverage. . As the proposed method, an adaptive heuristic algorithm can provide a simple solution and reach better performances.
		Performance assessment of ad hoc routing protocols, such as GPRS, OLSR, and AODV, in the context of swarms of UAVs, also considering the relative location of destination nodes [68].	. Maximize the throughput with a minimum number of neighbors into the swarms to ensure connectivity. . Minimize power consumption and optimize loiter time to prevent cross interference and redundant transmissions through spatial multiplexing technique.
		Ad hoc UAS-Ground Network (AUGNet) solution, where an Unmanned Aircraft provides additional connectivity for ground nodes driving into shorter routes with better throughput [69].	. Improve connectivity at the network coverage boundary. . Introduce the net-centric UAS operation concept, a tight coupling between communications, mobility, and task fulfillment.

Table 4. *Cont.*

Phase	Approaches	Strategies Models	Advantages/Findings
Medium: 2011–2016	Connectivity/ Coverage Enhancement	Mobility strategy for UAV-compound MANET to support communication data flow between ground nodes in a dynamic topology network [70].	. Provide the most appropriate air nodes position that maximizes network performance. . UAV nodes can flexibly communicate with ground nodes in the LOS, covering a greatly extended area. . Ground nodes periodically grant their communication status to the air-backbone to find the best mobile strategy.
		Analysis of the coverage problem to address several issues in UAV-FANETs, expecting to extend their operational scope and range and a reliable response time [50,51,71].	. The solid construction of FANET networking standards will result in scalable, reliable real-time peer-to-peer, new-form MANETs. . Aimed at robustness of the coverage algorithms, considering the several constraints in these kinds of networks, especially for UAVs fleets. . Cooperating UAV form aims to increase reliability for aerial missions, ensuring the connectivity of non-LOS systems.
		Neural-based cost function approach to improve coverage and boost capacity into geographical areas subject to high traffic demands [72].	. Provide reliable multi-connectivity using UAS overview as relays between a disconnected network and enhance connectivity. . The model can provide better capacity, reliability, and prolonged connectivity to tackle the inefficiency in handling macrocellular network traffic demands.
		The connectivity-based mobility model (CBMM) compares coverage and connectivity performance, looking for an optimal tracing and sense of a given area [73].	. Monitor inaccessible or dangerous areas to deliver information with lack-of-infrastructure regions. . CBMM allows adapting air node direction to maintain steady links to ground stations or their neighbors. . Reduce the overlap between covered areas by using an efficient and limited number of UAVs with a specific spatial density.
		Efficient 3D deployment of multiple UAVs as portable base stations, seeking downlink coverage performance's maximization in using a minimum transmit power and directional antennas [74].	. Aerial Base Stations have a higher chance of LOS links to ground users. . UAVs can readily move and have a flexible deployment to provide rapid, on-demand communications. . Using directional antennas, the model may enhance UAV-based networks because of effective beamforming schemes.
	Deployment Focusing	Low Altitude Small UAVs (SUAV) pilot provides a micro-scale mobile communication relay, attempting at a superior propagation model and increasing bandwidth reuse for emerging traffic hotspots [75].	. The model achieves an improvement of mean throughput (>22%) and QoS (>70%) in both rural and urban environments. . Offer new possibilities for addressing local traffic imbalances and providing great local coverage.
		Deployment of Drone Small Cells (DSCs) or aerial wireless base station to optimize the covered area. In the presence of D2D users, new challenges—such as coverage performance—should be tackled [76,77].	. The optimal UAVs' altitude results in maximum coverage and system sum rate simultaneously when introduced into underlaid D2D communications links. . In the case of two or more DSCs, an optimal separation distance will grant maximum coverage for a given target area.
	Civil Applications	QoS requirements ranking of UAV networks marked into a practical choice for commercial applications. These aims will outline the design of emerging aerial networks [57].	. Delimitation of the missions into four categories: Search and Rescue, Coverage Expanding, Delivery/Transport, Construction. . SUAVs have turned into handy but inexpensive options for commercial aims due to their their ease of deployment, low maintenance costs, high-maneuverability, and ability to hover. . Wi-Fi technology can support several prior categories whether each application requires a few number of hops amongst the nodes.
		UAV-aided Wireless Communication may be a promising solution for scenarios without coverage infrastructure [78].	. UAV systems are more cost-effective than other solutions—such as HAPs and satellites—providing performance enhancement and adaptive communications. . UAV-based networks involve three typical use cases: *ubiquitous coverage, relaying*, and *information dissemination and data collection*.

Table 5. Phases of UAV-based network models (continuation).

Phase	Approaches	Strategies/Models	Advantages/Findings
Novel: >2016	**Rural Panorama Addressing**	Energy consumption optimization aims to improve aerial node missions and connectivity in the countryside by using a graph-based structure [8], besides an optimized model called RURALPLAN [79,80].	. The multi-period graph approach derives into Genetic Algorithms. It guarantees the coverage and efficient management of UAV consumed energy. . RURALPLAN can reduce energy consumption by up to 60%. . The deployment of UAV-based networks can adopt a short-distance LOS, decreasing installation costs. . By considering a set of optical fiber links to support the backhaul network, the capital and operation expenditures can be compensated, simplifying the stated model.
		Analysis of joined-architecture networks, mixing UAVs and GEO/LEO satellites, to increase radius coverage and state the usability of aerial nodes to assist fixed-infrastructure networks in the countryside [81,82].	. The use of aerial nodes, acting as relays, can cover vast rural extensions, addressing further mobile network generations—such as 5G—to implement steady-links IoT devices. . Bearing in mind the optimizing cellular networks aim in the countryside, heritage functionalities of LTE can achieve prominent coverage radius in the sub-1 GHz bands, raising RF propagation. . Since Non-Terrestrial Networks may be an integral part of 5G infrastructure, UAVs become the bedrock of a mixed-architecture network, especially in collecting data in massive MTC types of application.
		LTE networks can provide coverage by UAV nodes in rural areas, chiefly to boost the Command and Control downlink channel, despite the raised interference due to height dependency [83,84].	. The dependency of the large-scale path loss on the drone's height may be challenging for achieving significant growth in coverage level, boosting the aerial-node's perceived interference level. . Applying network diversity, it is possible to improve the network coverage level and its reliability, since SINR would be better than the achieved −6 dB index under the full-load assumption. . The interference conditions—because drastically changed UAV height— will determine channel characterization to assess wireless remote controls for the aerial nodes.
		Boosting aerial coverage of rural area network deployment to clear limitations by interference mitigation techniques [85].	. Interference canceling and antenna beam selection are strategies to improve overall—aerial and terrestrial— system performance. . The abovementioned schemes will gain a 30% of throughput and achieve a 99% reliability increase. . Downlink and Uplink radio interference trigger poor performance within aerial traffic.
		A Non-Orthogonal Multiple Access (NOMA) layout for UAV-assisted networks to provide emergency services in rural areas [86].	. The proposal carry out the performance of terrestrial users enhancement, resulting in a by-device that is consumed in energy minimizing. . The proposed user-centric strategy follows stochastic geometry approaches for terrestrial users—placed into Voronoi cells—served by UAVs, achieving the location model of both nodes and UEs. . In the case of the NOMA-assisted multi-UAV framework, the analysis of coverage probability can aim to properly set up the network's power allocation factors and targeted rates.

Table 5. *Cont.*

Phase	Approaches	Strategies/Models	Advantages/Findings
		Optimization of the UAV-mounted base stations (MBSs) placement, setting forth a Geometric Disk Cover (GDC) algorithmic solution, which coats with all ground terminals (GTs) in an inward spiral manner [87].	. The correct deployment of MBSs can cover a set of k nodes with a minimum number of disks of a given circular surface with radius *r*. . The computational complexity may be significantly reduced when coverage starts from the perimeter of the area boundary.
		The Path Loss (PL) Characterization for urban, suburban, and rural environments enhances the access technologies for low-altitude aerial networks, considering UAV height effects on the channel [88,89].	. By introducing a Correction Factor (CF), which relies on the UAV altitude, the large-scale fading and the PL of the A2G channel will be accurately characterized. . In urban contexts, PL increases with horizontal distance. In the case of rural zones, PL is irrelevant to UAV heights, albeit it approximates to free-space propagation models at heights around 100 m. . UAV-based networks face a large amount of neighboring interference due to the down-tilted antenna pattern of cellular networks. Moreover, the coverage behavior will be affected beneath this scheme.
	Cellular Network Advance	Improvement of coverage and capacity for future 5G configurations of aerial networks beneath two algorithmic approaches, entropy-based network formation [90] and latency-minimal 3D cell association scheme [91].	. By correctly selecting the UAV controller and then performing network bargaining, the aerial base station could top off a more remarkable improvement on its throughput, SINR per UE capacity in the order of 6.3% and minimal delays and error rates. . With the increase in simultaneous requests within the next-generation heterogeneous wireless network, entropy approaches appear to be suitable for overcoming UAV allocation and Macro Base Station decision problems. . Lifting 3D configuration for aerial cellular networks, a yield of reducing up to 46% in the average total latency would enhance spectral efficiency.
		Optimal design of aerial nodes trajectory in cellular-enabled UAV communication with Ground-BS (GBS) subject to quality-of-connectivity constraints about the link GBS-UAV [92].	. The optimization problem converges in a non-convex approach to find high-quality approximate trajectory solutions. . Channel's delay-sensitive rates and SNR requirements restrict the target communication performance. . UAV's mission completion time may guarantee an efficient method for checking the strategy's feasibility.
		Cooperation of small and mini drones can further enhance the performance of the coverage area of FANETs—even other aerial-kind networks—by establishing a hierarchical structure of efficient collaboration of drones [93,94].	. In the case of ultra-dense networks, the approach efficiently broadens the common issues such as sparse and low-quality coverage and the non-steady aerial links. . The rapidly unfolding of UAV carries out in the non-dependency of geographical constraints and implies system performance lifting by establishing LOS communication links in most scenarios. . Among other advantages—at the top of cooperative distributed UAV networks—are the distributed gateway-selection algorithms used and stability-control regimes.

3.1.1. The A2G Channel Modeling

W. Khawaja et al. comprehensively addressed the available A2G channel models—also including the timeliness to be extended even for Air-to-Air (A2A)—in the study given by [56]. Throughout the paper, they stated the most relevant UAV's A2G schemes and thereby identified the limitations and future research directions for UAS-based communications scenarios. An outstanding classification of the analyzed literature includes deterministic methods, stochastic modeling, and their fusion.

Even though the performed analysis provided by Khawaja et al. remarkably spurred the development of a foremost deep-study on the A2G channel technical model, it is relevant to include a third slope into the classification: Machine Learning (ML) approach, which is scarcely investigated in the article [56]. To shed light on this end, Table 6 attempts to state some leading research studies to sketch out the aerial channel beneath two main strategies, ML and Geometry-based (GM) scheme, in addition to setting forth whether these strategies lie in stochastic (St) or non-stochastic (N-St) approaches.

Table 6. Some efforts addressed in A2G modeling.

Cite	Approach		Scenario			Method		Aim	Contributions
	ML	GM	UMa	UMi	RMa	St	N-St		
[95]	X		X	X	X	X		PL and Delay Spread prediction for mmWave channels.	• Low computational complexity. • Full feature selection scheme. • Frequency/scene-based transfer learning model.
[96]	X		X	X	X	X		PL and Shadowing effects analysis in 3D- LOS/NLOS Channel.	Unsupervised learning clustering technique to derive a 3D temporary channel.
[97]	X		X			X		PL empirical prediction with environmental parameters.	• Location-based method by using 3D-GPS coordinates. • Learning phase includes atmospheric conditions.
[98]	X		X	X	X		X	Collaborative algorithm to solve communication overload by achieving 1.5x throughput.	Optimization of Multi-UAV user deployment based in modified K-means distribution and POO.
[99]		X	X			X		3D non-stationary geometry-based stochastic channel model for A2G.	• 3D arbitrary trajectories. • 3D antenna arrays for 5G. • Computational Methods for time-variant channel parameters.
[100]		X	X			X		A MIMO wideband truncated ellipsoidal-shaped method with scatterer consideration.	Statistical derivation of space-time-correlation function and Dopler power spectrum density.
[101]		X		X		X		Geometrical model for UAV flight's Multi-Path Components evolution.	• Geometrical parametrization for the main MPCs. • Simulation under non-intuitive effects of propagation.
[102]		X	X	X	X		X	Spatial-temporal correlation in function of UAV's hover radius, flight altitude, and elevation angle.	• Numerical approach of PL, Multi-shadow fading, Doppler shift, and channel correlation. • Fixed-Wings UAV-BS Model.

3.1.2. Regulation

The 3GPP Association mainly tackles the protocols and regulations for UAS-FANET communication beneath the addressed need of the quickly maturing sector [103]. Consequently, in the eagerness to state new studies and new features for safe operations, there has been joint work with GUTMA, even involving the novel 5G framework use cases. To our best knowledge, Figure 6 introduces the areas that are being addressed in the latest 3GPP Releases, from Release 15 to Release 17.

There are other institutions concerned with developing UAS standards, such as GUTMA/GSMA, ASTM International, IEEE, ISO, EUROCAE, IETF, and JARUS [47]. For a handy insight on network safety, by avoiding a loss of service due to their proximity, we have briefly recapped the 3GPP-suggested edges [103] in the Listing 1, as long as new releases emerge in supporting the LTE aim [48].

Figure 6. UAS addressing in 3GPP standards.

Listing 1. 3GPP Releases Outline involved in UAS Communications.

- Release 15 addressed the research studies about the ability for UAVs to be served using LTE networks in addition to a comprehensive analysis of potential interferences between eNodeB and UAS.
- Release 16 has an overview of the potential requirements and use cases to enable the necessary connectivity between UAS and UTM.
- Release 17 approaches the use cases and requirements for UAS identification and tracking beneath the application layer. It also gathers the 5G connectivity needs of drones in new KPIs into a 3GPP subscription.

3.2. Other Engaging Solutions

We have thoroughly reviewed the implications in assisting rural networks by employing UAVs; moreover, other engaging solutions can enable broad coverage in the countryside and in shedding light on its connectivity. On the onset of the first decade of 2000s, SCPs appeared to be a prominent answer for fixed and mobile applications. These devices remarkably outpaced the unprofitable gap since they have arisen as a cost-effective solution for urban, suburban, and rural areas [21].

Aside from dedicated area coverage independence, the authors in [21] pointed out that Sky Station platforms may provide higher capacity—by higher frequency reuse—than other wireless systems, the possibility of grant enhanced roaming, and the possibility of choosing their stationary point. Another seamless option for rural connectivity includes satellites, namely LEO configuration. The following subsections will deepen the strategies mentioned earlier, and UAVs also fall into this category.

3.2.1. Altitude Platforms

The altitude platforms are grouped into LAPs (Low-altitude platforms) and HAPs (High-altitude platforms). Song et al. in [104] granted the main difference about the prior categories. LAPs gather the aerial platforms at an altitude of 20 km. UAVs, drones, and blimps fall into this group since they cannot support higher payload capacities, and their autonomy relies on SWAP constraints [105]. As Section 3.1 discussed, UAVs can perform far-flung coverage, increase redundancy, and increase survivability, leveraging the swarm FANET architecture [104].

LAPs have lent dynamic and scalable networks that can quickly cover broader regions, although there are by-payloads stuck. In this case, there are two method to limit this situation: First, developing a suitable propagation model that includes the elevation angle—deployed at several altitudes—along with the MIMO output antenna diversity gain, especially for the last mile connectivity [106]. In the UAV case, the strategy may contain a formulation of statistical assessment of A2G propagation by either using Ray Launching or Ray Tracing geometrical optics models [107]. Second, Drone-to-Drone communication arises as a reliable collision avoidance system [105].

On the other hand, HAPs operate in a quasi-stationary position at an altitude of 20 to 50 km, becoming a viable option to furnish capacity and coverage enhancement [104]. The authors in [108] have envisioned these platforms as a super macro BS (HAPs-SMBS) to unfold high-traffic-volume networks in a metropolitan area in bargaining with smart city paradigms. Facing the LEO constellation shortcomings, HAPs-SMBS can mask the high path loss and high mobility effects.

The potential uses of HAPs—to tackle rural not-spots—shed light on dynamically managed radio resources and mitigate the crossed interference [109]. The rural environment has admitted more prevalence to network coverage instead of higher capacity density. The reason for this is that HAPS requires lower investment and provides high quality—even providing higher terrestrial QoS—and this alternative has been carried out to cover rural and remote areas [110]. At this point, the likely exploitation of radio environment maps and artificial intelligence in the ongoing infrastructure may allow a radius coverage area of more than 30 km, as Chukwuebuka highlighted in [110].

3.2.2. Satellites

Satellite-based architecture has furnished an outstanding architecture to hook up the highly dispersed and remote rural nodes due to their scalability and flexibility in reaching vast geographical areas. In function of the developed network scope, the satellites' orbit relies on defined classification [111]: LEO (altitude between 500 km and 2000 km), MEO (altitude into the range 5000 and 20,000 km), and GEO (altitude of 35,800 km).

Underneath the condition of service-as-primary-resource, LEO architecture, on the one hand, solves latency issues [112]; on the other hand, it has added remarkable bit rate capacities by multi-beam technology [113]. In contrast, e.g., GEO holds limited these parameters. Heading to the best alternative for rural not-spots, LEO has become the best complementary structure of terrestrial networks in the countryside, figuring out several shouldered challenges, such as routing problems and raining attenuation [114].

In order to provide seamless and continuous service by LEO satellite networks, these approaches have adopted constellation shapes whereas QoS is guaranteed, fueled by novel routing protocols regarding UE location and exploiting deterministic LEO topology. Therefore, the route bottlenecks should be foreseen in any pair of end-users, as the authors said in [115]. By avoiding design planning deficiencies, the system's user capacity becomes greater, and the covered geographical zone becomes larger [114].

4. Discussion

At this height, the rural zones have struggled to embrace fully fledged connectivity. Regarding Latin America's situation, three considerable constraints are in conflict with the ubiquitous rural coverage aim: First, MNOs do not furnish a suitable telecom infrastructure outside of urban environments. Secondly, rural settlements are concentrated but geographically sparse, occupying common hot spots. Finally, the studied strategies should be based on bespoke hardware requirements since uneven relief and ecosystem variation hamper the static estimation of channel parameters—the latter demands higher investment cost—which seems unprofitable for ISPs and MNOs.

The not-spots affect directly rural inhabitants, especially those who attempt to foster rural businesses, which include mainly agricultural and new industrial activities in the countryside. Hence, Table 7 states the advantages and shortcomings of the studies solutions—in Section 3—while we spur ongoing research of UAV-assisted networks deployment driven by mobility, cost-effective, and the other leverages outlined in Table 7 that can bring networks to the uncovered regions. Further works include the analysis of dynamical propagation model and simulations of LTE—aiming for 5G-NR deployment—at incoming experimental stages.

Table 7. Comparison among the analyzed solutions for rural coverage.

Solution (Section)	Advantages	Disadvantages
UAVs [Section 3.1]	• Easily deployable and portable. • Reliable infrastructure to enhance coverage. • New security standards by new routing protocols. • Compatible with others as terrestrial and aerial network's platforms.	• Static-channel-modeling intermittent connectivity. • Energy constraints and limited effective payload. • Uncertainty on legislation. • Inefficient obstacle awareness rollout.
HAPs [Section 3.2]	• Commit to cover immensely inaccessible areas. • Allows adaptable resource allocation. • Low roll-out costs. • Guarantee connectivity by a single platform. • Agile deployment. • Payload upgrading.	• Few protocol standardization. • Unfit design of traffic aggregation. • Poor raters of interference mitigation in shared spectrum.
LEOs [Section 3.2]	• Enable higher QoS than terrestrial. • Reach a latency issue standard. • Add significant bit rate capacity. • Provide high capacity backhaul.	• Insufficient coverage time assessment. • Higher cost of deployment and maintenance. • Most affected by fading effects. • Unreliable communication at low elevation angles.

Future Research Opportunities

In the prior section, we have introduced three achievable solutions to strive against the countryside's not-spots. There remain shortcomings stuck in the fully fledged method of granting connectivity to pursue endurance in the deployed system. UAV-based networks seem to be an attractive option due to their commercial affordability, as we pointed out in Section 3.1. However, for now, both UAS and Altitude Platforms have factors opposed to large scale use, such as payload capability—to shoulder the network equipment—and insufficient MNO interest, behind higher returning investment rates.

Consequently, three categories claim further analysis in the case of deploying aerial networks. Firstly, A2G channel modeling needs to be supplied by a realistic propagation model, since most of them are still limited to a single device or particular environments. Moreover, there is a considerable need to characterize the by-mobility Doppler effect, in addition to the channel's captured time variation addressing more precision and accuracy. Fueled by the efforts to sketch the A2G channel—as it was analyzed in Section —the need to address research in this field has arisen and becomes more decisive as consumer demands grow.

Secondly, aerial platforms lacking an optimal 3D placement are of concern. A matchless location bestows the coexistence with the terrestrial cellular networks and avoids mutual interference with GBS. In the case of UAS, an optimal arrangement of UAV-BS can yield a minimum downlink transmission latency, setting up previously the drone-BS location and transmission bandwidth [52]. This approach can reduce the total flight time while also enhancing energy consumption.

With respect to concern on energy issues of SCP—which may be the foremost challenge to assure connectivity's significant periods—the utilization of peer-to-peer energy sharing has been investigated since energy is a limited resource in mobile networks because they are jammed in non-renewable sources such as batteries. According to the application, an attractive solution to outpace the excessive consumption of energy, mainly focused on renewable sources, appears to be a significant research field for guaranteeing communication availability [116].

Thirdly, outstanding cellular network planning foresees the minimum number of required aerial nodes to cover a given geographical area, either partially connected or entirely disconnected. Hence, to maximize the total covered zones, there should be a previous identification of users and obstacles. Beneath this regard, prior frequency planning and signaling overhead analysis can assure greater network throughput, especially for high-frequency bands.

Finally, an in-depth design of the bespoke-solution construction affords countless advantages to aim for in a fully connected countryside. In this case, embracing an expected radius of 30 km [109]—or even a greater value—assures the coexistence of either LAPs or HAPs with terrestrial systems, and sharing the same spectrum can extend the coverage in rural and remote areas. Other strategies involve a novel antenna array design and aerial swarms or constellations, which are expected to be further researched in broader 5G investigations.

5. Conclusions

Nowadays, complete rural internet access may be an incongruous reality due to the lack of efforts to deploy a suitable mobile networks infrastructure. However, data demand has grown recently since many rural inhabitants consider using technology to improve their quality of life by implementing trending technologies, such as IoT. Although Latin American countries have recently envisaged closing the connectivity gap, there are remote geographical zones where not-spots are a significant challenge to governments because they strive to outpace inequality under the insight of fully fledged coverage.

Bearing in mind our study cases, in Mexico and Colombia, which have economically and technologically developed in the last decade, the connectivity gaps are noticeable. Therefore, implementing alternative and efficient solutions—as listed in Section 3—will involve hooking the peripheral population up with reliable deployments. The COVID-19 pandemic has accelerated the reshaping of a noteworthy need for connectivity since most of our performed activities leverage digitalization growth to partake in affordability and access. Although several rural populations remain fully offline, the recent efforts to stimulate new steady links have triggered new opportunities to access online education, employment, or critical health and sanitation advice.

We have summarized some strategies to strengthen connectivity in rural environments, especially for Latin American countries. By establishing statistics that best draw the mentioned phenomena, we encourage further access to ICT and encourage the target of providing affordable access to the internet in developed countries, which in turn considers rural and geographically remote populations. Hence, solutions such as UAVs, HAPs/LAPs, and LEO satellites have arisen for most cost-effective bargaining. However, we have comprehensively studied UAS scope in communication because its efficient maneuverability can solve the coverage problem through a solid construction of either GBS or FANET approaches.

Author Contributions: D.F.C.-C. carried out the most literature search, outlined, drafted, and wrote the manuscript. Also, D.F.C.-C. sketched all the figures and tables by collecting the information from references. A.A.-Z. and G.C.-A. provided the main insights of the Discussion, reviewed the draft and furnished substantial comments. All authors have read and agreed to the published version of the manuscript.

Funding: This research received no external funding.

Institutional Review Board Statement: Not applicable.

Informed Consent Statement: Not applicable.

Data Availability Statement: This article has retrieved information from GSM Association Intelligence databases and reports. Both global MCI and RDP may be consulted through 2020-GSMA Mobile Connectivity Index. © GSMA Intelligence 2004–2021.

Acknowledgments: D.F.C.-C. would like to acknowledge to the Consejo Nacional de Ciencia y Tecnología (CONACyT) since this work has been partially supported by national scholarship with id. 1013611.

Conflicts of Interest: The authors declare no conflict of interest.

Abbreviations

The following abbreviations are used in this manuscript:

A2G	Air-to-Ground
AMPS	Advanced Mobile Phone System
APs	Altitude Platforms
BS	Base Station
CAPEX	Capital Expenditure
CDMA	Code Division Multiple Access
D2D	Device-To-Device
FANET	Flying Ad-Hoc Network
FFTx	Fiber-To-The-x
GBS	Ground BS
GEO	Geostationary Orbit Satellite
GT	Ground Terminal
HAPs	High-altitude Platform
ISP	Internet Service Provider
KPI	Key Performance Indicator
LAPs	Low-Altitude Platform
LDC	Least-Developed Country
LEO	Low-altitude Earth Orbit Satellite
LOS	Line-of-Sight
LTE-NR	LTE-New Radio
MANET	Mobile Ad-Hoc Network
MCI	Mobile Connectivity Index
MEO	Medium-altitude Earth Orbit Satellite
ML	Machine Learning
MNO	Mobile Network Operator
MTC	Machine-Type Communication
NLOS	Non-Line-of-Sight
NOMA	Non-Orthogonal Multiple Access
PL	Path Loss
QoS	Quality Of Service
RDP	Rural Population Density
SNR	Signal-to-Noise Ratio
SCPs	Stratospheric Communication Platforms
SUAV	Small UAV
SWAP	Size, Weight, and Power constraint
TBD	To-Be-Developed
TVWS	Television White Space
UAS	Unmanned Aerial System
UAV	Unmanned Aerial Vehicle
UE	User Equipment
UHF	Ultra High Frequency
UTM	UAS Traffic Management
VANET	Vehicular Ad-Hoc Network
VDSL	Very-high-bit-rate Digital Subscriber Line

References

1. Kim, H.; Kim, Y.; Park, P.; Ahn, S.; Jo, J.; Park, S.; Park, C. Network coverage expansion in radio access network sharing. In Proceedings of the 2017 Sixth International Conference on Future Generation Communication Technologies (FGCT), Dublin, Ireland, 21–23 August 2017; pp. 1–5. [CrossRef]

2. Paliwal, K.K.; Singh, S.; Bajaj, S. Rural communication enhancement using mobile ad-hoc network. In Proceedings of the 2017 7th International Conference on Cloud Computing, Data Science Engineering—Confluence, Noida, India, 12–13 January 2017; pp. 387–391. [CrossRef]
3. Office of Communications of United Kingdom. *Connected Nations Report 2016*; Technical Report; OfCom: London, UK, 2016.
4. Kalvin Bahia, A.D. *The State of Mobile Internet Connectivity 2020*; Technical Report; GSMA Intelligence: London, UK, 2020.
5. Secretary-General. *Progress towards the Sustainable Development Goals*; Technical Report; United Nations: New York, NY, USA, 2020.
6. United Nations. Goal 9: Build Resilient Infrastructure, Promote Sustainable Industrialization and Foster Innovation. 2019. Available online: www.un.org/sustainabledevelopment/infrastructure-industrialization/ (accessed on 06 May 2021).
7. GSMA Intelligence. GSMA Mobile Connectivity Index. 2019. Available online: www.mobileconnectivityindex.com (accessed on 27 March 2021).
8. Jimenez, J.G.; Chiaraviglio, L.; Amorosi, L.; Blefari-Melazzi, N. Multi-Period Mission Planning of UAVs for 5G Coverage in Rural Areas: A Heuristic Approach. In Proceedings of the 2018 9th International Conference on the Network of the Future, NOF 2018, Poznan, Poland, 19–21 November 2018; pp. 52–59. [CrossRef]
9. Gatwaza, W.W.; Afullo, T.J.; Sewsunker, R. Characterization of rural traffic and evaluation of cellular protocols for fixed-mobile rural application. In Proceedings of the 2004 IEEE Africon. 7th Africon Conference in Africa (IEEE Cat. No. 04CH37590), Gaborone, Botswana, 15–17 September 2004; Volume 1, pp. 245–250 . [CrossRef]
10. Rudd, R.F. Indoor coverage considerations for high-elevation angle systems. In Proceedings of the Second International Conference on 3G Mobile Communication Technologies, London, UK, 26–28 March 2001; pp. 171–174. [CrossRef]
11. Chen, J. Network Analysis Based on 2/4G Common Coverage Research in Rural Area. In Proceedings of the 2018 IEEE International Conference on Internet of Things (iThings) and IEEE Green Computing and Communications (GreenCom) and IEEE Cyber, Physical and Social Computing (CPSCom) and IEEE Smart Data (SmartData), Halifax, NS, Canada, 30 July–3 August 2018; pp. 295–300. [CrossRef]
12. Ding, Y.; Liu, H.; Cao, X. An Intelligent Computation of Coverage and Capacity of Base Station in 3G Mobile Communications Network. In Proceedings of the 2008 International Conference on Computational Intelligence and Security, Suzhou, China, 13–17 December 2008; Volume 2, pp. 494–497. [CrossRef]
13. Ganesh, R.; Ning Yang.; Regitz, S.; Giordano, A.; Fye, D. Coverage considerations for CDMA overlay of rural analog sites. In Proceedings of the 2000 IEEE International Conference on Personal Wireless Communications. Conference Proceedings (Cat. No. 00TH8488), Hyderabad, India, 17–20 December 2000; pp. 288–292. [CrossRef]
14. Manna, G.C.; Jharia, B. Mobile WiMAX coverage evaluation for rural areas of India. In Proceedings of the 13th International Conference on Advanced Communication Technology (ICACT2011), Gangwon, Korea (South), 13–16 February 2011; pp. 942–947.
15. Amine, L.M.; Adil, E.B.; Aawatif, P.H. Toward enhancing connectivity through TVWS in outdoor rural isolated environment. In Proceedings of the 2017 Sensors Networks Smart and Emerging Technologies (SENSET), Beiriut, Lebanon, 12–14 September 2017; pp. 1–4. [CrossRef]
16. Araújo, M.; Ekenberg, L.; Confraria, J. Rural networks cost comparison between 5G (mobile) and FTTx (fixed) scenarios. In Proceedings of the 2018 IEEE 29th Annual International Symposium on Personal, Indoor and Mobile Radio Communications (PIMRC), Bologna, Italy, 9–12 September 2018; pp. 259–264. [CrossRef]
17. Rendon Schneir, J.; Xiong, Y. A cost study of fixed broadband access networks for rural areas. *Telecommun. Policy* **2016**, *40*, 755–773. [CrossRef]
18. Smirnov, N.I.; Eremichev, V.I.; Tsyrenova, L.A. Increase in the efficiency of the coverage area in the cellular communication systems. In Proceedings of the 2017 Systems of Signal Synchronization, Generating and Processing in Telecommunications (SINKHROINFO), Kazan, Russia, 3–4 July 2017; pp. 1–4. [CrossRef]
19. Prieger, J.E. The broadband digital divide and the economic benefits of mobile broadband for rural areas. *Telecommun. Policy* **2013**, *37*, 483–502. [CrossRef]
20. Kelif, J. Coverage and Performance of Stratospheric Balloons Wireless Networks. In Proceedings of the 2016 IEEE 27th Annual International Symposium on Personal, Indoor, and Mobile Radio Communications (PIMRC), Valencia, Spain, 4–8 September 2016; pp. 1–6. [CrossRef]
21. Ilcev, S.D.; Singh, A. Development of stratospheric communication platforms (SCP) for rural applications. In Proceedings of the 2004 IEEE Africon. 7th Africon Conference in Africa (IEEE Cat. No. 04CH37590), Gaborone, Botswana, 15–17 September 2004. [CrossRef]
22. Huang, L.; Zhou, Y.; Wang, Y.; Han, X.; Shi, J.; Chen, X. Advanced coverage optimization techniques for small cell clusters. *China Commun.* **2015**, *12*, 111–122. [CrossRef]
23. Loon LLC. The Loon Project. 2020. Available online: www.loon.com/ (accessed on 5 April 2021).
24. Alastair Westgarth. Saying Goodbye to Loon. 2021. Available online: https://medium.com/loon-for-all/loon-draft-c3fcebc11f3f/ (accessed on 16 April 2021).
25. Yael Maguire. High Altitude Connectivity: The Next Chapter. 2018. Available online: www.engineering.fb.com/2018/06/27/connectivity/high-altitude-connectivity-the-next-chapter/ (accessed on 1 May 2021).
26. Mucalo, A.K.; Simac, G.; Tekovic, A. Criteria for mobile coverage obligations—Overview in Europe. In Proceedings of the 2015 57th International Symposium ELMAR (ELMAR), Zadar, Croatia, 28–30 September 2015; pp. 181–184. [CrossRef]

27. Ilcev, D.S. *Introduction to Stratospheric Communication Platforms (SCP)*; University for Business and Technology: Pristina, Kosovo, 2015. [CrossRef]
28. Espinoza, D.; Reed, D. Wireless technologies and policies for connecting rural areas in emerging countries: A case study in rural Peru. *Digit. Policy Regul. Gov.* **2018**, *20*, 479–511. [CrossRef]
29. Ye, J.; Dang, S.; Shihada, B.; Alouini, M.S. Space-Air-Ground Integrated Networks: Outage Performance Analysis. *IEEE Trans. Wirel. Commun.* **2020**, *19*, 7897–7912. [CrossRef]
30. Talgat, A.; Kishk, M.A.; Alouini, M.S. Stochastic Geometry-Based Analysis of LEO Satellite Communication Systems. *IEEE Commun. Lett.* **2021**, *25*, 2458–2462. [CrossRef]
31. Tang, Y.F.; Zhang, Y.X.; Kan, S.L. Indoor RF signal coverage solution. In Proceedings of the 2017 IEEE Asia Pacific Microwave Conference (APMC), Kuala Lumpur, Malaysia, 13–16 November 2017; pp. 233–236. [CrossRef]
32. Lauridsen, M.; Nguyen, H.; Vejlgaard, B.; Kovacs, I.Z.; Mogensen, P.; Sorensen, M. Coverage Comparison of GPRS, NB-IoT, LoRa, and SigFox in a 7800 km^2 Area. In Proceedings of the 2017 IEEE 85th Vehicular Technology Conference (VTC Spring), Sydney, Australia, 4–7 June 2017. [CrossRef]
33. Lauridsen, M.; Kovacs, I.Z.; Mogensen, P.; Sorensen, M.; Holst, S. Coverage and Capacity Analysis of LTE-M and NB-IoT in a Rural Area. In Proceedings of the 2016 IEEE 84th Vehicular Technology Conference (VTC-Fall), Montreal, QC, Canada, 18–21 September 2016; pp. 1–5. [CrossRef]
34. Aldossari, S.M.; Chen, K. Machine Learning for Wireless Communication Channel Modeling: An Overview. *Wirel. Pers. Commun.* **2019**, *106*, 41–70. [CrossRef]
35. Akpakwu, G.A.; Silva, B.J.; Hancke, G.P.; Abu-Mahfouz, A.M. A Survey on 5G Networks for the Internet of Things: Communication Technologies and Challenges. *IEEE Access* **2017**, *6*, 3619–3647. [CrossRef]
36. Naddafzadeh-Shirazi, G.; Lampe, L.; Vos, G.; Bennett, S. Coverage enhancement techniques for machine-to-machine communications over LTE. *IEEE Commun. Mag.* **2015**, *53*, 192–200. [CrossRef]
37. Yazdani, O.; MirJalili, G. A survey of distributed resource allocation for device-to-device communication in cellular networks. In Proceedings of the 19th CSI International Symposium on Artificial Intelligence and Signal Processing (AISP 2017), Shiraz, Iran, 25–27 October 2017, pp. 236–239. [CrossRef]
38. Ioannou, N.; Katsianis, D.; Varoutas, D. Comparative techno-economic evaluation of LTE fixed wireless access, FTTdp G.fast and FTTC VDSL network deployment for providing 30Mbps broadband services in rural areas. *Telecommun. Policy* **2020**, *44*. [CrossRef]
39. Yilmaz, O.N.C.; Teyeb, O.; Orsino, A. Overview of LTE-NR Dual Connectivity. *IEEE Commun. Mag.* **2019**, *57*, 138–144. [CrossRef]
40. Villapol, M.E.; Liu, W.; Gutierrez, J.; Qadir, J.; Gordon, S.; Tan, J.; Chiaraviglio, L.; Wu, J.; Zhang, W. A Sustainable Connectivity Model of the Internet Access Technologies in Rural and Low-Income Areas. In *Smart Grid and Innovative Frontiers in Telecommunications*; Chong, P.H.J., Seet, B.C., Chai, M., Rehman, S.U., Eds.; Springer International Publishing: Cham, Switzerland, 2018; pp. 93–102.
41. Sundquist, M.; Markendahl, J. A case study cost modelling of regulatory alternatives to mitigate the mobile network coverage and capacity problems in rural areas. In Proceedings of the 26th European Regional Conference of the International Telecommunications Society (ITS): "What Next for European Telecommunications?", Madrid, Spain, 24–27 June 2015.
42. Handforth, C. *Closing the Coverage Gap: How Innovation Can Drive Rural Connectivity*; Technical Report; GSMA Association: London, UK, 2019.
43. Cruz, G.; Touchard, G.; Buckwell, M.; Liberatore, F. *Enabling Rural Coverage: Regulatory and Policy Recommendations to Foster Mobile Broadband Coverage in Developing Countries*; Technical Report; GSMA Association, Squire Patton Boggs: London, UK, 2018.
44. Hasan, S.; Heimerl, K.; Harrison, K.; Ali, K.; Roberts, S.; Sahai, A.; Brewer, E. GSM whitespaces: An opportunity for rural cellular service. In Proceedings of the 2014 IEEE International Symposium on Dynamic Spectrum Access Networks (DYSPAN), McLean, VA, USA, 1–4 April 2014; pp. 271–282. [CrossRef]
45. Niessen, C.W. Battlefield connectivity via airborne communications nodes. In Proceedings of the SPIE—The International Society for Optical Engineering, Orlando, FL, USA, 26 June 1997; Volume 3080, pp. 72–78. [CrossRef]
46. *UAS Traffic Management Architecture*; Technical Report; GUTMA: Lausanne, Switzerland, 2017.
47. *Designing UTM for Global Success*; Technical Report; GUTMA: Lausanne, Switzerland, 2020.
48. Muruganathan, S.D.; Lin, X.; Maattanen, H.L.; Sedin, J.; Zou, Z.; Hapsari, W.A.; Yasukawa, S. An Overview of 3GPP Release-15 Study on Enhanced LTE Support for Connected Drones. *arXiv* **2018**, arXiv:1805.00826.
49. *LTE Aerial Profile v1.00*; Technical Report; GSMA, GUTMA: Lausanne, Switzerland, 2020.
50. Sahingoz, O.K. Networking models in flying Ad-hoc networks (FANETs): Concepts and challenges. *J. Intell. Robot. Syst. Theory Appl.* **2014**, *74*, 513–527. [CrossRef]
51. Sahingoz, O.K. Mobile networking with UAVs: Opportunities and challenges. In Proceedings of the 2013 International Conference on Unmanned Aircraft Systems, Atlanta, GA, USA, 28–31 May 2013; pp. 933–941. [CrossRef]
52. Mozaffari, M.; Saad, W.; Bennis, M.; Nam, Y.; Debbah, M. A Tutorial on UAVs for Wireless Networks: Applications, Challenges, and Open Problems. *IEEE Commun. Surv. Tutor.* **2019**, *21*, 2334–2360. [CrossRef]
53. Li, B.; Fei, Z.; Zhang, Y. UAV communications for 5G and beyond: Recent advances and future trends. *IEEE Internet Things J.* **2019**, *6*, 2241–2263. [CrossRef]

54. Fotouhi, A.; Qiang, H.; Ding, M.; Hassan, M.; Giordano, L.G.; Garcia-Rodriguez, A.; Yuan, J. Survey on UAV Cellular Communications: Practical Aspects, Standardization Advancements, Regulation, and Security Challenges. *IEEE Commun. Surv. Tutor.* **2019**, *21*, 3417–3442. [CrossRef]
55. Shakhatreh, H.; Sawalmeh, A.H.; Al-Fuqaha, A.; Dou, Z.; Almaita, E.; Khalil, I.; Othman, N.S.; Khreishah, A.; Guizani, M. Unmanned Aerial Vehicles (UAVs): A Survey on Civil Applications and Key Research Challenges. *IEEE Access* **2019**, *7*, 48572–48634. [CrossRef]
56. Khawaja, W.; Guvenc, I.; Matolak, D.W.; Fiebig, U.; Schneckenburger, N. A Survey of Air-to-Ground Propagation Channel Modeling for Unmanned Aerial Vehicles. *IEEE Commun. Surv. Tutor.* **2019**, *21*, 2361–2391. [CrossRef]
57. Hayat, S.; Yanmaz, E.; Muzaffar, R. Survey on Unmanned Aerial Vehicle Networks for Civil Applications: A Communications Viewpoint. *IEEE Commun. Surv. Tutor.* **2016**, *18*, 2624–2661. [CrossRef]
58. Basu, P.; Redi, J.; Shurbanov, V. Coordinated flocking of UAVS for improved connectivity of mobile ground nodes. In Proceedings of the—IEEE Military Communications Conference MILCOM, Monterey, CA, USA, 31 October–3 November 2004; Volume 3, pp. 1628–1634. [CrossRef]
59. Chandrashekar, K.; Dekhordi, M.R.; Baras, J.S. Providing full connectivity in large ad-hoc networks by dynamic placement of aerial platforms. In Proceedings of the—IEEE Military Communications Conference MILCOM, Monterey, CA, USA, 31 October–3 November 2004; Volume 3, pp. 1429–1436. [CrossRef]
60. Palazzi, C.E.; Roseti, C.; Luglio, M.; Gerla, M.; Sanadidi, M.Y.; Stepanek, J. Satellite coverage in urban areas using unmanned airborne vehicles (UAVs). In Proceedings of the 2004 IEEE 59th Vehicular Technology Conference. VTC 2004-Spring (IEEE Cat. No. 04CH37514), Milan, Italy, 17–19 May 2004; Volume 5, pp. 2886–2890. [CrossRef]
61. Palazzi, C.E.; Roseti, C.; Luglio, M.; Gerla, M.; Sanadidi, M.Y.; Stepanek, J. Enhancing transport layer capability in HAPS-satellite integrated architecture. *Wirel. Pers. Commun.* **2005**, *32*, 339–356. [CrossRef]
62. Dovis, F.; Sellone, F. Increasing the capacity of existing terrestrial outdoor radio mobile systems by means of UAV-HALE platforms. In Proceedings of the 2000 IEEE 51st Vehicular Technology Conference Proceedings (Cat. No. 00CH37026), Tokyo, Japan, 15–18 May 2000; Volume 3, pp. 2365–2369.
63. Saad, W.; Han, Z.; Başar, T.; Debbah, M.; Hjørungnes, A. A selfish approach to coalition formation among unmanned air vehicles in wireless networks. In Proceedings of the 2009 International Conference on Game Theory for Networks, GameNets '09, Istanbul, Turkey, 13–15 May 2009; pp. 259–267. [CrossRef]
64. Goddemeier, N.; Daniel, K.; Wietfeld, C. Coverage evaluation of wireless networks for unmanned aerial systems. In Proceedings of the 2010 IEEE Globecom Workshops, GC'10, Miami, FL, USA, 6–10 December 2010; pp. 1760–1765. [CrossRef]
65. Cerasoli, C. An analysis of unmanned airborne vehicle relay coverage in urban environments. In Proceedings of the—IEEE Military Communications Conference MILCOM, Orlando, FL, USA, 29–31 October 2007; pp. 1–7. [CrossRef]
66. Han, Z.; Swindlehurst, A.L.; Liu, K.J.R. Optimization of MANET connectivity via smart deployment/movement of unmanned air vehicles. *IEEE Trans. Veh. Technol.* **2009**, *58*, 3533–3546. [CrossRef]
67. Han, Z.; Swindlehurst, A.L.; Liu, K.J.R. Smart deployment/movement of unmanned air vehicle to improve connectivity in MANET. In Proceedings of the IEEE Wireless Communications and Networking Conference, WCNC, Las Vegas, NV, USA, 3–6 April 2006; Volume 1, pp. 252–257. [CrossRef]
68. Hyland, M.T.; Mullins, B.E.; Baldwin, R.O.; Temple, M.A. Simulation-based performance evaluation of mobile ad hoc routing protocols in a swarm of unmanned aerial vehicles. In Proceedings of the—21st International Conference on Advanced Information Networking and Applications Workshops/Symposia, AINAW'07, Niagara Falls, ON, Canada, 21–23 May 2007; Volume 1, pp. 249–256. [CrossRef]
69. Brown, T.; Argrow, B.; Frew, E.; Dixon, C.; Henkel, D.; Elston, J.; Gates, H. Experiments using small unmanned aircraft to augment a mobile ad hoc network. In *Emerging Technologies in Wireless LANs: Theory, Design, and Deployment*; Cambridge University Press: Cambridge, UK, 2007; pp. 695–718. ISBN 9780521895842.
70. Huo, J.; Xu, Z.; Zhang, Y.; Shan, X. A UAV mobile strategy in mobile ad hoc networks. In Proceedings of the 2011 International Conference on Electronics, Communications and Control, Ningbo, China, 9–11 September 2011; pp. 686–690. [CrossRef]
71. Chen, Y.; Zhang, H.; Xu, M. The coverage problem in UAV network: A survey. In Proceedings of the 5th International Conference on Computing Communication and Networking Technologies, ICCCNT 2014, Hefei, China, 11–13 July 2014; pp. 1–5. [CrossRef]
72. Sharma, V.; Bennis, M.; Kumar, R. UAV-Assisted Heterogeneous Networks for Capacity Enhancement. *IEEE Commun. Lett.* **2016**, *20*, 1207–1210. [CrossRef]
73. Yanmaz, E. Connectivity versus area coverage in unmanned aerial vehicle networks. In Proceedings of the IEEE International Conference on Communications, Ottawa, ON, Canada, 10–15 June 2012; pp. 719–723. [CrossRef]
74. Mozaffari, M.; Saad, W.; Bennis, M.; Debbah, M. Efficient Deployment of Multiple Unmanned Aerial Vehicles for Optimal Wireless Coverage. *IEEE Commun. Lett.* **2016**, *20*, 1647–1650. [CrossRef]
75. Guo, W.; Devine, C.; Wang, S. Performance analysis of micro unmanned airborne communication relays for cellular networks. In Proceedings of the 2014 9th International Symposium on Communication Systems, Networks and Digital Signal Processing, CSNDSP 2014, Manchester, UK, 23–25 July 2014; pp. 658–663. [CrossRef]
76. Mozaffari, M.; Saad, W.; Bennis, M.; Debbah, M. Unmanned Aerial Vehicle with Underlaid Device-to-Device Communications: Performance and Tradeoffs. *IEEE Trans. Wirel. Commun.* **2016**, *15*, 3949–3963. [CrossRef]

77. Mozaffari, M.; Saad, W.; Bennis, M.; Debbah, M. Drone small cells in the clouds: Design, deployment and performance analysis. In Proceedings of the 2015 IEEE Global Communications Conference—GLOBECOM 2015, San Diego, CA, USA, 6–10 December 2015; pp. 1–6. [CrossRef]
78. Zeng, Y.; Zhang, R.; Lim, T.J. Wireless communications with unmanned aerial vehicles: Opportunities and challenges. *IEEE Commun. Mag.* **2016**, *54*, 36–42. [CrossRef]
79. Chiaraviglio, L.; Amorosi, L.; Blefari-Melazzi, N.; Dell'olmo, P.; Lo Mastro, A.; Natalino, C.; Monti, P. Minimum Cost Design of Cellular Networks in Rural Areas with UAVs, Optical Rings, Solar Panels, and Batteries. *IEEE Trans. Green Commun. Netw.* **2019**, *3*, 901–918. [CrossRef]
80. Amorosi, L.; Chiaraviglio, L.; D'Andreagiovanni, F.; Blefari-Melazzi, N. Energy-efficient mission planning of UAVs for 5G coverage in rural zones. In Proceedings of the 2018 IEEE International Conference on Environmental Engineering, Milan, Italy, 12–14 March 2018; pp. 1–9. [CrossRef]
81. Giambene, G.; Addo, E.O.; Kota, S. 5G aerial component for IoT support in remote rural areas. In Proceedings of the IEEE 5G World Forum, Dresden, Germany, 30 September–2 October 2019; pp. 572–577. [CrossRef]
82. Carrillo, D.; Seki, J. Rural area deployment of internet of things connectivity: LTE and LoRaWAN case study. In Proceedings of the 2017 IEEE 24th International Congress on Electronics, Electrical Engineering and Computing—INTERCON 2017, Cusco, Peru, 15–18 August 2017; pp. 1–4. [CrossRef]
83. Kovács, I.Z.; Amorim, R.; Nguyen, H.C.; Wigard, J.; Mogensen, P. Interference analysis for UAV connectivity over LTE using aerial radio measurements. In Proceedings of the 2017 IEEE 86th Vehicular Technology Conference (VTC-Fall), Toronto, ON, Canada, 24–27 September 2017; pp. 1–6. [CrossRef]
84. Nguyen, H.C.; Amorim, R.; Wigard, J.; Kovács, I.Z.; Mogensen, P. Using LTE networks for UAV command and control link: A rural-area coverage analysis. In Proceedings of the 2017 IEEE 86th Vehicular Technology Conference (VTC-Fall), Toronto, ON, Canada, 24–27 September 2017; pp. 1–6. [CrossRef]
85. Nguyen, H.C.; Amorim, R.; Wigard, J.; Kovacs, I.Z.; Sorensen, T.B.; Mogensen, P.E. How to Ensure Reliable Connectivity for Aerial Vehicles over Cellular Networks. *IEEE Access* **2018**, *6*, 12304–12317. [CrossRef]
86. Hou, T.; Liu, Y.; Sun, X.; Song, Z.; Chen, Y. Non-orthogonal multiple access in multi-UAV networks. In Proceedings of the IEEE Vehicular Technology Conference, Honolulu, HI, USA, 22–25 September 2019; pp. 1–5. [CrossRef]
87. Lyu, J.; Zeng, Y.; Zhang, R.; Lim, T.J. Placement Optimization of UAV-Mounted Mobile Base Stations. *IEEE Commun. Lett.* **2017**, *21*, 604–607. [CrossRef]
88. Al-Hourani, A.; Gomez, K. Modeling Cellular-to-UAV Path-Loss for Suburban Environments. *IEEE Wirel. Commun. Lett.* **2018**, *7*, 82–85. [CrossRef]
89. Wang, K.; Zhang, R.; Wu, L.; Zhong, Z.; He, L.; Liu, J.; Pang, X. Path loss measurement and modeling for low-altitude UAV access channels. In Proceedings of the 2017 IEEE 86th Vehicular Technology Conference (VTC-Fall), Toronto, ON, Canada, 24–27 September 2017; pp. 1–5. [CrossRef]
90. Sharma, V.; Srinivasan, K.; Chao, H.; Hua, K..; Cheng, W. Intelligent deployment of UAVs in 5G heterogeneous communication environment for improved coverage. *J. Netw. Comput. Appl.* **2017**, *85*, 94–105. [CrossRef]
91. Mozaffari, M.; Kasgari, A.T.Z.; Saad, W.; Bennis, M.; Debbah, M. Beyond 5G with UAVs: Foundations of a 3D Wireless Cellular Network. *IEEE Trans. Wirel. Commun.* **2019**, *18*, 357–372. [CrossRef]
92. Zhang, S.; Zeng, Y.; Zhang, R. Cellular-enabled UAV communication: A connectivity-constrained trajectory optimization perspective. *IEEE Trans. Commun.* **2019**, *67*, 2580–2604. [CrossRef]
93. Wang, J.; Jiang, C.; Han, Z.; Ren, Y.; Maunder, R.G.; Hanzo, L. Taking Drones to the Next Level: Cooperative Distributed Unmanned-Aerial-Vehicular Networks for Small and Mini Drones. *IEEE Veh. Technol. Mag.* **2017**, *12*, 73–82. [CrossRef]
94. Wang, H.; Ding, G.; Gao, F.; Chen, J.; Wang, J.; Wang, L. Power Control in UAV-Supported Ultra Dense Networks: Communications, Caching, and Energy Transfer. *IEEE Commun. Mag.* **2018**, *56*, 28–34. [CrossRef]
95. Yang, G.; Zhang, Y.; He, Z.; Wen, J.; Ji, Z.; Li, Y. Machine-learning-based prediction methods for path loss and delay spread in air-to-ground millimetre-wave channels. *IET Microwaves Antennas Propag.* **2019**. *13*, 1113–1121. [CrossRef]
96. Wang, J.L.; Li, Y.R.; Adege, A.B.; Wang, L.C.; Jeng, S.S.; Chen, J.Y. Machine Learning Based Rapid 3D Channel Modeling for UAV Communication Networks. In Proceedings of the 2019 16th IEEE Annual Consumer Communications Networking Conference (CCNC), Las Vegas, NV, USA, 11–14 January 2019. [CrossRef]
97. Tahat, A.; Edwan, T.; Al-Sawwaf, H.; Al-Baw, J.; Amayreh, M. Simplistic Machine Learning-Based Air-to-Ground Path Loss Modeling in an Urban Environment. In Proceedings of the 2020 Fifth International Conference on Fog and Mobile Edge Computing (FMEC), Paris, France, 20–23 April 2020. [CrossRef]
98. Park, Y.M.; Lee, M.; Hong, C.S. Multi-UAVs Collaboration System based on Machine Learning for Throughput Maximization. In Proceedings of the 2019 20th Asia-Pacific Network Operations and Management Symposium (APNOMS), Matsue, Japan, 18–20 September 2019. [CrossRef]
99. Zhu, Q.; Wang, Y.; Jiang, K.; Chen, X.; Zhong, W.; Ahmed, N. 3D non-stationary geometry-based multi-input multi-output channel model for UAV-ground communication systems. *IET Microwaves Antennas Propag.* **2019**, *13*, 1104–1112. [CrossRef]
100. Chang, H.; Bian, J.; Wang, C.X.; Bai, Z.; Sun, J.; Gao, X. A 3D Wideband Geometry-Based Stochastic Model for UAV Air-to-Ground Channels. In Proceedings of the 2018 IEEE Global Communications Conference (GLOBECOM), Abu Dhabi, United Arab Emirates, 9–13 December 2018. [CrossRef]

101. Rodríguez-Piñeiro, J.; Huang, Z.; Cai, X.; Domínguez-Bolaño, T.; Yin, X. Geometry-Based MPC Tracking and Modeling Algorithm for Time-Varying UAV Channels. *IEEE Trans. Wirel. Commun.* **2021**, *20*, 2700–2715. [CrossRef]
102. Liu, J.; Zhang, H.; Sheng, M.; Su, Y.; Chen, S.; Li, J. High Altitude Air-to-Ground Channel Modeling for Fixed-Wing UAV Mounted Aerial Base Stations. *IEEE Wirel. Commun. Lett.* **2021**, *10*, 330–334. [CrossRef]
103. 3GPP. 3GPP Regulation for UAS-UAV. 2019. Available online: https://www.3gpp.org/uas-uav (accessed on 24 March 2021).
104. Song, S.; Choi, M.; Goh, Y.; Yun, J.; Yoo, W.; Yang, W.; Jung, J.; Chung, J. Analysis of wireless backhaul networks based on aerial platform technology for 6G systems. *Comput. Mater. Contin.* **2020**, *62*, 473–494. [CrossRef]
105. Becker, D.; Fiebig, U.; Schalk, L. Wideband Channel Measurements and First Findings for Low Altitude Drone-to-Drone Links in an Urban Scenario. In Proceedings of the 2020 14th European Conference on Antennas and Propagation (EuCAP), Copenhagen, Denmark, 15–20 March 2020; pp. 1–5. [CrossRef]
106. Almalki, F.A.; Angelides, M.C. A machine learning approach to evolving an optimal propagation model for last mile connectivity using low altitude platforms. *Comput. Commun.* **2019**, *142–143*, 9–33. [CrossRef]
107. Arpaio, M.J.; Vitucci, E.M.; Barbiroli, M.; Degli-Esposti, V.; Masotti, D.; Fuschini, F. Narrowband Characteristics of Air-to-Ground Propagation for UAV Assisted Networks in Urban Environments By Means of Fast Ray-Launching Simulations. In Proceedings of the 2020 IEEE 91st Vehicular Technology Conference (VTC2020-Spring), Antwerp, Belgium, 25–28 May 2020; pp. 1–5. [CrossRef]
108. Alam, M.S.; Kurt, G.K.; Yanikomeroglu, H.; Zhu, P.; Đào, N.D. High Altitude Platform Station Based Super Macro Base Station Constellations. *IEEE Commun. Mag.* **2021**, *59*, 103–109. [CrossRef]
109. Arum, S.C.; Grace, D.; Mitchell, P.D. A review of wireless communication using high-altitude platforms for extended coverage and capacity. *Comput. Commun.* **2020**, *157*, 232–256. [CrossRef]
110. Setiawan, E. The Potential Use of High Altitude Platform Station in Rural Telecommunication Infrastructure. In Proceedings of the 2018 International Conference on ICT for Rural Development (IC-ICTRuDev), Badung, Indonesia, 17–18 October 2018. [CrossRef]
111. Vatalaro, F.; Corazza, G.E.; Caini, C.; Ferrarelli, C. Analysis of LEO, MEO, and GEO global mobile satellite systems in the presence of interference and fading. *IEEE J. Sel. Areas Commun.* **1995**, *13*, 291–300. [CrossRef]
112. Seyedi, Y.; Safavi, S.M. On the Analysis of Random Coverage Time in Mobile LEO Satellite Communications . *IEEE Commun. Lett.* **2012**, *16*, 612–615. [CrossRef]
113. Araujo, M.; Ekenberg, L.; Confraria, J. Satellite backhaul for macro-cells, as an alternative to optical fibre, to close the digital divide. In Proceedings of the IEEE Wireless Communications and Networking Conference, Marrakesh, Morocco, 15–18 April 2019; pp. 1–6. [CrossRef]
114. Arifin, M.A.; Khamsah, N.M.N. A Case Study in User Capacity Planning for Low Earth Orbit Communication Satellite. In Proceedings of the ICARES 2018—Proceedings of the 2018 IEEE International Conference on Aerospace Electronics and Remote Sensing Technology, Bali, Indonesia, 20–21 September 2018; pp. 1–6. [CrossRef]
115. Özgür Erçetin.; Krishnamurthy, S.; Dao, S.; Tassiulas, L. Provision of guaranteed services in broadband LEO satellite networks. *Comput. Netw.* **2002**, *39*, 61–77. [CrossRef]
116. Dhungana, A.; Bulut, E. Peer-to-peer energy sharing in mobile networks: Applications, challenges, and open problems. *Ad. Hoc. Netw.* **2020**, *97*, 102029. [CrossRef]

sensors

MDPI

Review

The Internet of Drones: Requirements, Taxonomy, Recent Advances, and Challenges of Research Trends

Abdelzahir Abdelmaboud

Department of Information Systems, King Khalid University, Muhayel Aseer 61913, Saudi Arabia; aelnour@kku.edu.sa

Abstract: The use of unmanned aerial vehicles or drones are a valuable technique in coping with issues related to life in the general public's daily routines. Given the growing number of drones in low-altitude airspace, linking drones to form the Internet of drones (IoD) is a highly desirable trend to improve the safety as well as the quality of flight. However, there remain security, privacy, and communication issues related to IoD. In this paper, we discuss the key requirements of security, privacy, and communication and we present a taxonomy of IoD based on the most relevant considerations. Furthermore, we present the most commonly used commercial case studies and address the latest advancements and solutions proposed for the IoD environments. Lastly, we discuss the challenges and future research directions of IoD.

Keywords: Internet of drones; communication; security; privacy

Citation: Abdelmaboud, A. The Internet of Drones: Requirements, Taxonomy, Recent Advances, and Challenges of Research Trends. *Sensors* **2021**, *21*, 5718. https://doi.org/10.3390/s21175718

Academic Editor: Margot Deruyck

Received: 20 July 2021
Accepted: 16 August 2021
Published: 25 August 2021

1. Introduction

The Internet of drones (IoD) can be described as an infrastructure designed to provide control and access over the Internet between drones and users. In reality, drones are rapidly becoming readily available commodity items, allowing any user to fly different missions in controlled airspace using multiple drones. Although technology helps mass-produce the onboard components of unmanned aerial vehicles (UAVs), including processors, sensors, storage and battery life, the performance limitations of these components impede and reduce expectations. IoD offers drones coupling vehicle as well as cloud mobility functions to allow remote drone access and control, as well as seamlessly scalable offloading and capabilities of remote cloud storage [1,2]. Figure 1 illustrates the IoD environment that includes base stations, signal link, and cloud environments.

The key advantage of a UAV with fixed wings compared to a UAV with rotary wings is that a less complex repair and maintenance process is required by simple structure, thereby giving the consumer more operating time at a reduced cost. The simple structure offers a high-speed aircraft that can have a longer flight length and can cover more ground. UAVs may make use of non-power supply techniques to make gliding more efficient. It is also worth noting that fixed-wing airplanes can carry a greater payload for longer distances when flying with less power giving them the capability to carry a combination of bigger more advanced sensors with a pair of complementary sensors [3].

Until recently, UAVs were operated individually, but today a higher number of coordinated drones may work together to accomplish complex missions. In these circumstances, drone communication is absolutely essential. In other words, it is vital for users to fully comprehend UAV communication systems. One additional kind of wireless channel and network protocol is utilized in drone communications, but on the other hand, several distinct types of wireless routes and network protocols are applied in drone communications. For this reason, the network design for UAVs is determined by their application. As a basic example, researchers have discovered that a point-to-point line-of-sight link between a drone and a gadget may maintain continuous data transmission even when transmission is extended. Drones that use satellite communications to talk to each other for surveillance, when employed for safety defense, or more broad outreach activities, satellite communication is a better option for drones. Alternatively, cellular communications systems are more commonly used in civic and personal applications. For example, indoor communication, in particular for the mesh network and WSN, P2P protocols such as Bluetooth have shown to be more efficient. When applied to drones, working with a multi-layered network can be a difficult and challenging procedure [4]. Examples of important problems are outlined below.

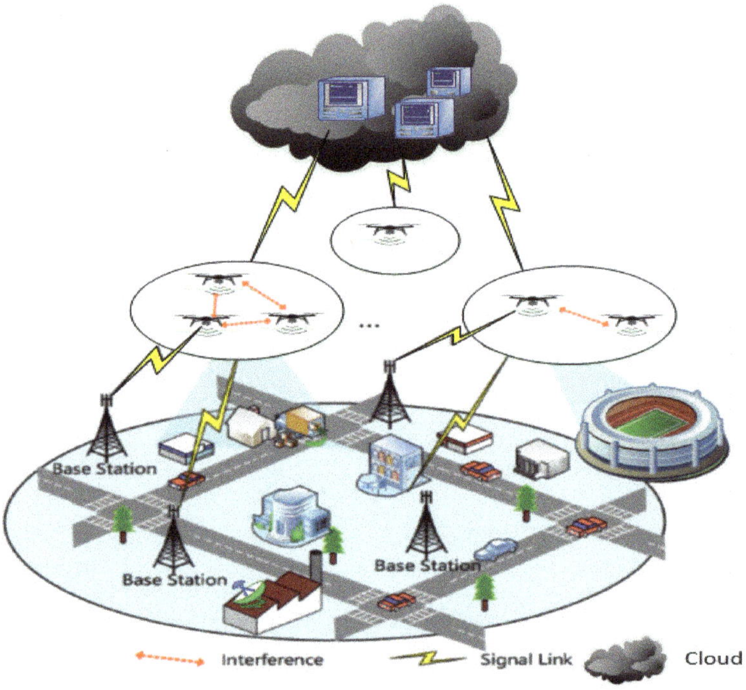

Figure 1. IoD environment.

A remote hijacking of the drones could be achieved by leveraging the vulnerability in the software of the UAVs that act as a sophisticated tool for military purposes. Global positioning system (GPS) signals are under the influence of malware programs on drones that can be controlled by malicious users for malicious objectives (attackers). By doing this, unreasonable attacks, such as dropping bombs, could be committed by the attacker, endangering lives. The control signal is a significant feature of IoD environments due to the different communications among entities and should not be disclosed or exposed in any circumstance. There is a need for robust security measures to avert harm from security attacks. Moreover, to facilitate personal and business drones for independent flight, a certain type of authentication and key exchange protocols are required between the two entities in the sky. Both the entities then create a symmetric security key for future data transmission [4–6].

The rise in the popularity of drones has increased the frequency of cyber assaults against UAV systems during the last decade. An adversary will target the radio connections of UAV systems in order to hinder the systems' ability to communicate with user equipment. This includes data required by the user's cellular equipment that control signals and GPS signals. This is an example of an interception of information giving enemies the ability to steal data that drones transmit and request, as well as foregoing direct control of the drones through control signals. To guarantee the security of the wireless communication channels, guaranteeing that both control and data signals are sent has become a critical element of the overall UAV system [7]. This review focuses on the recent advancements in IoD development. The key contributions to this paper, which are presented from Sections 3–8, are to:

- Investigate the main requirements of IoD (Section 3).
- Provide the main parameters-based thematic taxonomy (Section 4).
- Discuss the most commonly used commercial case studies of IoD (Section 5).
- Investigate the progress recently described in the literature (Section 6).
- Discuss the IoD open challenges and prospective study directions (Section 7).

2. Related Works

In order to examine the possible solutions involving the use of the Internet of drones, several survey studies have been published in the literature. The specific surveys are briefly summarized in this section and given in Table 1, which also covers the substance of this article.

Boccadoro et al. [8] conducted an overview which focused on the consequences of widespread drone adoption for the economy and society. They covered many aspects of the IoD, including a comprehensive analysis of privacy and security issues. Additionally, they explored on what are considered to be the primary research problems and projects that are of high interest to the IoD for prospective future developments.

Rejeb et al. [9] investigated drones to discover their skills, achievements, and challenges in logistics for humanitarian use. They provided an in-depth overview of the capabilities, challenges, and results associated with humanitarian drones in logistical operations, management, and governance. Furthermore, they examined possible humanitarian uses for drones and established a road map for more extensive research on the issue. Yahuza et al. [10] conducted a review to seek and evaluate the factors that impact the IoD network security and privacy. They examined the variety of drone danger levels and security and privacy vulnerabilities. Additionally, they analyzed the recently developed IoD security controls. In addition, they provided the methods for determining the performance, such as assessment and measurement procedures, that are used by the approaches.

Ayamga et al. [11] reviewed the advancements in agricultural, military, and medical drones to showcase the strengths, weaknesses, opportunities, and threats of the current technologies. In addition, the study suggested a need for further research investigation to better integrate drones into current transportation and supply chain infrastructure, such as increasing payload and flight length, and addressing the unique aspects of a country's culture in which drones are easily accepted and embraced.

Merkert and Bushell [12] conducted a review of systematic literature related to historical concerns such as privacy, acceptance, and protection that have been gradually replaced by operational considerations, including contact with and impacts on other users of airspace, in order to recognize critical issues and research gaps. The study demonstrates that unrestricted use of drones can cause problems for other users of the airspace, such as airports and emergency services. In addition, analysis of current regulatory approaches illustrates the need for more policy and management response to both handle rapid and efficient growth in drone usage and promote innovation, with systems of the airspace management for all cases of drone use being one promising strategic response.

Zaidi et al. [13] reviewed the state of the art with an emphasis on the Internet of flying things (IoFT). They proposed taxonomy literature, including description, classification, and comparative analysis of various IoFT works. In addition, the study presents an applications problem for IoFT and future perspectives. Moreover, the survey provides scientific researchers with basic concepts and a full overview of recent IoFT studies.

Table 1. Related works.

Survey Paper	Year of Publication	Highlight	Comparison of Current Research Works in Terms of	Domain
This work	2021	Comprehensive study covering requirements, taxonomy, recent advances, and challenges of future research trends	Drone applications Commercial case studies Recent advances in IoD	IoD
Boccadoro et al. [8]	2021	Drone adoption for the economy and society	Drone applications	IoD
Rejeb et al. [9]	2021	Humanitarian drones	The humanitarian drone's capabilities and performance	Drones
Yahuza et al. [10]	2021	Evaluate IoD network security and privacy	Drones classification	IoD
Ayamga et al. [11]	2021	Review the advancements in agricultural, military, and medical drones	No comparison table	Drones
Merkert and Bushell et al. [12]	2020	Need for more policy and management response to drone usage	Countries and regions contributing	Airborne drones
Zaidi et al. [13]	2020	State of the art of IoFT	IoT, FANET, and IoFT MANET, VANET, UANET, and FANET UAV system, multi-UAV system, and FANET	IoFT
Yaacoub et al. [14]	2020	Analysis to assist ethical hackers in recognizing the current vulnerabilities of UAVs	Frequencies of drones communication Drones cyber-attacks Analytical drone review	Unmanned Aerial Vehicles
Al-Turjman et al. [15]	2020	Review the performance assessment techniques and cybersecurity applications	Flight stacks Objective, constraint, area, and assessment approach Mass simulators Testbeds	IoT networks
Alsamhi et al. [16]	2019	Improve smart cities' real-time implementation	Surveys and current work Drone and IoT for smart cities	Smart cities
Fotouhi et al. [7]	2019	Overview of advancements supporting compatibility of UAVs into mobile networks	Characteristics of drones Comparison of aerial placement	UAV Cellular
Wazid et al. [4]	2018	Define the strengths and limitations of user authentication of IoD	Authentication schemes Communication overheads Security features and functionality	IoD
Bagloee et al. [17]	2016	Vehicles' connection and infrastructure of future growth of AV	No comparison table	Autonomous vehicles
Gharibi et al. [1]	2016	Investigate and extract key concepts of networks	No comparison table	IoD

Yaacoub et al. [14] conducted an overview and analysis of the use of drones or unmanned aerial vehicles (UAV) in multiple domains such as civilian and military, terrorism, and for other various purposes. They describe how a simulated attack on a given drone was carried out by the authors following the hacking period. The study analysis will significantly assist ethical hackers in recognizing the current vulnerabilities of UAVs in both military and civilian realms. In addition, their proposed future study will enable implementation of new techniques and technologies for the identification and defense of enhanced UAV attacks.

Al-Turjman et al. [15] highlighted an overview of drone applications, emergency networks, and surveillance monitoring in software-defined networking-enabled drone-base stations (DBS). Furthermore, they reviewed the performance assessment techniques and related aspects of cybersecurity applications. In addition, the study presents the necessity to advance in IoT-enabled spaces by developing an innovative and multi-faceted drone performance assessment framework with primary concerns, meeting user-defined requirements, and providing safe and reliable services.

Alsamhi et al. [16] presented a survey of the possible collaborative drone and IoT techniques and technologies recently proposed to increase the intelligence of smart cities. Their study focused on data collection, privacy and protection, public safety, disaster management, energy use and quality of life in smart cities, and enhancing smart cities' smartness.

Fotouhi et al. [7] presented an overview of advancements supporting compatibility of UAVs into mobile networks. Their study focuses on discovering what types of consumer UAVs are available to purchase off-the-shelf, and what other people and organizations have addressed when it comes to standardization for serving aerial users with the existing terrestrial base stations. In addition, the study investigates the potential issues and solutions that are being addressed by standardization bodies for aerial users with respect to interference.

Wazid et al. [4] discussed certain security issues and IoD environment specifications. They also discuss a taxonomy of different security protocols in the IoD setting. The study emphasizes the analysis of some of the user authentication systems recently proposed for IoD communication. In addition, the study defines the strengths and limitations of user authentication through a systematic comparative analysis of existing systems with IoD contact systems.

Bagloee et al. [17] discussed an overview of literature related to computer ethics and safety including vehicles' connection and infrastructure of future growth of autonomous vehicles (AV). The study illustrates that a significant information gap exists in AV technology with regards to route activities. In addition, the study shows a tremendous opportunity to incorporate a routing system that offers an effective and intelligent connected vehicle technology. Gharibi et al. [1] presented a conceptual model of how such an architecture can be designed and defined the features that should be implemented by an IoD framework. They investigated and extracted key concepts of current large-scale networks such as cellular network, air traffic control network, and the Internet to link with drone traffic management architecture.

3. Requirements of Internet of Drones (IoD)

The expected increase in the use of drones in a variety of applications could expose operators to a whole new array of risks, particularly damages to third parties and liability. Several of the requirements for potential drones are presented and classified in the following sub-sections. Figure 2 shows the key requirements of IoD.

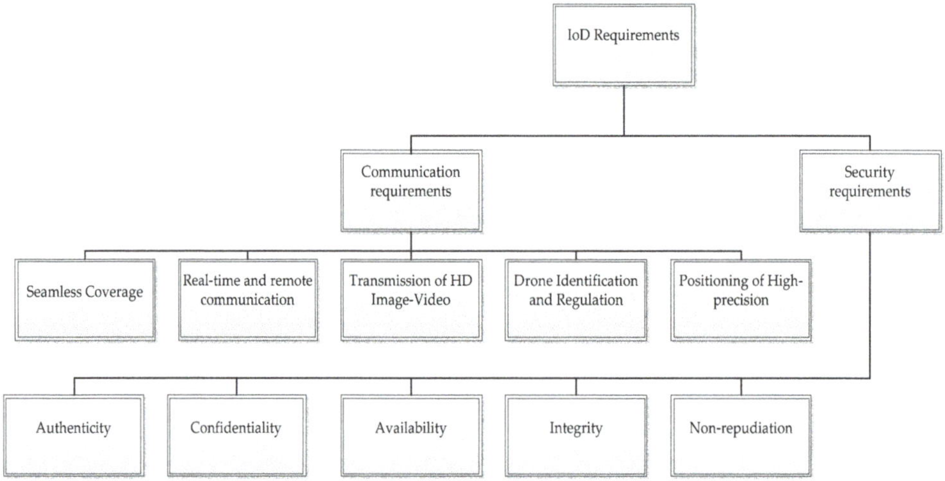

Figure 2. The key requirements of IoD.

3.1. *Communication Requirements*

The impact of IoD communication vulnerabilities is receiving increasing attention from researchers. Many remote locations would be hard to reach were it not for the usage of drones. As a result, drones are commonly employed for important tasks such as rescuing victims, providing surveillance, transportation, and helping with conserving and protection of the environment [18]. Therefore, critical communication requirements to support the different drone applications are discussed as follows.

3.1.1. Seamless Coverage

Hot-spot coverage (stages, tourist areas, and industrial areas) is appropriate for aerial entertainment. Widespread coverage in suburban, urban, and rural areas is required for the inspection and logistics of power or base stations. In the future, seamless drone coverage will become more essential for network planning. Unlike conventional network coverage serving mostly land users, enhanced sky coverage is required to support drone users flying at different heights.

Coverage of up to 10 m altitude is appropriate for plant protection (e.g., spraying of agricultural chemicals). Coverage of up to 50 to 100 m altitude is required for power line inspection. Coverage of up to 200 to 300 m altitude is sufficient for mapping of agricultural lands, while coverage of the upper air pipeline of up to 300 to 3000 m altitude may be needed. It is difficult for networks to serve this large spectrum of coverage scenarios at varying altitudes [19].

3.1.2. Real-Time and Remote Communication

Real-time and remote communication capabilities permit remote controllers to issue a time-based command and control instructions on the basis of the drone flight status report in real-time, such as space co-orders and equipment status. Real-time and remote controls are primarily used in the monitoring of flight conditions, drone task, and equipment and emergency control. The latency and rate of certain data requirements should be met to allow remote control for drones. The downlink (from the base station to drone) data rates in many application scenarios are about 300–600 kbps, and the existing 4G+ networks will fulfill this requirement. For potential implementations, such as remote real-time operations, the latency criterion is strict to guarantee the precision and experience of service [19].

3.1.3. Transmission of HD Image/Video

Networks should be able to provide a high uplink (from the drone to the base station) data rate for drones to permit the transmission of HD image/video. The data rate required is calculated mainly by picture/video size and quality.

In the future, the demand for higher resolution images/videos in vertical industries needing 4K/8K HD video support would need a higher Gbps-level data rate. The 5G networks are well equipped to support such services with a data rate requirement of multi-Gbps. Transmission of HD image/video will dramatically extend the drones' application scenarios including energy and power line inspection, agricultural exploration, control and rescue, entertainment and monitoring. With high transmission rates, drones linked to networks are able to transmit HD images/videos to enable the immersive experience of augmented reality and virtual reality [19].

3.1.4. Drone Identification and Regulation

Mobile networks may help identify and control drones by supporting drone registration, tracking, provision, and coordination [19,20].

- Registration: Identifying but standardizing the drone equipment number, serial number, and flight control serial number helps track the whole process orderly from initial drone production to in-use. Through standardizing registration of drone users, owners and mobile networks, drone users and owners can be legally monitored.
- Monitoring: Drone connections and data communications can be detected and monitored through mobile networks. Drone implementations can be completely tracked in real-time with additional regulatory protocols.
- Forecast: Flight situations can be dynamically evaluated and early warning of possible risks can be achieved by tracking drone positions and monitoring the flight traffic and path.
- Coordination: Knowledge exchange between industries and different companies can be carried out by approved oversight of all vertical industries involved.

3.1.5. Positioning of High-Precision

For numerous drone applications, positioning is critical. In several drone applications, vertical positioning also is important, in addition to traditional positioning on the horizontal plane. The requirement for positioning accuracy will increase from tens of meters to sub-meters with the drone applications' development. Fifty meters positioning precision is adequate for regular monitoring activities. Applications such as agricultural land mapping and automated loading involve high precision positioning at the sub-meter [19].

3.2. Security Requirements

Researchers have created several security and privacy approaches to secure the Internet of drones (IoD) network to protect the location of unmanned aerial vehicles (UAVs) and the privacy and security problems that come with using the IoD network. Due to these localization errors, drone positioning was previously unreliable, which had disastrous implications for the whole IoD network. Another critical aim of the IoD network is to increase the level of security and privacy to a point where it cannot be compromised [21]. Therefore, the main requirements of security and privacy of the IoD network include authenticity, confidentiality, availability, integrity, and non-repudiation [4,11,22]:

3.2.1. Authenticity

Authentication is required for sensing devices, users, and portal nodes before access to a limited resource is enabled or essential information is disclosed [23]. In addition, for communicating entities to have mutual authentication, two of the communicating entities have to be a monitoring drone and a ground-control station. To guarantee full forward secrecy, it is essential to employ a secure key exchange using a method that produces session keys that are impossible to recover [24].

3.2.2. Confidentiality

Confidentiality or privacy of the wireless communication channel protects from the unauthorized disclosure of information [24]. Another significant barrier to IoD implementation is making data available, and controlling access to that data (data confidentiality). For instance, when a group of drones collects road traffic data from several places, there is a continuing problem in sharing this data safely and effectively [20].

3.2.3. Availability

Registered users should also be granted access to appropriate network services in conjunction with system denial-of-service attacks. Both the mechanism and the system are capable of recognizing if a drone is engaged in combat and keeping track of the battle limit, which governs whether the flight management system can pick up on a malfunctioning drone and determine if the availability criteria are compromised [10].

3.2.4. Integrity

Integrity is essential to guarantee the trustworthiness of the information (for example, that it has not been altered in transit, and the source of the information is genuine) [24].

3.2.5. Non-Repudiation

The goal is to ensure that a criminal organization does not conceal its actions. In fact, when there are several parties conducting an action, one of the necessary security measures is to make sure that the action cannot be rejected without the others' knowledge [24].

4. Taxonomy of IoD

This section presents a thematic taxonomy for the Internet of drones that consists of the elements of IoD, notably architecture, middleware, data fusion and sharing, security and various applications, as shown in Figure 3.

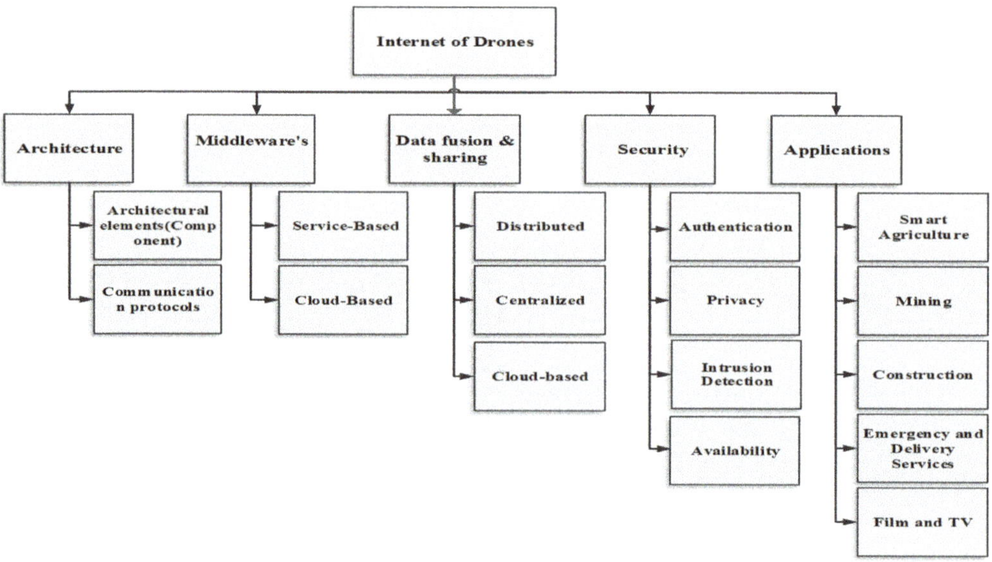

Figure 3. Taxonomy of IoD.

4.1. Architecture

IoD architecture is divided into two major elements: Architecture components and communication protocols. IoD architecture plays a vital role in controlling and administrating unmanned aerial vehicles to perform their operation more efficiently [25].

4.1.1. Architecture Elements (Component)

There are numbers of the Internet of things components present in the architecture of IoD. The role and operations of these components depend on interaction with the IoD architecture. Furthermore, stability, data acquisition, and communication methods of IoD are determined by its architecture components [26]. In addition, IoD architecture components are intended to perform operations mainly related to deciding and controlling the drones and ensuring that the data reach the correct destination from the source nodes [1,27]. Lastly, in the development of IoD architecture, we have to consider resources such as airspace, intersections, and nodes [1].

4.1.2. Communication Protocols

IoD enables multiple communication protocols to support and transfer data between the nodes. Authors propose a drone planner which assists in two communication protocols, namely the MAVLink protocol and ROSLink protocol. The MAVLink is a lightweight message marshaling protocol and ROSLink is integrated with the robot operating system which provides robots into the IoT [28]. The efficiency of the communication protocol is decided by the node movements and behavior in the network [29]. Furthermore, the communication protocol for IoD has to accelerate in routing competencies where data communication between the source node to target nodes is crucial [30]. Due to the above reasons, selection of the communication protocol for IoD should be analyzed thoroughly.

4.2. Middlewares

In a connected IoT devices environment, the middleware layer plays an essential role functioning as a mediator between various nodes and applications [31]. Moreover, the middleware layer ensures abstraction between the various interfaces of the IoD, namely programming language, operating system, networks, and architecture [32]. Two major middlewares, service-based and cloud-based are discussed in the sub-section.

4.2.1. Service-Based

Service-based middleware facilitates in offering network access, local message delivery, caching and name resolution to the IoD architecture [33]. Service-based middleware can achieve robust connectivity for the entire IoD architecture. Furthermore, it can integrate well with other network layers to provide efficient collaboration, performance, and is adaptive to the architecture. Authors propose service-based middleware that enables effective data transmission and extended connectivity time between the network communications [34]. In IoD architecture, simultaneous communication between the ground network and drone is a fundamental principle. The authors propose concurrent communication between ground ad hoc networks and multi-UAV network for on-time communication [1].

4.2.2. Cloud-Based

Cloud-based middleware for IoD service facilitates diverse applications to integrate their operations into the network architecture. The most commonly used robotics application middleware is the robot operating system (ROS) which utilizes the benefits of its elastic resources by offloading computation and processing to the cloud for efficient resource usage. This process is named cloud robotics [1]. The cloud-based middleware provides reliable communication between the ground network and UAV [35]. The cloud-based middleware delivers a very rapid response to the requested service. It further benefits the IoD to actively communicate with the other networks.

4.3. Data Fusion and Sharing

Currently, multiple drones are connected to perform various operations simultaneously. Incorporating data fusion and sharing for IoD allows the processing and merging of several data sources to generate correct information for decision making. Moreover, data fusion based algorithms for multi-UAV further support more efficient performance in the smart environment. In this sub-section, we outline the three different types of data fusion and sharing for the IoD, namely distributed, centralized, and cloud-based.

4.3.1. Distributed

The distributed data fusion and sharing are also called meta-architecture. This decentralizes the data into local interaction and no components are important to any other operation and results in several benefits for the IoD operation, namely scalability, being fault-tolerant, interoperability, and ease of redesign [36]. These benefits help reconfigure or remove the particular unavailable or disconnected drone from the network. Moreover, the distributed environment is more convenient to make a decision independently and in the case of emergency can work as collaborative sharing.

4.3.2. Centralized

The centralized data fusion and sharing mechanism needs high-power communication equipment to perform the operation smoothly. Furthermore, it shares all the information through the fusion center with other devices in the network and leads to provision of more accurate information about the operation [37]. At the early stage of deployment and operation for the shorter distance, centralized sharing performs more efficiently. However, when it comes to longer distances and later part of the operations, it will be much more complicated to reposition or control the device [38].

4.3.3. Cloud-Based

Cloud-based data fusion and sharing are completely controlled by cloud interfaces that consolidate various services to support the client to make an effective decision. Furthermore, services such as analytics and intelligent operation can be performed in the architecture. The developed framework to integrate various sources to operate traffic management of smart cities applications use cloud-based data fusion. In addition, these frameworks offer services to safety, collision avoidance, and risk-aware navigation [39].

4.4. Security

IoD transport and convey by more sensitive and confidential information about the operation. Correspondingly, unauthorized access or control to drones leads to the destruction of any services or physical attack. In this sub-section, an overview of four different areas in security is discussed.

4.4.1. Authentication

Authentication is one of the major threats to any drone, whereby information disclosure results in leakage of identity, location, flying routes, and other meta-data information about the device. To improve the authentication, the authors propose a fast modular arithmetic operation for the drone that helps control the signature key generation and computing capability of any UAV resources. In addition, preserving authentication further secures the authentication performance for UAV-to-UAV in real-time [2]. Moreover, authentication should be given higher priority when it comes to accessing information in real-time, otherwise the attacker can change the original message while operations are performing in real-time. Most of the communication channels are protected by encryption algorithms, which help protect the unauthorized authentication of the IoD [40].

4.4.2. Privacy

The privacy issues of the IoD arise from three major areas: Sensors, communications, and multi-UAV [41]. Eaves-dropping and keylogging are the popular methods in data interception which is aimed to affect the privacy of the flowing information between the drone and communication center [42]. Furthermore, the IoD utilizes the open-access communication environment which is vulnerable to many security and privacy issues such as message integrity, and disclosure of the UAV identity [2]. The recent advancement in drone usage by civilian and business entities further increases the possible harm to the right of the individual or organization to privacy [43].

4.4.3. Intrusion Detection

An intrusion detection system is classified into two main categories, namely host and network-based. The authors propose a hybrid method for the UAV network intrusion detection system, which comprises spectral traffic analysis and a robust controller to observe the abnormal behavior in the UAV environment. Moreover, it performs a statistical signature to explore the threat. Furthermore, the wireless network faces several threats, namely overload, flash crowds, worms, port scans, jamming attacks, etc. [44].

To handle the above-mentioned modern threats, integrating deep learning and big data technologies for intrusion detection will provide a more efficient method to identify the threats [45].

4.4.4. Availability

The specialty of the IoD is to have high availability of the data flows in real-time to the network. The IoD streams the high rate of data in the emergency response operation [46]. Moreover, in an emergency, the rivals may alter the incoming drone data or stop the availability of the data to threaten the current situation. This would affect the evacuation operation in the area [47]. Apart from this, these atomic attacks, namely battery exhaustion attack, fuzzing attack, and physical component warning suppression would be performed to disrupt the availability of drone data [48]. Lastly, any modern security system should protect the collected drone data to be available in the network without interruption for further analysis.

4.5. Applications

In this section, we discuss the most frequently used IoD applications, namely smart agriculture, mining, construction, emergency and delivery services, and film and TV. Table 2 illustrates the summary of drone applications and its modern usage areas.

4.5.1. Smart Agriculture

IoD based smart agriculture is being pursued to satisfy the growing demand emerging from an environmental change with an ever-increasing population and limitation to natural resources. In general, due to over-reliance on machinery, many farmers have left behind farming jobs leading to a shortage of workers for agriculture industries.

Innovations such as IoD, cloud, and IoT will help them monitor their fields in real-time to make an efficient decision. In particular, adopting the IoD and wireless sensor network for smart agriculture will add more value and efficiency in monitoring and growing crops for better yields. Furthermore, IoD eases the process by aerial monitoring and smart spraying in crop affected areas [49]. In the era of precision agriculture, remote sensing with IoD and GPS technologies further helps increase the various types of detection in the farming sector, such as drought stress, nutrient status, and weed detection [50].

4.5.2. Mining

The usage of the IoD in mining applications protects human safety and reduces the cost of mining operations. Moreover, the disruptive technologies help in indoor and outdoor visual inspections, pit planning management, stockpile surveying in the mining industries. The collected IoD data can be further processed to make efficient decisions in mining industries. The authors propose that the framework of UAV applications in mining areas is divided into three major categories. Firstly, basic data from digital cameras, spectral, lidar, thermal infrared, and gas sensors. Secondly, monitoring objects such as dump, surface subsidence, coal gangue heap, open mining pit, and industrial site. Thirdly, different applications in mining areas [51].

4.5.3. Construction

IoD makes it easier to access the visual data of the construction site to the client's view simultaneously. Furthermore, connected sensors, GPS, and the high-quality camera provide various details to different stakeholders in the project. With sophisticated technologies, the project manager has real-time data access to the site to monitor and avoid waste in the construction process.

In addition, worker smart helmets in the construction site provide data about the safety of an individual person to the drone. However, usage of IoD for the construction industry is quite new compared to other application areas which need clear regulatory approval and availability of experts [52].

Table 2. Summary of drone applications and their modern usage areas.

Applications	Modern Usage Areas	Reference
Smart Agriculture	Soil and Field Analysis Crop Monitoring Irrigation Pathogen detection Drought stress Weed detection Nutrient status Yield prediction	https://www.microdrones.com/en/content/drones-and-precision-agriculture-the-future-of-farming/ (accessed on 10 December 2020). [47,49]
Mining	Field measurements Water sampling Stockpile management Site description Mine or quarry monitoring Sediment flow Hazard identification	https://wingtra.com/drone-mapping-applications/mining-and-aggregates/ (accessed on 10 December 2020). [51,53]
Construction	Construction Safety Autonomous operation Planning and analysis process Site Inspections	[52]
Emergency and Delivery Services	Rescue Operations Blood and medicine Transportation Patient Monitoring Disaster relief and Management Find Missing Persons Public safety	[53–55]
Film and TV	Scouting and planning Location managers Logistics and safety Multiple sensors shooting Visual effects	https://variety.com/2019/artisans/production/film-and-tv-shoots-drones-1203223079/ (accessed on 11 December 2020). [32,56]

4.5.4. Emergency and Delivery Services

IoD plays a significant role in forest search and emergency medicine delivery to remote areas. Furthermore, it is used for transporting blood, for disaster relief, missing individuals finding lost hiker in the hill station, and many more areas of emergency services.

Tech giants, such as Google and Amazon use the sky to provide medical care and food delivery. IoD is equipped with a unique signature to track individual missions to verify the place to which they traveled for disaster assistance. This helps coordinate the search and rescue more efficiently. IoD is even used in the assistance of underwater search and rescue operations. However, the greatest challenge relates to the bandwidth of IoD devices that need to assist with intelligent algorithms [53].

The use of drones for emergency response services, especially in medical situations, opens up new possibilities for life-saving measures. Using drones to provide "eyes" on a dangerous situation or to deliver medical supplies to stranded patients may improve emergency response physicians' ability to provide treatment in dangerous situations. IoD offers many emergency response services that impact daily life, discussed as follows.

Disaster management: Sensor data are a major problem in a broad geographical area. In the case of satellite remote sensing, the identification of different parameters such as forest vegetation, the tracking of the bottom of rivers, and the desert is very difficult. For these situations, most devices are specifically designed and have very low time sensitivity. In addition, environmental instability such as large clouds, cyclones, and volcanic ash can cause disturbances. A certain situation can occur when acquiring large-scale sensor data.

As is the case in most situations, the sensors are located in fixed positions and the network architecture is not suitable for rapid monitoring and management. In addition, the sensor nodes themselves are susceptible to failure and low transmission capacity. Thus, a dynamic node collection is highly important, which links the ground station and data processing sites by utilizing the utility of the drone ecosystem Internet. Furthermore, the drone functions as a highly complex data mule and sensing nodes, and sometimes the nodes are disconnected from the network. That is, the intrinsic phenomenon of flying nodes in dissipated networks and ad-hoc flying networks. Therefore, clearly, the nodes interconnect in an opportunistic way [54].

Public safety: Whether a delivery site or a specific restaurant, and being more popular with COVID-19, we can see that replacing personal delivery with a drone can boost protection (no chance of transmission) but there are other possible benefits of drone usage. Fewer cars will minimize CO_2 emissions and alleviate congested streets. Drones may reduce delivery times in non-urban areas or enable delivery to previously unreachable areas. Cost savings will also potentially occur, the average cost of owning and using a vehicle appears to be greater than a drone. Despite the labor pool, many restaurants and supermarkets had trouble filling open vacancies prior to COVID-19. The use of drones can allow the redeployment of their labor by these establishments [55].

4.5.5. Films and TV

Films and TV production are going through revolutionary changes from human intervention to automation. Algorithms, technologies, and expert knowledge systems provide more efficiency to produce a high quality of creative content into cinematography industries. In the existing approach, labor workers are heavily used to position, and target the camera position for adjusting the camera motion. These problems are easily addressed by a drone-based intelligent shooting system, which further integrates with machine learning and computer vision to assist in more convenient automation.

IoD contains more than one high clarity camera to track the visual object for the production of films and TV. Moreover, IoD provides high precision and coordination to capture the shot for cinematography applications. However, the standardization of drones for shot types and camera trajectories need to be addressed [56]. Films and TV had comprehensively addressed the various existing elements in IoD, in particular, architecture, middleware, data fusion and sharing, security, and various applications.

5. Commercial Case Studies

A broad variety of Internet of drone applications in the fields of mining and industrial automation services are being opened up by drones acting as sensor devices. Drones are now being designed for a variety of business applications that are an increasing using IoD. The following section discusses the most key commercial case studies of IoD, summarized in Table 3.

Table 3. Commercial case studies.

Case Study	Middleware Strategy	Data Sharing	Supported Applications and Services
Matternet	Cloud-based	Cloud-based	Healthcare and logistics
Aria Insights	Service-based	Distributed	Oil and Gas
DroneSmartX	Cloud-based	Cloud-based	Agriculture Energy Manufacturing Oil and Gas
H3 Dynamics	Cloud-based	Cloud-based	Real Estate Mining Oil and Gas

5.1. Matternet

Matternet is the leading technology company for independent urban drone logistics systems, offering the technology platform as a service for healthcare and logistics organizations for the transport of medical supplies through dispersed healthcare networks. Matternet has successfully deployed drone networks all over the world and has been licensed by the Swiss Federal Office of Civil Aviation (FOCA) and the United States for regulatory purposes. The Federal Aviation Administration (FAA) carries out fully autonomous operations beyond the line of sight and flights over people to carry essential, important medical supplies. To date, over 10,000 commercial revenue drone flights have been facilitated by Matternet technology [57,58].

In order for Matternet to operate drones in Zurich, local regulators required a flight permit. Drone reliability and safety were crucial regulatory requirements. Regulators required that human operators access the status and location of the drone in real-time and had the ability to manually interact with and monitor the drone, if necessary. Moreover, Matternet had to incorporate data on air traffic control in real-time since helicopters frequently landed near hospitals. The built-in Matternet solution includes self-employed drones and landing stations, as well as a logistics cloud platform. The operations platform is in charge of flight planning and drone surveillance in real-time. A flight director can track the flight status and send orders to drones to come home or wait at certain places [59].

5.2. ARIA

Autonomous roadless intelligent array (ARIA) insights use artificial intelligence (AI) and drones to eliminate people from unpredictable circumstances. Smarter collection of data and machine learning allow decision-makers to solve a problem or conduct a task rapidly and efficiently, while ensuring that human lives are not placed at risk. Decision makers no longer have to spend hours recording and watching a video with sophisticated analytics. Instead, the robots from Aria insights can recognise information that is relevant, sound an alarm when new data are detected, and eventually link all the data to a digital 3D map [60].

The new machine learning technologies of Aria insights require consumers in the public safety, oil and gas, and other commercial sectors to gather information and then transform it into actionable plans. Aria's drones can conduct fully autonomous missions fitted by AI capabilities to make it easier to recognize characteristics of interest in multiple use cases. For example, drones may help maintain enclosed spaces, such as oil tankers or pipelines, allowing operators to locate areas that need to be repaired without human exploration into potentially unsafe working conditions. Aria insights drones can also track the progress of a natural disaster and inform first responders about increased activity in a given area, as well as warn event organizers when sections of the crowd may require more supervision [60].

5.3. DroneSmartX

DroneSmartX, a UAV service provider, has created the UAV SmartHub, a system that allows users to connect to smart sensors on the ground and send data directly to the cloud platform. According to the company, headquartered at the University of Central Florida's business center in Kissimmee, Fla., the UAV SmartHub is connected to several types of commercial-grade drones. DroneSmartX claims that the device integrates UAV as a service with the Internet of things to make drones smarter [61].

The platform gathers data from sources such as passive and active radio frequency identification (RFID), Bluetooth, GPS, and other sensors; processes the data; and delivers it back to the user through the cloud platform in real-time. Ideal applications include agriculture, energy, manufacturing, oil and gas, and inventory of warehouses. According to DroneSmartX, the SmartHub will add tags to thousands of items for inventory checks in minutes [61].

5.4. H3 Dynamics

H3 dynamics is developing the future of business drone services through the combination of machine learning, remote tele-robot, and off-grid power to ensure the deployment and management of condition control, surveillance, and security solutions from anywhere. H3 dynamics provide huge volumes of smart, professional-validated inspection reports into complex images, videos, and other sensor data, a cloud-based service for owners, airlines, contractors, and financial companies [62].

H3 dynamics have developed a station named "DBX" that can easily link all its functions with third-party networks, including particular security and communications networks, using an open application program interface (API) architecture. The DBX can respond in any mobile or fixed sensor to geo-located signals, such as an intelligent camera in a CCTV network, mobile phone, fire sensor or radar coordinate after an earthquake or movement of some sort. To facilitate the fight against COVID-19, Rooftop DBX drone facilities are being re-used as rooftop drone distribution and delivery networks for test kits, blood samples, and provision of supplies. H3 dynamics provide services to many sectors such as real estate, mining, and oil and gas [62].

6. Recent Advancements

This section explores recent advancements in IoD research activities (models, scheme frameworks, protocols, mechanisms, methods, architectures, and algorithms). Table 4 lists the ideas that have been put forward and they are addressed below.

Almulhem [63] proposed a model for a security threat, known as threat trees, to evaluate and enumerate threats to the architecture of the IoD. The suggested threat tree is intended to provide a holistic view of threats affecting an IoD scheme. However, the threat tree that has been proposed does not necessarily include all potential threats.

Zhang et al. [64] suggested a scheme to enable drones and users to authenticate each other. The suggested scheme protected from security attacks. In addition, the proposed scheme provides better protection and better functionality characteristics in terms of connectivity and processing costs but it still does not address certain types of attack.

Nikooghadam et al. [23] presented a scheme for developing a stable authentication based on an elliptical curve for drones to ensure surveillance of a smart city. The scheme demonstrates supporting security criteria and resisting established attacks, while incurring low computational and communication costs.

Table 4. Recent advancements in IoD.

References	Architectural Organization	Middleware Strategies	Solution Type	Proposed Solution	Supported Applications and Services
Almulhem [63]	Communication protocol	Cloud-based	Model	Provide a holistic view of threats affecting an IoD scheme	IoD system
Zhang et al. [64]	Component	Service-based	Scheme	Attain security and withstand various attacks	IoD architecture
Nikooghadam et al. [23]	Communication protocol	Service-based	Scheme	Develop authentication for drones to ensure surveillance of smart city	Smart City Surveillance
Deebak and Al-Turjman [65]	Communication protocol	Service-based	Scheme	Reduce the computation expense of the authentication protocol	IoD infrastructure
Bera et al. [22]	Communication protocol	Service-based	Scheme	Secure data management system between entities of IoD communication	IoD environment
Wang et al. [66]	Communication protocol	Service-based	Scheme	Improve the accuracy of object segmentation	IoD
Tian [2]	Communication protocol	Service-based	Framework	Secure authentication to protect privacy of UAV	IoD communication
Ever [3]	Communication protocol	Service-based	Framework	Secure authentication of UAV environment	IoD applications
Choudhary et al. [25]	Communication protocol	Service-based	Framework	Enhance IoD security aspects	Military IoD
Nouacer et al. [67]	Component	Cloud-based	Framework	Secure architectures based drone for software and hardware	IoD architecture
Sharma et al. [68]	Communication protocol	Service-based	Protocol	Minimize packet drops, to maintain fairness and boost energy quality	Wireless Sensor Network
Mukherjee et al. [54]	Communication protocol	Cloud-based	Mechanism	Enhance protocol	EdgeDrone
Bera et al. [69]	Communication protocol	Cloud-based	Mechanism	Access control for unauthorized UAV	IoD environment
Weng et al. [70]	Communication protocol	Service-based	Method	Improve mobility compensation of drones	IoT drone
Dawaliby et al. [71]	Component	Cloud-based	Architecture	Handle IoT drone operations	IoT drone
Gallego-Madrid et al. [72]	Component	Cloud-based	Architecture	Provide an intermediate processing layer running (VNF)	Drone Gateways
Koubâa et al. [28]	Communication protocol	Cloud-based	Architecture	Control, monitor, and communicate over Internet with drones	IoD
Rehman et al. [73]	Communication protocol	Service-based	Algorithm	Enable information search to explore the ranking of services	Drone network
Chang et al. [74]	Communication protocol	Service-based	Algorithm	Improve cost estimation for multiple in-flight rerouting	IoD
Huang et al. [75]	Communication protocol	Cloud-based	Algorithm	Support analysis of computational complexity	IoD
Yao et al. [76]	Communication protocol	Service-based	Algorithm	Minimize the energy consumption of the drone	IoD

Deebak and Al-Turjman [65] suggested a scheme of IoD to gather sensible knowledge independently. The scheme is implemented to reduce the computation expense of the authentication protocol. The suggested scheme is constructed to provide a valid authentication time for robustness between the IoT devices. However, the proposed scheme has not addressed privacy leakage. Bera et al. [22] discussed a scheme to secure a data management system between entities of IoD communication based on blockchain. The proposed scheme has the ability to resist multiple possible attacks. In addition, comparative analysis shows that the proposed scheme offers better conditions for security and reliability and also has fewer overheads for computational and communication compared to similar schemes. However, the proposed scheme required more investigation of blockchain technologies to support deployment in a real IoD environment.

Wang et al. [66] presented a scheme of joint learning segmentation combined with a model of conditional random-field and designed model based-enhanced U-net to improve the accuracy of object segmentation and to solve the problem of inappropriate edge recognition. The findings of the presented scheme indicated that the precision of the segmentation of the ground object increased up to 86.1%, which is a promising development.

Tian et al. [2] introduced a framework for authentication to protect privacy. The framework guarantees the efficiency of authentication when deployed on resource-constrained small-scale UAVs using the lightweight online/offline signature design. The presented framework is enabled to predict authentication by exploring mobile edge computing (MEC) in view of the high mobility of UAVs to further reduce the cost of authentication for future authentication activities. Moreover, the framework allowed privacy security in terms of the UAV identity, location, and flying routes by developing a username buffer and updated strategy of the public key. However, the study is missing the use of formal analysis to compare current authentication schemes in detail.

Ever [3] designed a framework for UAV environment authentication using elliptic-curve crypto-systems. The proposed framework was tested to ensure that it is immune to substantial well-known possible data confidentiality, shared authentication, password guessing, and key impersonation related attacks. However, the study has not covered privacy issues.

Choudhary et al. [25] proposed a framework for a security based-neural network, which implements MAC protocol managed by the length of Macaulay. The proposed framework enhances the sufficient duration of comprehensive control over ties to enhance the security aspects of IoD and introduces countermeasures against identified cyberattacks. However, the proposed framework does not support the combination of a security protocol with channel authorization and real-time performance.

Nouacer et al. [67] designed a framework to secure the architectures based drone for software and hardware. The framework holds a holistically built ecosystem ranging from electronic components to applications. Moreover, the framework enabled a tightly integrated solution for multi-vendor and scalable drone embedded architecture and an appropriate tool-chain. The problems of latency and computing capacity, on the other hand, are not fully discussed.

Sharma et al. [68] suggested a protocol that depends on congestion, which offers both symmetric reliability and congestion control based on rate adjustment. The protocol aims to minimize packet drops, maintain fairness, and boost energy quality. In addition, the protocol offers greater energy effectiveness and efficiency. However, the proposed protocol does not cover security and privacy issues.

Mukherjee et al. [54] presented a mechanism to enhance protocol solution as an amalgam with the opportunistic routing mechanism.

The mechanism kept monitoring the history of the encounter and the transitivity of encounter time. The analysis of the improved mechanism has been further checked and a substantially improved performance in terms of publisher bandwidth, latency, and run time has been observed. The findings of the proposed mechanism showed efficiency improvement for both memory usage and energy dissipation. However, although the problems of latency and computing power have been mitigated, there are concerns of bandwidth that remain unresolved.

Bera et al. [69] suggested a mechanism of access control for unauthorized UAV detection and mitigation. The proposed mechanism is enabled to protect data from a drone (UAV) to the server of the ground station and the abnormal data for the identification of unauthorized UAVs. The results of the proposed mechanism illustrated the possibility to conduct big data analytics on authenticated transactional data recorded on the blockchain. However, the study is missing the cover privacy and attack issues during the transfer of data form the UAV to the server.

Weng et al. [70] suggested a method to improve mobility compensation of drone based IoT. The method involves phases of frequency offset approximation and relative velocity measurement. The proposed method shows beneficial gains of mobility compensation of drones via the simulations of Monte Carlo, but the study has not considered security and privacy issues of drones' mobility.

Dawaliby et al. [71] designed an architecture-based blockchain to handle IoT drone operations while retaining and security. The results of the introduced architecture obtained in the case of practical agricultural usage illustrate the usefulness in reducing signaling and operating time, improving the percentage of effective maintenance operations, and providing trust and protection in autonomous drone management. The study needs to develop a platform to better meet the demands of IoT devices and use different placement strategies.

Gallego-Madrid et al. [72] introduced an architecture based on the ability of multi-access edge-computing (MEC) to host a drone-board virtualization platform to provide an intermediate processing layer running virtualized networking functions (VNF). The results of the presented architecture have shown significant communication improvements using LoRa-drone gateways in terms of link availability and covered areas, particularly in vast controlled extensions or at points with difficult access, such as rough zones. The study needs to conceptually expand the architecture with software-defined networking and artificial intelligence to dynamically pre-configure the networks according to the current context.

Koubâa et al. [28] proposed an architecture-based cloud for drone management to control, monitor, and communicate over the Internet. The architecture allowed smooth contact with the drones over the Internet, enabling them to be operated anywhere and at any time without distance limitations. In addition, the architecture provided drones with access to cloud computing services to offload heavy computations. Furthermore, the proposed architecture demonstrated the efficacy performance assessment analysis using a real drone for a real-time monitoring application. However, the study has not covered privacy issues of data movement between drones and clouds.

Rehman et al. [73] suggested an algorithm to support information searching that exploits the ranking of services in a drone network. The results of the proposed algorithm have shown that, with its average execution time not exceeding 2.5 ms on two separate machines, the proposed knowledge searching algorithm proved to be successful. The study has not defined in detail the services, classes, and mathematical models to rating parameter, in order to offer better optimization design and performance.

Chang et al. [74] presented an algorithm to improve cost estimation for multiple in-flight rerouting. The experimental flight test of the proposed algorithm was conducted ten times on different routes. In addition, the findings indicate that the adjusted cost estimation algorithm has a rerouting cost estimate of more than 92% accuracy under multiple rerouting path-points. However, the study did not use AI and IoT to develop the autonomous flight algorithm's foundation.

Huang et al. [75] introduced an algorithm to support the analysis of the computational complexity of IoD. They demonstrated the efficacy of the proposed algorithm, detailed computer simulations are carried out, and a comparison with identified objectives is given to determine performance gains. This study did not disprove assumption and identify potential locations for drone-BSs to provide service to mobile users. Another flaw is that environmental variables such as rain and wind were not taken into account in this study.

Yao et al. [76] presented an algorithm to minimize the energy consumption of the drone. The suggested algorithm obtained the optimal solution and transformed the convex optimization problem, and then updated the Lagrangian parameters with an updated Newton method. The results of the simulation showed that the proposed algorithm performs better than the current algorithms. However, the issues of latency and computational power are not completely addressed.

In conclusion, there have been few recent studies that have made advances to contribute to the research fields. For example, only a few studies addressed schemes, frameworks, architectures, and mechanisms, as well as systems for enhancing and improving the security and privacy of IoD services. Therefore, further research and investigation of issues are required for IoD environments.

7. Challenges and Future Research Trends

This section highlights the most crucial challenges and future research trends on the Internet of drones.

7.1. Privacy and Security-Related Challenges

Since government agencies grant licenses for drone usage in civilian and commercial applications, the modern IoD is equipped with numerous connected sensor devices. These devices are vulnerable to various threats including hijack, human error, and loss. These issues should be given higher priority in the design of drone applications.

Integrating AI-based security detection with blockchain to protect the devices and a more sophisticated defense system to control and monitor the application will help reduce the security threat to IoD [77]. In fact, when referring to UAVs one may have to stress the "temporary" relation since these devices could be "untethered" from land. Communications networks are vulnerable to jamming attacks that could catch new vehicles or shut down those vehicles by manipulating their controls. To ensure this goal is met, high integrity protected data links must be used between the flight and ground controllers, and the Federal Aviation Authority (FAA). Technological advancement such as augmented telecommunication and data protection technologies can help avoid the security breaches of the future. Nevertheless, anti-jamming using high powered signal encoding is possible with modern high technology. Communications protection depends on the communication channel frequency, communications media, technology, and the relationships between each other. Usually, encryption algorithms that operate at low bit rates are a bigger problem than those that run at high bit rates as there are more costly categories. Therefore, lowering costs are often followed by a reduction in security or in the number of operations [78].

Additionally, owing to the UAVs autonomous nature, safety is a major concern. The UAVs are noisy and not easy to monitor, hence more safeguards should be established to maintain data protection. There will be many possibilities, implications, and events from the analysis of UAV. To secure UAV systems, using many techniques including redundant sensory devices and security systems that are linked to the UAV ground station and others, should be addressed. Other methods of preventing UAVs from being hacked have been reported. The recommendations from the defense department (DoD) discuss taking steps to counter hijacking. Thus, there are very few details on anti-terror strategies for coping with hijacking. The potential danger of hacking in the air can look as malware in the software created, spoofing attacks on communications-relay, and manipulation hardware or software. One of the possibilities to counter the attack is for the control station to manipulate the electromagnetic field of the carrier signal with hardware on the malicious attacker side. Other forms of attacks include stealing data and gaining control of device functions. To minimize the implications of these conditions, enabling the use of risk reduction standards that could be exercised in court is essential. Which risks are compatible to achieve the highest utility? Based on the available details, these decisions will lead to potential choices to be made. A ground radar device that detects vehicle hijacking is potentially useful by telling the driver that the vehicle has been electronically changed. The only drawback will be the expense of developing, equipping, evaluating, and implementing the product since they had a dollar for dollar return on investment [79,80].

7.2. Global Resource Management-Related Challenges

Resource distribution is critical in serving productivity and reducing cost and can be divided into two groups, the global allocation of resources and local allocation. Global concerns concentrate on the expenditure spent on the global resources of time, energy, and equipment. In addition, the maximum global productivity can be deployed by various equipment in the Internet of everything (IoE), such as edge computers, cloud servers, and UAVs. Under digital media transmission in IoE, global efficiency can be increased by networking algorithms and video coding. Energy and power are allocated for every role based on committed performance. For instance, in a smart building scenario (such as in a local network), an efficient data collection algorithm is good for increasing the user's data rate. In addition, for the flying terminals, the energy management system is essential for making use of limited energy. This is the concept of allocating resources at various nodes [81].

In fact, a challenge for UAV applications is to preserve efficiency while facing cost issues. First, developing protocols for the provisioning of network services based on information rate and requirements of computation is a challenge. The problems to be addressed are for systems that are clustered, use of cloud resources, and autonomous services. In addition, the most popular economic activity is "resource allocation" since it includes competing interests of individuals and companies or services to receive particular resources. The next challenge is discovering reliable and cost-effective ways to assign consumer demands. Another difficulty in resource utilization centers on determining how best to specify the priorities and paths to the minimization of activity energy expenditure. Moreover, resource mapping aims to offer an equal amount of services to buyers and suppliers. Resource mapping may face problems such as mapping the physical components and the rational distribution of resources to fulfill the requirement, which is often hindered by physical limitations and obstacles. Designing algorithms that can use generic methods to quickly obtain a mapping process fast is critical. The application's compatibility with the UAV platform needs to be checked by installing corresponding hardware prerequisites. Yet another difficulty to be resolved would be to build models that can correctly measure the performance of multi-core CPUs, PC storage medium, communication channels, and information center data stores. In addition, this consideration may be a problem when mapping resources [79].

7.3. Sensor Communication-Related Challenges

The IoD sensor's communication protocols are designed with lightweight and highly sensible objects which have a substantial chance for data loss or receive the wrong data from other nodes. Furthermore, sensors face routing issues when communicating with multiple drones with a network center. Some of the drone manufacturers use cheap hardware components which increase the chance of other network communication issues namely, high throughput, latency, and delay between the device and center [82]. To address these issues, next-generation network technologies such as 5G, intelligent routing, narrowband-Internet of things, LoRa-based IoT systems, Sigfox, NB-IoT, and LTE-M need to support connection choices in the future. Lastly, the standardized policy should be developed to use the authorized component for drones' communication protocol.

7.4. Coordination and Tasks Scheduling-Related Challenges

Cloud computing and edge computing are going to be combined in the future to handle computationally intensive IoT applications. These intelligent external intelligent network applications need an appropriate centralized AI analysis and individual big data analysis for coordination. Regarding the situation-target scheduling, the applied intelligence first examines the gathered information on the necessary computing tasks and then decides whether the tasks are required to be submitted to the remote cloud. The object of migration is different from the requirement of resources for a real-world data or application. Computer architecture and networking will be adapted to international performance. This method can be changed at any time for possible changes and improvements. This will certainly become a regular theme [83–85].

7.5. Drones Distribution and Deployment-Related Challenges

Apart from data confidentiality, data sharing and access control are challenges that face IoD deployment. For instance, in the application where a set of drones can collaborate to collect road traffic data of different regions, how to securely and efficiently share these collected data (e.g., in the sense that only authorized entities have access to the data) remain an ongoing challenge [20,22].

8. Conclusions

Unmanned aerial vehicles or drones are helpful in raising efficiency in everyday life. With the number of drones in low altitude air territory, linking drones into the Internet of drones (IoDs) will help enhance the safety and performance of the drone aircraft. However, privacy, security, and communications are still a big concern of the governments. We have addressed key prospects for the association of IoD. This article provides a taxonomy of the IoD based on the most important perspectives. We have included commercial cases that are the most common and illustrate the most recent techniques used for IoD. Finally, we have addressed the obstacles IoD faces and provide some direction for future studies.

Funding: The authors extend their appreciation to the Deanship of Scientific Research at King Khalid University for funding this work through General Research Project, under grant number (GRP-319-40).

Institutional Review Board Statement: Not applicable.

Informed Consent Statement: Not applicable.

Data Availability Statement: Not applicable.

Conflicts of Interest: The authors declare no conflict of interest. The funders had no role in the design of the study; in the collection, analyses or interpretation of data; in the writing of the manuscript or in the decision to publish the results.

References

1. Gharibi, M.; Boutaba, R.; Waslander, S.L. Internet of drones. *IEEE Access* **2016**, *4*, 1148–1162. [CrossRef]
2. Tian, Y.; Yuan, J.; Song, H. Efficient privacy-preserving authentication framework for edge-assisted Internet of Drones. *J. Inf. Secur. Appl.* **2019**, *48*, 102354. [CrossRef]
3. Ever, Y.K. A secure authentication scheme framework for mobile-sinks used in the Internet of Drones applications. *Comput. Commun.* **2020**, *155*, 143–149. [CrossRef]
4. Wazid, M.; Das, A.K.; Lee, J.-H. Authentication protocols for the internet of drones: Taxonomy, analysis and future directions. *J. Ambient Intell. Humaniz. Comput.* **2018**, 1–10. [CrossRef]
5. Motlagh, N.H.; Taleb, T.; Arouk, O. Low-altitude unmanned aerial vehicles-based internet of things services: Comprehensive survey and future perspectives. *IEEE Internet Things J.* **2016**, *3*, 899–922. [CrossRef]
6. Won, J.; Seo, S.-H.; Bertino, E. Certificateless cryptographic protocols for efficient drone-based smart city applications. *IEEE Access* **2017**, *5*, 3721–3749. [CrossRef]
7. Fotouhi, A.; Qiang, H.; Ding, M.; Hassan, M.; Giordano, L.G.; Garcia-Rodriguez, A.; Yuan, J. Survey on UAV cellular communications: Practical aspects, standardization advancements, regulation, and security challenges. *IEEE Commun. Surv. Tutor.* **2019**, *21*, 3417–3442. [CrossRef]
8. Boccadoro, P.; Striccoli, D.; Grieco, L.A. An extensive survey on the Internet of Drones. *Ad Hoc Netw.* **2021**, *122*, 102600. [CrossRef]
9. Rejeb, A.; Rejeb, K.; Simske, S.; Treiblmaier, H. Humanitarian Drones: A Review and Research Agenda. *Internet Things* **2021**, *16*, 100434. [CrossRef]
10. Yahuza, M.; Idris, M.Y.I.; Ahmedy, I.B.; Wahab, A.W.A.; Nandy, T.; Noor, N.M.; Bala, A. Internet of Drones Security and Privacy Issues: Taxonomy and Open Challenges. *IEEE Access* **2021**, *9*, 57243–57270.

11. Ayamga, M.; Akaba, S.; Nyaaba, A.A. Multifaceted applicability of drones: A review. *Technol. Forecast. Soc. Chang.* **2021**, *167*, 120677. [CrossRef]
12. Merkert, R.; Bushell, J. Managing the drone revolution: A systematic literature review into the current use of airborne drones and future strategic directions for their effective control. *J. Air Transp. Manag.* **2020**, *89*, 101929. [CrossRef] [PubMed]
13. Zaidi, S.; Atiquzzaman, M.; Calafate, C.T. Internet of Flying Things (IoFT): A survey. *Comput. Commun.* **2020**, *165*, 53–74. [CrossRef]
14. Yaacoub, J.-P.; Salman, O. Security Analysis of Drones Systems: Attacks, Limitations, and Recommendations. *Internet Things* **2020**, *11*, 100218. [CrossRef]
15. Al-Turjman, F.; Abujubbeh, M.; Malekloo, A.; Mostarda, L. UAVs assessment in software-defined IoT networks: An overview. *Comput. Commun.* **2020**, *150*, 519–536. [CrossRef]
16. Alsamhi, S.H.; Ma, O.; Ansari, M.S.; Almalki, F.A. Survey on collaborative smart drones and internet of things for improving smartness of smart cities. *IEEE Access* **2019**, *7*, 128125–128152. [CrossRef]
17. Bagloee, S.A.; Tavana, M.; Asadi, M.; Oliver, T. Autonomous vehicles: Challenges, opportunities, and future implications for transportation policies. *J. Mod. Transp.* **2016**, *24*, 284–303. [CrossRef]
18. Sharma, A.; Vanjani, P.; Paliwal, N.; Basnayaka, C.M.W.; Jayakody, D.N.K.; Wang, H.-C.; Muthuchidambaranathan, P. Communication and networking technologies for UAVs: A survey. *J. Netw. Comput. Appl.* **2020**, *168*, 102739. [CrossRef]
19. Yang, G.; Lin, X.; Li, Y.; Cui, H.; Xu, M.; Wu, D.; Rydén, H.; Redhwan, S.B. A telecom perspective on the internet of drones: From LTE-advanced to 5G. *arXiv* **2018**, arXiv:1803.11048.
20. Lin, C.; He, D.; Kumar, N.; Choo, K.-K.R.; Vinel, A.; Huang, X. Security and privacy for the internet of drones: Challenges and solutions. *IEEE Commun. Mag.* **2018**, *56*, 64–69. [CrossRef]
21. Rahman, M.F.B.A. *Smart CCTVS for Secure Cities: Potentials and Challenges, S. Rajaratnam School of International Studies*; Rajaratnam School of International Studies: Singapore, 2017.
22. Bera, B.; Saha, S.; Das, A.K.; Kumar, N.; Lorenz, P.; Alazab, M. Blockchain-envisioned secure data delivery and collection scheme for 5g-based iot-enabled internet of drones environment. *IEEE Trans. Veh. Technol.* **2020**, *69*, 9097–9111. [CrossRef]
23. Nikooghadam, M.; Amintoosi, H.; Islam, S.H.; Moghadam, M.F. A provably secure and lightweight authentication scheme for Internet of Drones for smart city surveillance. *J. Syst. Archit.* **2021**, *115*, 101955. [CrossRef]
24. Ko, Y.; Kim, J.; Duguma, D.G.; Astillo, P.V.; You, I.; Pau, G. Drone Secure Communication Protocol for Future Sensitive Applications in Military Zone. *Sensors* **2021**, *21*, 2057. [CrossRef] [PubMed]
25. Choudhary, G.; Sharma, V.; You, I. Sustainable and secure trajectories for the military Internet of Drones (IoD) through an efficient Medium Access Control (MAC) protocol. *Comput. Electr. Eng.* **2019**, *74*, 59–73. [CrossRef]
26. Lagkas, T.; Argyriou, V.; Bibi, S.; Sarigiannidis, P. UAV IoT framework views and challenges: Towards protecting drones as "Things". *Sensors* **2018**, *18*, 4015. [CrossRef]
27. Zhang, P.; Wang, C.; Qin, Z.; Cao, H. A multidomain virtual network embedding algorithm based on multiobjective optimization for Internet of Drones architecture in Industry 4.0. *Softw. Pract. Exp.* **2020**. [CrossRef]
28. Koubâa, A.; Qureshi, B.; Sriti, M.-F.; Allouch, A.; Javed, Y.; Alajlan, M.; Cheikhrouhou, O.; Khalgui, M.; Tovar, E. Dronemap planner: A service-oriented cloud-based management system for the internet-of-drones. *Ad Hoc Netw.* **2019**, *86*, 46–62. [CrossRef]
29. Sánchez-García, J.; García-Campos, J.; Arzamendia, M.; Reina, D.G.; Toral, S.; Gregor, D. A survey on unmanned aerial and aquatic vehicle multi-hop networks: Wireless communications, evaluation tools and applications. *Comput. Commun.* **2018**, *119*, 43–65. [CrossRef]
30. Bousbaa, F.Z.; Kerrache, C.A.; Mahi, Z.; Tahari, A.E.K.; Lagraa, N.; Yagoubi, M.B. GeoUAVs: A new geocast routing protocol for fleet of UAVs. *Comput. Commun.* **2020**, *149*, 259–269. [CrossRef]
31. Ahmed, A.I.A.; Gani, A.; Ab Hamid, S.H.; Abdelmaboud, A.; Syed, H.J.; Mohamed, R.A.A.H.; Ali, I. Service Management for IoT: Requirements, Taxonomy, Recent Advances and Open Research Challenges. *IEEE Access* **2019**, *7*, 155472–155488. [CrossRef]
32. Vallejo, D.; Castro-Schez, J.; Glez-Morcillo, C.; Albusac, J. Multi-agent architecture for information retrieval and intelligent monitoring by UAVs in known environments affected by catastrophes. *Eng. Appl. Artif. Intell.* **2020**, *87*, 103243. [CrossRef]
33. López, J.; Royo, P.; Pastor, E.; Barrado, C.; Santamaria, E. A middleware architecture for unmanned aircraft avionics. In Proceedings of the 2007 ACM/IFIP/USENIX International Conference on Middleware Companion, Newport Beach, CA, USA, 26–30 November 2007; pp. 1–6.
34. Sharma, V.; Kumar, R.; Patiala, P. Service-oriented middleware for multi-UAV guided ad hoc networks. *IT Converg. Pract.* **2014**, *2*, 24–33.
35. Ribeiro, J.P.; Fontes, H.; Lopes, M.; Silva, H.; Campos, R.; Almeida, J.M.; Silva, E. UAV cooperative perception based on DDS communications network. In Proceedings of the OCEANS 2017-Anchorage, Anchorage, AK, USA, 18–27 September 2017; pp. 1–8.
36. Aggarwal, S.; Shojafar, M.; Kumar, N.; Conti, M. A new secure data dissemination model in internet of drones. In Proceedings of the ICC 2019–2019 IEEE International Conference on Communications (ICC), Shanghai, China, 20–24 May 2019; pp. 1–6.
37. Xu, S.; Doğançay, K.; Hmam, H. Distributed pseudolinear estimation and UAV path optimization for 3D AOA target tracking. *Signal. Process.* **2017**, *133*, 64–78. [CrossRef]
38. Patra, A.N.; Regis, P.A.; Sengupta, S. Distributed allocation and dynamic reassignment of channels in UAV networks for wireless coverage. *Pervasive Mob. Comput.* **2019**, *54*, 58–70. [CrossRef]

39. Mohamed, N.; Al-Jaroodi, J.; Jawhar, I.; Idries, A.; Mohammed, F. Unmanned aerial vehicles applications in future smart cities. *Technol. Forecast. Soc. Chang.* **2018**, *153*, 119293. [CrossRef]
40. Fotohi, R. Securing of Unmanned Aerial Systems (UAS) against security threats using human immune system. *Reliab. Eng. Syst. Saf.* **2020**, *193*, 106675. [CrossRef]
41. Zhi, Y.; Fu, Z.; Sun, X.; Yu, J. Security and Privacy Issues of UAV: A Survey. *Mob. Netw. Appl.* **2019**, *25*, 1–7. [CrossRef]
42. Manesh, M.R.; Kaabouch, N. Cyber Attacks on Unmanned Aerial System Networks: Detection, Countermeasure, and Future Research Directions. *Comput. Secur.* **2019**, *85*, 386–401. [CrossRef]
43. Volovelsky, U. Civilian uses of unmanned aerial vehicles and the threat to the right to privacy–An Israeli case study. *Comput. Law Secur. Rev.* **2014**, *30*, 306–320. [CrossRef]
44. Condomines, J.-P.; Zhang, R.; Larrieu, N. Network intrusion detection system for UAV ad-hoc communication: From methodology design to real test validation. *Ad Hoc Netw.* **2019**, *90*, 101759. [CrossRef]
45. Amanullah, M.A.; Habeeb, R.A.A.; Nasaruddin, F.H.; Gani, A.; Ahmed, E.; Nainar, A.S.M.; Akim, N.M.; Imran, M. Deep learning and big data technologies for IoT security. *Comput. Commun.* **2020**, *151*, 495–517. [CrossRef]
46. Besada, J.A.; Bernardos, A.M.; Bergesio, L.; Vaquero, D.; Campaña, I.; Casar, J.R. Drones-as-a-service: A management architecture to provide mission planning, resource brokerage and operation support for fleets of drones. In Proceedings of the 2019 IEEE International Conference on Pervasive Computing and Communications Workshops (PerCom Workshops), Kyoto, Japan, 11–15 March 2019; pp. 931–936.
47. Abdallah, A.; Ali, M.Z.; Mišić, J.; Mišić, V.B. Efficient Security Scheme for Disaster Surveillance UAV Communication Networks. *Information* **2019**, *10*, 43. [CrossRef]
48. Eng-Wilmot, G.L. *Universal People: Audiovisual Experiments in Black Science Fiction, 1980–1986*; University of California: Irvine, CA, USA, 2017.
49. Radoglou-Grammatikis, P.; Sarigiannidis, P.; Lagkas, T.; Moscholios, I. A Compilation of UAV Applications for Precision Agriculture. *Comput. Netw.* **2020**, *172*, 107148. [CrossRef]
50. Maes, W.H.; Steppe, K. Perspectives for remote sensing with unmanned aerial vehicles in precision agriculture. *Trends Plant Sci.* **2019**, *24*, 152–164. [CrossRef] [PubMed]
51. Ren, H.; Zhao, Y.; Xiao, W.; Hu, Z. A review of UAV monitoring in mining areas: Current status and future perspectives. *Int. J. Coal Sci. Technol.* **2019**, *6*, 1–14. [CrossRef]
52. Gheisari, M.; Esmaeili, B. Applications and requirements of unmanned aerial systems (UASs) for construction safety. *Saf. Sci.* **2019**, *118*, 230–240. [CrossRef]
53. Manickavasagam, B.; Amutha, B. Analysis of efficient unmanned aerial vehicles to handle medical emergency data transmission surveillance system by using wireless body area network. *Comput. Commun.* **2020**, *152*, 19–33. [CrossRef]
54. Mukherjee, A.; Dey, N.; De, D. EdgeDrone: QoS aware MQTT middleware for mobile edge computing in opportunistic Internet of Drone Things. *Comput. Commun.* **2020**, *152*, 93–108. [CrossRef]
55. Chamola, V.; Hassija, V.; Gupta, V.; Guizani, M. A comprehensive review of the COVID-19 pandemic and the role of IoT, drones, AI, blockchain, and 5G in managing its impact. *IEEE Access* **2020**, *8*, 90225–90265. [CrossRef]
56. Karakostas, I.; Mademlis, I.; Nikolaidis, N.; Pitas, I. Shot type constraints in UAV cinematography for autonomous target tracking. *Inf. Sci.* **2020**, *506*, 273–294. [CrossRef]
57. Matternet Launches Operations in Berlin. Available online: https://mttr.net/ (accessed on 22 December 2020).
58. Lacey, J. Technology for Good: Stratus Information Systems Helps Matternet Drones Deliver Critical Medical Supplies. Available online: https://blogs.cisco.com/partner/stratus-information-systems-helps-matternet-drones-deliver-critical-medical-supplies (accessed on 22 December 2020).
59. HiveMQ's Reliable IoT Communication Enables Real-Time Monitoring of Matternet's Autonomous Drones. Available online: https://www.hivemq.com/case-studies/matternet/ (accessed on 22 December 2020).
60. Insights, A. New Company 'Aria Insights' Turns Drone Collected Data into Actionable Intelligence. Available online: https://www.officer.com/command-hq/technology/security-surveillance/uav-uas/press-release/21041425/aria-insights-formerly-known-as-cyphy-works-inc-persistent-aerial-solutions-aria-insights-uses-ai-to-turn-drone-collected-data-into-actionable-intelligence (accessed on 22 December 2020).
61. Lillian, B. DroneSmartX Develops Cloud-Based UAV Attachment for Data Collection. Available online: https://unmanned-aerial.com/dronesmartx-develops-cloud-based-uav-attachment-for-data-collection (accessed on 20 December 2020).
62. Dynamics, H. DBX Automates Various Professional Drones. Available online: https://www.h3dynamics.com/security (accessed on 23 December 2020).
63. Almulhem, A. Threat modeling of a multi-UAV system. *Transp. Res. Part A Policy Pract.* **2020**, *142*, 290–295. [CrossRef]
64. Zhang, Y.; He, D.; Li, L.; Chen, B. A lightweight authentication and key agreement scheme for Internet of Drones. *Comput. Commun.* **2020**, *154*, 455–464. [CrossRef]
65. Deebak, B.; Al-Turjman, F. A smart lightweight privacy preservation scheme for IoT-based UAV communication systems. *Comput. Commun.* **2020**, *162*, 102–117. [CrossRef]
66. Wang, E.K.; Chen, C.-M.; Wang, F.; Khan, M.K.; Kumari, S. Joint-learning segmentation in Internet of drones (IoD)-based monitor systems. *Comput. Commun.* **2020**, *152*, 54–62. [CrossRef]

67. Nouacer, R.; Hussein, M.; Espinoza, H.; Ouhammou, Y.; Ladeira, M.; Castiñeira, R. Towards a Framework of Key Technologies for Drones. *Microprocess. Microsyst.* **2020**, *77*, 103142. [CrossRef]
68. Sharma, B.; Srivastava, G.; Lin, J.C.-W. A bidirectional congestion control transport protocol for the internet of drones. *Comput. Commun.* **2020**, *153*, 102–116. [CrossRef]
69. Bera, B.; Das, A.K.; Sutrala, A.K. Private blockchain-based access control mechanism for unauthorized UAV detection and mitigation in Internet of Drones environment. *Comput. Commun.* **2020**, *166*, 91–109. [CrossRef]
70. Weng, L.; Zhang, Y.; Yang, Y.; Fang, M.; Yu, Z. A mobility compensation method for drones in SG-eIoT. *Digit. Commun. Netw.* **2020**, *7*, 196–200. [CrossRef]
71. Dawaliby, S.; Aberkane, A.; Bradai, A. Blockchain-based IoT platform for autonomous drone operations management. In Proceedings of the 2nd ACM MobiCom Workshop on Drone Assisted Wireless Communications for 5G and Beyond, London, UK, 25 September 2020; pp. 31–36.
72. Gallego-Madrid, J.; Molina-Zarca, A.; Sanchez-Iborra, R.; Bernal-Bernabe, J.; Santa, J.; Ruiz, P.M.; Skarmeta-Gómez, A.F. Enhancing Extensive and Remote LoRa Deployments through MEC-Powered Drone Gateways. *Sensors* **2020**, *20*, 4109. [CrossRef] [PubMed]
73. Rehman, A.; Paul, A.; Ahmad, A.; Jeon, G. A novel class based searching algorithm in small world internet of drone network. *Comput. Commun.* **2020**, *157*, 329–335. [CrossRef]
74. Chang, Y. An enhanced rerouting cost estimation algorithm towards internet of drone. *J. Supercomput.* **2020**, *76*, 1–14. [CrossRef]
75. Huang, H.; Savkin, A.V.; Ding, M.; Kaafar, M.A. Optimized deployment of drone base station to improve user experience in cellular networks. *J. Netw. Comput. Appl.* **2019**, *144*, 49–58. [CrossRef]
76. Yao, J.; Ansari, N. QoS-aware power control in internet of drones for data collection service. *IEEE Trans. Veh. Technol.* **2019**, *68*, 6649–6656. [CrossRef]
77. Alladi, T.; Chamola, V.; Sahu, N.; Guizani, M. Applications of blockchain in unmanned aerial vehicles: A review. *Veh. Commun.* **2020**, *23*, 100249. [CrossRef]
78. Faughnan, M.S.; Hourican, B.J.; MacDonald, G.C.; Srivastava, M.; Wright, J.-P.A.; Haimes, Y.Y.; Andrijcic, E.; Guo, Z.; White, J.C. Risk analysis of unmanned aerial vehicle hijacking and methods of its detection. In Proceedings of the 2013 IEEE Systems and Information Engineering Design Symposium, Charlottesville, VA, USA, 26 April 2013; pp. 145–150.
79. Idries, A.; Mohamed, N.; Jawhar, I.; Mohamed, F.; Al-Jaroodi, J. Challenges of developing UAV applications: A project management view. In Proceedings of the 2015 International Conference on Industrial Engineering and Operations Management (IEOM), Dubai, United Arab Emirates, 3–5 March 2015; pp. 1–10.
80. Nikolic, J.; Burri, M.; Rehder, J.; Leutenegger, S.; Huerzeler, C.; Siegwart, R. A UAV system for inspection of industrial facilities. In Proceedings of the 2013 IEEE Aerospace Conference, Big Sky, MT, USA, 6 January 2013; pp. 1–8.
81. Mohamed, N.; Al-Jaroodi, J. Service-oriented middleware for collaborative UAVs. In Proceedings of the 2013 IEEE 14th International Conference on Information Reuse & Integration (IRI), San Francisco, CA, USA, 14–16 August 2013; pp. 185–192.
82. Kriz, V.; Gabrlik, P. Uranuslink-communication protocol for uav with small overhead and encryption ability. *IFAC-PapersOnLine* **2015**, *48*, 474–479. [CrossRef]
83. Liu, Y.; Dai, H.-N.; Wang, Q.; Shukla, M.K.; Imran, M. Unmanned aerial vehicle for internet of everything: Opportunities and challenges. *Comput. Commun.* **2020**, *155*, 66–83. [CrossRef]
84. Tran, T.X.; Hajisami, A.; Pandey, P.; Pompili, D. Collaborative mobile edge computing in 5G networks: New paradigms, scenarios, and challenges. *IEEE Commun. Mag.* **2017**, *55*, 54–61. [CrossRef]
85. Stergiou, C.; Psannis, K.E.; Kim, B.-G.; Gupta, B. Secure integration of IoT and cloud computing. *Future Gener. Comput. Syst.* **2018**, *78*, 964–975. [CrossRef]

MDPI
St. Alban-Anlage 66
4052 Basel
Switzerland
Tel. +41 61 683 77 34
Fax +41 61 302 89 18
www.mdpi.com

Sensors Editorial Office
E-mail: sensors@mdpi.com
www.mdpi.com/journal/sensors

www.ingramcontent.com/pod-product-compliance
Ingram Content Group UK Ltd.
Pitfield, Milton Keynes, MK11 3LW, UK
UKHW060124010826
14089UKWH00006B/134